MW00759971

Smart Flow Control Processes in Micro Scale

Smart Flow Control Processes in Micro Scale

Volume 2

Special Issue Editors

Bengt Sunden
Jin-yuan Qian
Junhui Zhang
Zan Wu

MDPI • Basel • Beijing • Wuhan • Barcelona • Belgrade • Manchester • Tokyo • Cluj • Tianjin

Special Issue Editors
Bengt Sunden
Lund University
Sweden

Jin-yuan Qian
Zhejiang University
China

Junhui Zhang
Zhejiang University
China

Zan Wu
Lund University
Sweden

Editorial Office
MDPI
St. Alban-Anlage 66
4052 Basel, Switzerland

This is a reprint of articles from the Special Issue published online in the open access journal *Processes* (ISSN 2227-9717) (available at: https://www.mdpi.com/journal/processes/special_issues/Flow_Micro_Scale).

For citation purposes, cite each article independently as indicated on the article page online and as indicated below:

LastName, A.A.; LastName, B.B.; LastName, C.C. Article Title. *Journal Name* **Year**, *Article Number*, Page Range.

Volume 2
ISBN 978-3-03936-511-1 (Hbk)
ISBN 978-3-03936-512-8 (PDF)

Volume 1-2
ISBN 978-3-03936-513-5 (Hbk)
ISBN 978-3-03936-514-2 (PDF)

Contents

About the Special Issue Editors

Bengt Sunden received his M.Sc. in 1973, Ph.D. in 1979, and was appointed Docent in 1980, all at Chalmers University of Technology, Gothenburg, Sweden. He was appointed Professor of Heat Transfer at Lund University, Lund, Sweden, in 1992. He has served as Professor Emeritus and Senior Professor since 2016. His main research interests include heat transfer enhancement techniques, gas turbine heat transfer, and computational modeling and analysis of multiphysics and multiscale transport phenomena for fuel cells. He serves as Guest Professor of numerous prestigious universities. He is a Fellow of ASME, regional editor for *Journal of Enhanced Heat Transfer* since 2007, and associate editor of *Heat Transfer Research* since 2011, the ASME J. Thermal Science, Engineering and Applications (2010–2016), and ASME *Journal of Electrochemical Energy Conversion and Storage* since 2017. He is a recipient of the ASME Heat Transfer Memorial Award 2011 and Donald Q. Kern Award 2016. He received the ASME HTD 75th Anniversary Medal 2013. He has edited 30 books and authored three textbooks. He has published over 400 papers in numerous journals, with a h-index of 39 and over 6400 citations.

Jin-yuan Qian is presently Lecturer at the Institute of Process Equipment, College of Energy Engineering, Zhejiang University, China, a position he has held since his appointment in 2018. He received his B.Sc. and Ph.D. degrees, both in Chemical Process Equipment, from Zhejiang University, China, in 2011 and 2016, respectively. He was a joint Ph.D. student at TU Bergakademie Freiberg, Germany, from 2013 to 2014 and a Postdoc Researcher in the Department of Energy Sciences, Lund University, Sweden, from 2016 to 2017. His research interests include heat transfer, multiphase flow, flow control, and computational fluid dynamics. He has co-authored around 50 papers in international journals and conference proceedings.

Junhui Zhang is presently Research Professor at College of Mechanical Engineering, Zhejiang University, China, a position he has held since 2013, and has been Deputy Director of the State Key Laboratory of Fluid Power and Mechatronic Systems since 2019. He received his B.Sc. and Ph.D. degrees, both in Mechatronic Engineering, from Zhejiang University, China, in 2007 and 2012, respectively, and won the Outstanding Youth Science Foundation in 2019. His research interests are focused on the design and measurement of hydraulic components, especially the high-power-density axial piston pump. He has co-authored about 60 papers in international journals and conference proceedings.

Zan Wu is presently Senior Lecturer in the Department of Energy Sciences, Lund University, Lund, Sweden. He received his B.Sc. in Energy and Environmental System Engineering in 2008, and Ph.D. in Energy Engineering, both from Zhejiang University, Hangzhou, China. His research interests include multiphase flow, phase-change heat transfer enhancement techniques, microfluidics, surface modification, nanofluids, thermophysical properties, compact heat exchangers, and proton exchange membrane fuel cells. He has co-authored around 70 papers in international journals and conference proceedings as well as four book chapters.

Preface to "Smart Flow Control Processes in Micro Scale"

In recent years, microfluidic devices with a large surface-to-volume ratio have witnessed rapid development, allowing them to be successfully utilized in many engineering applications. Within microfluidic devices, the fluid flow at microscale shows obvious differences and unique flow characteristics compared to that at the common macroscale. Thus, the flow behaviors at microscale have attracted many researchers for the purpose of innovative heat and mass transfer enhancement. A smart control process has been proposed for many years, while many new innovations and enabling technologies have been developed for smart flow control, especially concerning "smart flow control" at the microscale. This Special Issue aims to highlight the current research trends related to this topic, presenting a collection of 33 papers from leading scholars in this field. Among these include studies and demonstrations of flow characteristics in pumps or valves as well as dynamic performance in roiling mill systems or jet systems to the optimal design of special components in smart control systems. We do think smart flow control at the microscale will continue to become more and more useful in the near future. To end, we would like to express our heartful gratitude to all the scientific contributors of the papers submitted to this Special Issue.

Bengt Sunden, Jin-yuan Qian, Junhui Zhang, Zan Wu
Special Issue Editors

Article

A Numerical Research on Vortex Street Flow Oscillation in the Double Flapper Nozzle Servo Valve

Liang Lu [1,*], Shirang Long [1] and Kangwu Zhu [2,3]

[1] School of Mechanical Engineering, Tongji University, Shanghai 200092, China; 1930201@tongji.edu.cn
[2] Shanghai Institute of Aerospace Control Technology, Shanghai 201109, China; zjuzkw@zju.edu.cn
[3] Shanghai Servo System Engineering Technology Research Center, Shanghai 201109, China
* Correspondence: luliang829@tongji.edu.cn

Received: 27 August 2019; Accepted: 8 October 2019; Published: 11 October 2019

Abstract: The oscillating flow field of the double nozzle flapper servo valve pre-stage is numerically analyzed through Large Eddy Simulation (LES) turbulent modeling with the previous grid independence verification. The vortex street flow phenomenon can be observed when the flow passes through the nozzle flapper channel, the vortex alternating in each side produces the periodical flow oscillation. The structural and flow parameter effects on the oscillating flow are emphasized, and it could be determined that the pressure on the flapper is nearly proportional to the flow velocity and inversely proportional to the actual distance between the flapper and the nozzle. On the other hand, the main frequency of oscillation decreases with the velocity and increases with the distance between the nozzle flapper. The main stage movement is further considered with a User Defined Function (UDF), and it could be determined that the influences of the structural and flow parameters on the flow oscillation are rarely changed, but the main frequencies drop, generally.

Keywords: double flapper nozzle servo valve; Karman vortex; self-sustained flow oscillation; computational fluid dynamics

---

1. Introduction

As the core component of hydraulic control systems, the electro-hydraulic servo valve has certain advantages of high performance and high reliability. Its first appearance was to the application in fighter planes during World War II, but the single-stage open-loop structure made it difficult to meet the control requirements at that time [1]. It was not until Massachusetts institute of technology (MIT) replaced the solenoid with a high-frequency permanent magnet torque motor that the servo valve ushered in its golden period of development. In 1953, the single nozzle flapper valve was firstly invented by Moog [2]. After four years, the single nozzle structure was further improved to have double nozzles by Howard [3]. In 1962, Atchley [4] invented the jet tube servo valve. Thanks to the development of electronic technology, Vanderlaan et al. [5] made the servo motors directly drive the spool movement in 1987. In 1993, Laux [6] improved and invented the rotary direct drive servo valve. However, the direct drive servo valves are still limited by insufficient motor power, leading to the frequency response not being quick enough. With the advantages of a high power density ratio and a high frequency response, jet servo valves are still popularly applied in crucial industrial applications, including aerospace, ship engineering, high-end robots, etc. At present, research on jet servo valves are still in progress.

There are many related aspects of servo valve research. For example, on the control algorithm, Samakwong et al. [7] found that a genetic algorithm (GA) could better optimize the parameters of the PID controller and control the performance of the servo valve than the Ziegler–Nichols adjustment method. With respect to mathematical modeling, Brito et al. [8] established a Hammerstein model for aerospace servo valves, the results showing that the identified model can represent the general non-linear behavior of servo valves. With respect to hydraulic power, Zohreh et al. [9] simulated the valve core pressure under unsteady conditions. It was found that in the two-stage flapper nozzle electro-hydraulic valve the external acceleration would change the fluid pressure leaving the nozzle and produce the same effect as the external force. Ye et al. [10] established the dynamic model of the plunger pump, simulated it by Computational Fluid Dynamics (CFD), and verified it experimentally. The results show that the vibration speed of the plunger pump on the X_F axis is higher than that on the Y_F axis. The excitation moment M_{CY} and M_{PY} on the Y_F axis contribute greatly to the vibration of the plunger pump. On the flow field characteristic, many scholars adopted CFD approaches to obtain the complex valve flow detail. Brito et al. [8] also carried out experimental and numerical studies to determine the mechanism of cavitation in the fluid region between the flapper and nozzle by using 3D models and CFD grids. Li et al. [11] observed the cavitation phenomena in the flow field from Reynolds numbers 630–2500 with the comparison of CFD simulation. They found that the computational results were in good agreement with the experimental observations and came to the conclusion that the position of the cavitation source is shown at the tip of the nozzle inner wall, the tip of the nozzle outer wall, and the front of the flapper. Chen et al. [12] revealed the effect of oil viscosity on the transient distribution of cavitation and small-size vortices, indicating the noise accompanied by the flow resonance in the nozzle. When the pressure fluctuation near the flowmeter is large enough in the two-stage servo valve, flow acoustic resonance and screaming may occur. Qian et al. [13] researched the forward and reverse flow of Al_2O_3-water nanofluids in micro T45-R Tesla valves at different flow rates, temperatures, and nanoparticle volume fractions by CFD on the basis of the verified numerical model, finding that the main flow percentage was proportional to the above three factors and the flow rate has the greatest influence on the polarity of the valve. Chao et al. [14] found that the inward inclined design of cylinder ports could effectively decrease the gaseous cavitation and increase the effective output flow of cylinder by using centrifugal effects of rotating fluid, which provided a new way to optimize the performance of (Electro-Hydrostatic Actuator) EHA. Qian et al. [15] used CFD to simulate the valve core diameter, single hole/porous diameter, hole diameter, and its arrangement at the bottom of the valve core steadily and instantaneously, and found that the pressure difference between the two sides increases with the increase of the diameter of the valve core and the decrease of the aperture. Meanwhile, the opening time of the main valve also increases with the increase of the diameter of the valve core. Zhang et al. [16] proposed a damping sleeve with a throttle hole. Through experiments and numerical calculation, it was found that the designed damper sleeve had a significant effect on the pressure distribution and jet direction on the surface of the cone, which can significantly reduce the flow force and the opening time of the valve.

Recently, with the improvement of working requirements, the jet flow velocity comes to a higher level with a larger Mach number. The flow compressible effect is more and more obvious. The authors acknowledge there are rarely any studies that have paid enough attention to the high-speed compressible flow oscillations in the jet servo valves. For the present paper content organization, the CFD approach is employed with a (Large Eddy Simulation) LES turbulent model to obtain the vortex flow oscillation conditions. After determining the independence of the grid, the flow field of the fluid in the servo valve with double nozzles and flappers is analyzed under the condition of changing the inlet oil flow rate and the deflection displacement of the flappers, while the force acting on the servo valve flappers under the coupling of the main valve is also discussed.

2. Flow Structure and Grid Independence Analysis

2.1. Operation Principle and Structural Parameters

The two-stage double nozzle flapper force feedback electro-hydraulic servo valve is taken as the research object. As shown in Figure 1a, when the servo valve is in the initial position, the coil is not electrified, the flapper is located in the middle of the nozzle without deflecting, the flow force acting on the flapper is offset each other, the pressure loss caused by the variable throttle hole is the same, the pressure at both ends of the main valve core is the same, and the main valve core is not moving. When the corresponding electric signal is input, the coil generates a magnetic field, which makes the torque motor produce a magnetic moment, and drives the flapper to produce the corresponding deflection angle, thus promoting the movement of the main valve core. As shown in Figure 1b, the simplified structure of the jet location, when the current flowing through the left and right coils is different, for example of $i_1 > i_2$, the electromagnetic moment produced by the left coil is larger than that of the right coil, which makes the coil rotate clockwise, and makes the flapper shift to the left, thus making the distance between the flapper of the jet flapper valve and the two nozzles different, the left side smaller and the right side large. The flow resistance of the hole changes, making the pressure loss on the left side small, the pressure large, the pressure loss on the right side large and the pressure small, so that the oil hydraulic pressure at the two ends of the main valve core is different, driving the main valve core to move to the right, generating load flow and driving the load operation. At the same time, the armature rotates, driving the feedback rod fixed on the armature to shift to the left. Deformation results in counter-clockwise feedback moment. The motion of the valve core makes the feedback rod more deformed and the feedback moment correspondingly larger. When the feedback moments generated by the two are superimposed on the flapper and balanced with the electromagnetic moments generated by the torque motor, the spool is in a predetermined position, and the flapper is in a balanced state. At this time, the required load flow and pressure are generated, and the servo valve is in a predetermined working state. When the load changes or external disturbance causes the spool to deviate from the balanced position, the feedback moment changes, which makes the flapper deviate from the balanced position, and the flow resistance of the variable throttle hole changes accordingly. The pressure difference between the two sides of the spool is generated again, so that the flapper moves in the direction of reducing the deviation until the spool reaches the balanced state [17].

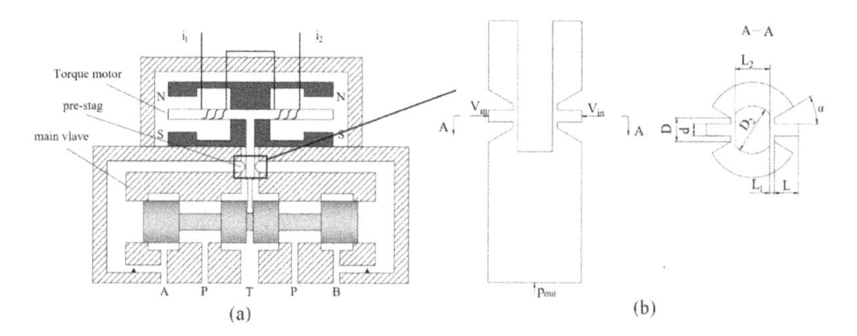

Figure 1. Principle diagram of jet flapper valve. (**a**) Structure schematic diagram of servo valve. (**b**) Pre-stage flow field and boundary conditions.

The structural parameters of the nozzle flapper chamber and the set working oil parameters are both as shown in Table 1. According to the working conditions, the maximum working pressure of the servo valve is less than 31 MPa, and the rated flow rate is 0.48–6.9 L/min. The inlet pressure is selected according to the inlet speed. The outlet pressure is set atmospheric pressure.

Table 1. Nozzle parameters of the flapper nozzle valve.

Parameter		Numerical Value
Reference pressure/MPa		0.1
Density/kg·m^{-3}		889
Dynamic Viscosity/Pa·s		0.04
Modulus of elasticity/MPa		1000
Nozzle	Internal diameter/d	0.5
	External diameter/D	1
	Spacing/L_1	0.2
	Length/L	1.0
	Angle/α	30
Flapper	Diameter/D_1	2.0
	Thickness/L_2	1.5

The pre-stage valve in this study is perpendicular to the nozzle and is mainly affected by the force in the x-axis, i.e., the horizontal direction. While the force in the y-axis, i.e., the direction perpendicular to the paper surface, counteracts each other, and has little influence on the servo valve, it can be simulated by two-dimensional flow chart, which saves on computational resources. Moreover, it is convenient to observe the flow field changes in the channel, and it has no effect on the final simulation results. The following grid and simulation are based on the 2D model shown in Figure 1b.

2.2. Boundary Conditions and Simulation Settings

As shown in Figure 1b, there are three boundary conditions in the two-dimensional model of the flow channel, inlet, outlet, and wall. Among them, the inlet boundary is where the oil enters, Since the intensity of the change of oil flow field at the nozzle is most directly affected by the flow velocity, the inlet oil flow velocity is set as the inlet boundary condition. The outlet oil tank is set as the pressure outlet. The other surface is set as the wall. As shown in Figure 1b, V_{in} represents the velocity inlet, p_{out} represents the pressure outlet, and the unspecified surfaces are all wall surfaces.

Fluent 19.0 (ANSYS 19.0, ANSYS, Inc., Canonsburg, PA, USA, 2018) is used for present numerical solution. Since there are vertical pipes in the channel model, the gravity factor should be taken into account and the gravity acceleration is 9.81 m/s^2. As mentioned above, flow oscillation is a wave–vorticity coupled flow phenomenon. Therefore, when setting the fluid as a turbulent flow, compressibility must be considered, fluid state as a transient state, and the LES model is suitable for solving model. The pressure–velocity coupling term is selected as the Semi-Implicit Method for Pressure-Linked Equations Consistent (SIMPLEC) algorithm, which is improved by the SIMPLE algorithm. The correction term in the velocity equation is not neglected in each iteration. Therefore, the pressure correction value obtained is generally appropriate, and an under-relaxation coefficient less than 1 can be selected according to the situation to accelerate the convergence of the solution in the iteration process [18]. When factors such as cavitation are not considered, the oil is a single-phase flow. Since the LES model belongs to direct numerical simulation to some extent, the discrete scheme of momentum and size terms are chosen as the second-order upwind scheme with second-order accuracy. Since most of the flow field grids are quadrilateral grids and mainly focus on wave–eddy coupling, it is necessary to observe the eddy effect in turbulence, so in order to avoid errors in the difference value and the eddy effect in turbulence, the discrete scheme is adopted as the second-order upwind scheme with second-order accuracy. The pressure gradient hypothesis on the boundary is advantageous to the flexible calculation, and the pressure term is chosen as PRESTO. In order to ensure continuity, the convergence condition is defined as the third-order residuals of each parameter, and the transient equation is a second-order implicit equation. The time step is 0.0001 s, as shown in Table 2.

Table 2. Solution strategy.

Emulation Items	Emulation Settings
Fluid state	Single-phase turbulent transient
Pressure algorithm	SIMPLEC
Discrete scheme	Second-order upwind
Pressure correction algorithm	PRESTO!
Transient equation	Second-order implicit equation
Simulation step size	0.0001s

2.3. Grid Generation and Independence Analysis

The runner belongs to the regular geometric model, so the grid module of ANSYS software, mesh (ANSYS 19.0, ANSYS, Inc., Canonsburg, PA, USA, 2018) is selected to divide the runner with the Quadrilateral Dominant method. Since there are a few trapezoidal areas in the runner, the type of grid division is Quad/Tri. At the same time, the nozzle part of the key research area has a larger grid density in order to capture as many transient details as possible. The partitioned grid is shown in Figure 2, and the grid parameters are shown in Table 3.

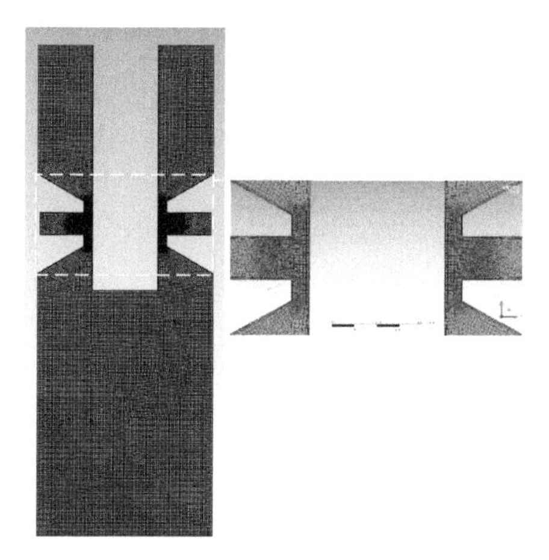

Figure 2. Grid model with a grid size of 0.012 and the flapper at the zero position when $L_1 = 0.02$.

Table 3. Grid parameters.

Grid Size/mm	Nodes	Elements	Element Quality	Skewness	Skewness (max)	Orthogonal Quality	Orthogonal Quality (min)
0.010	26,090	25,411	0.835	0.078	0.643	0.982	0.544
0.012	22,795	22,150	0.839	0.068	0.609	0.985	0.544
0.015	20,048	19,440	0.839	0.066	0.759	0.985	0.312
0.020	17,815	17,243	0.841	0.066	0.788	0.985	0.272

For a servo valve, flow and pressure are the ultimate indicators, so pressure and flow can be used to determine whether the grid size is appropriate. The outlet flow and average pressure on flapper at different grid sizes are shown in Figure 3.

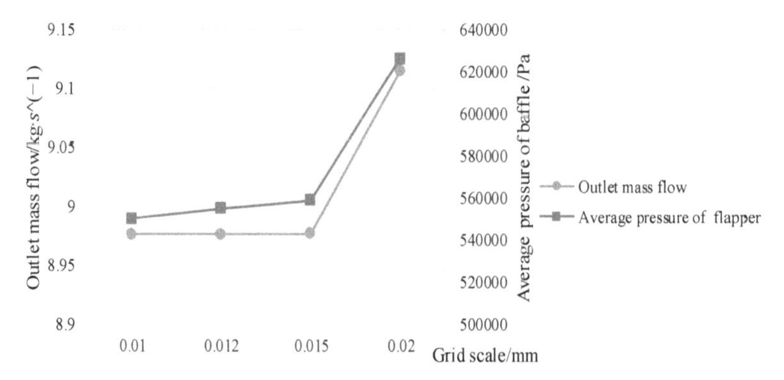

Figure 3. Flow and pressure at different grid sizes.

It can be seen that when the grid size is 0.010/0.012 and 0.015, the average pressure of the flapper is not much different, only within 0.005 MPa, while the mass flow of the outlet is almost the same, the difference being only 0.000048 kg/s. Therefore, when the grid size is in the range of 0.010–0.015, the grid quality has no effect on the simulation results. The median value of range of 0.012 is taken as the grid size when grid is independent.

After finding the optimal grid size, the channel model with different flappers are meshed. Since the distance between the flapper and the two nozzles is 0.2 mm when the flapper is located at the zero position, the number and quality parameters of the grid at different positions are obtained as shown in Table 4. The displacement of the flapper is in the right direction, i.e., the X direction shown in Figure 1 is in the positive direction.

Table 4. Grid model parameters of different spacing between nozzle and flapper.

Flapper Displacement x/mm	Nodes	Elements	Element Quality	Skewness	Skewness (max)	Orthogonal Quality
$x = 0.00$	22,795	22,150	0.839	0.068	0.609	0.985
$x = 0.05$	20,922	20,291	0.899	0.071	0.669	0.986
$x = 0.10$	22,953	22,292	0.836	0.069	0.747	0.983
$x = 0.15$	20,967	20,321	0.909	0.061	0.676	0.989

3. Results and Discussion

3.1. Flow Field Characteristics with Constant Velocity and Flapper Displacement

The nozzle area is selected as the observation object, and the observation points on the flow channel model are shown in Figure 4a, while Figure 4b shows that the wall Y plus number along the wall position. The results show that nearly all the wall Y plus number is less than 1 as suggested by Fluent user's guide, except for the leftmost and rightmost boundaries; however, the Y plus numbers are still less than 2. Analogously, the simulation obtained Y plus numbers in 5 m/s, 25 m/s, and 50 m/s conditions are also less than 1, mostly, including the wall areas near points 1–6, in which the research is much concerned.

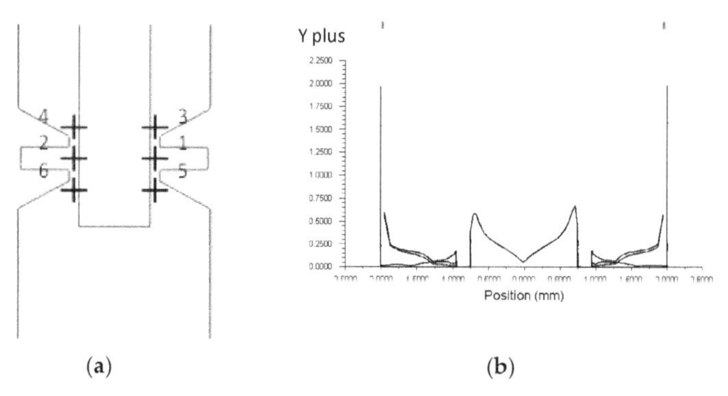

(a) (b)

Figure 4. Nozzle structure with observation points and the Y plus distribution. (**a**) Nozzle structure with observation points. (**b**) The Y plus distribution when velocity of inlet is 10 m/s.

The flow field model with an inlet velocity of 25 m/s and the flapper at the zero position is used for overall analysis. Since the flapper is in the zero position, the distance between the nozzles on both sides and the flapper is equal, making the flow resistance the same. According to the theoretical calculation, the pressure loss of the nozzles on both sides is the same, and the flow velocity through the nozzles should be equal. In addition, because the flow state of the oil injection to the two sides of the flapper is exactly the same, without considering the wear of the nozzles, the pressure fluctuation curves generated by the flappers on both sides should also be the same. By observing Figure 5, it is found that the oil velocity on both sides of the nozzle is the same, which is in accordance with the theory. However, in the wake region, the oil continuously makes periodic whirlpool motions, which makes the flapper subject to periodic pressure oscillations through the disturbance of the oil cycle, and in order to observe the pressure oscillation on the left and right sides of the flapper, point 1 and point 2 of Figure 4 are selected as the comparison, and the pressure situation at the two points is measured as shown in Figure 6a–d. Pressure oscillation phenomena have been observed at both points. It can be seen that there exists a fluid self-excitation oscillation between the nozzle and flapper. Moreover, the pressure oscillation curves at both points are approximately the same, the average pressure is about 0.57 MPa, and the oscillation amplitude is 0.12 MPa. Observing the power spectrum, it is found that the pressure fluctuation is mainly high frequency, and the peak frequency appears at 3000 Hz.

It is known from the foregoing that when the flapper of the jet flapper valve is located at the zero position, the average pressure, velocity and pressure fluctuation of the flapper on both sides of the flapper are almost the same, so in order to observe the pressure situation at different positions of the flapper, according to Figure 4, only points 1, 3, and 5 on the right side are selected as the observation objects. The pressure fluctuation curves are shown in Figure 6a,b,e–h respectively. From the graph analysis, it can be seen that the pressure pulsation at point 1 and point 3 oscillates around 0.57 MPa and has positive pressure, while point 5 oscillates near −0.045 MPa and the pressure is negative. This is because point 1 is located in the central region of the oil injection, so its pressure corresponds to the highest, while point 5 is located in the wake region, so the pressure is low; besides point 1, point 3, and point 5, the main frequency is low. The performance is not obvious because point 1 is the main area to be impacted by the jet, so the pressure oscillation here is also the most significant. At the same time, the pressure fluctuations in the three locations are mainly high frequency, generally over 3000 Hz, and the maximum power size appears at 3000 Hz, which indicates that the main frequency of pressure oscillation is 3000 Hz. In addition, the pressure amplitudes of the three locations are very stable in the time domain, and there is no sudden change point, which indicates that pressure oscillation exists. It is not caused by accidental factors, but by the high-speed impact of oil on the flapper.

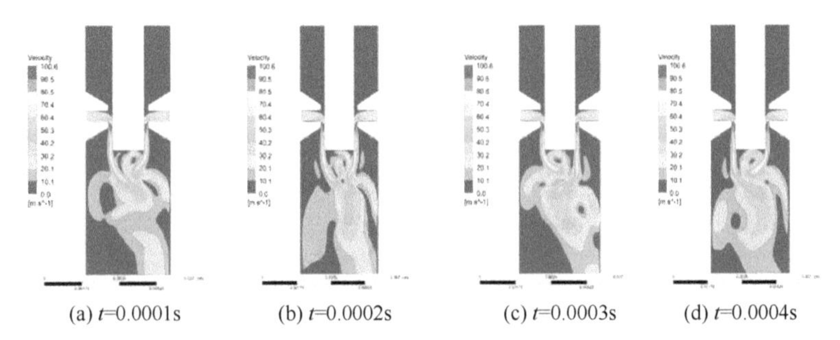

Figure 5. Velocity nephogram in one oscillating period.

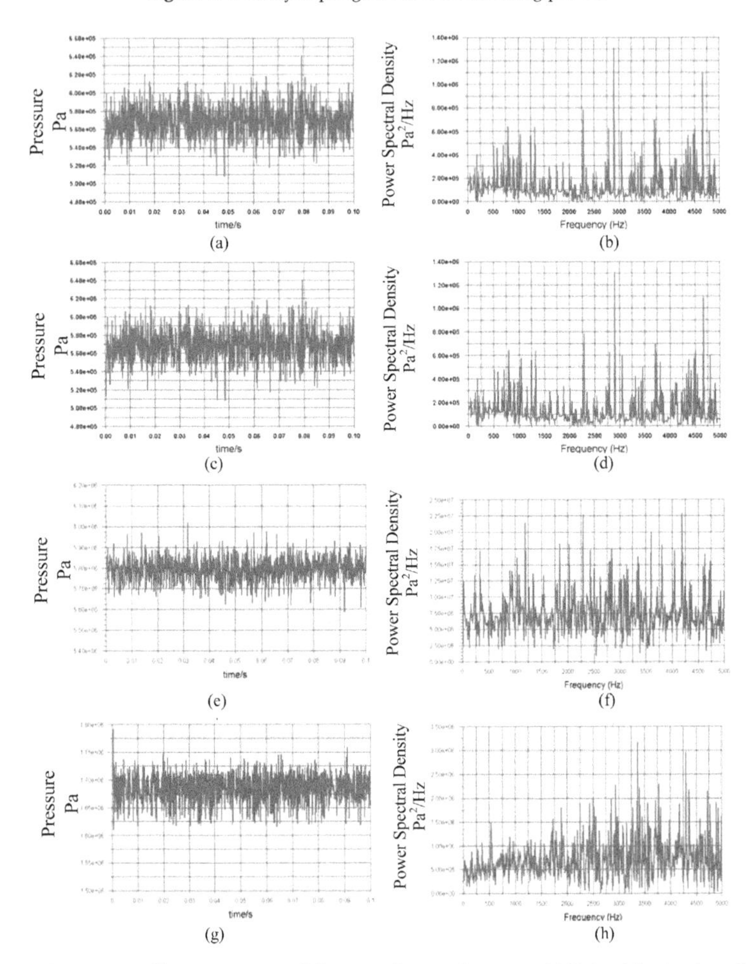

Figure 6. Pressure oscillation curves at different points on flappers. (**a**) Point 1 in the time domain. (**b**) Point 1 in the frequency domain. (**c**) Point 2 in the time domain. (**d**) Point 2 in the frequency domain. (**e**) Point 3 in the time domain. (**f**) Point 3 in the frequency domain. (**g**) Point 5 in the time domain. (**h**) Point 5 in the frequency domain.

The simulation time step is 0.0001 s, and each step is saved automatically. As a result, a flow cycle is obtained as shown in Figure 5. It can be found that the fluid disturbance is very intense when it enters the nozzle flapper area. In Figure 5a, after the fluid is injected into the nozzle flapper, it intersects at the bottom of the flapper, and the phenomenon of reflux is intensified in Figure 5b. Then a small eddy current is initially formed in Figure 5c, which is generated in Figure 5d and completes a cycle in 0.0003 s. Figure 6b shows that the main frequency is about 2850 Hz and the calculated period is about 0.00035 s, which are in good agreement with each other, indicating that the intercepted period is correct. This is because the oil continuously forms eddy currents on the flapper, producing repetitive pulses on the flapper, and thus forming a pressure oscillation on the flapper.

3.2. Variation of Main Frequency and Amplitude of Oscillation at Different Velocities

It can be seen from the above section that the pressure oscillation at point 1 and point 2 of the nozzles are the most obvious. Since the flapper is in the zero position and the pressure on both sides of the flapper is the same when the velocity of investigation is affected, point 1 is chosen as the research object. The results are shown in Figures 7–10, where the figures of 20 times of l *g* "power spectrum density" are displayed, which show the "−5/3" slope of the frequency, much accorded with the k41 theory spectra distribution characteristic, which insures the simualtion results are turbulent, not numerical errors. After comparison, it can be found that the pressure fluctuation increases with the increase of velocity, and there is a positive correlation between them. However, with the increase in velocity, the relative oscillation amplitude of the pressure does not follow the positive correlation, but decreases with the increase of velocity. Thus, the higher the velocity of the nozzle, the pressure oscillation phase oscillation will occur. The lower the amplitude, it can be inferred that when the velocity is high enough, the pressure oscillation of the flapper will gradually disappear, and its oscillation will only occur in a certain velocity range. In addition, combined with the power spectrum, it can be determined that the peak frequency of the pressure decreases gradually with the increase of velocity. This is because, with the increase of velocity, the stronger the ability of the oil to maintain its original motion state before injecting the nozzle into the flapper, the smaller the tendency to develop into turbulence, and the smaller the flow field dissipation, and the smaller the oscillation frequency generated along with it. The relationship between pressure and the main frequency and velocity is shown in Figure 11.

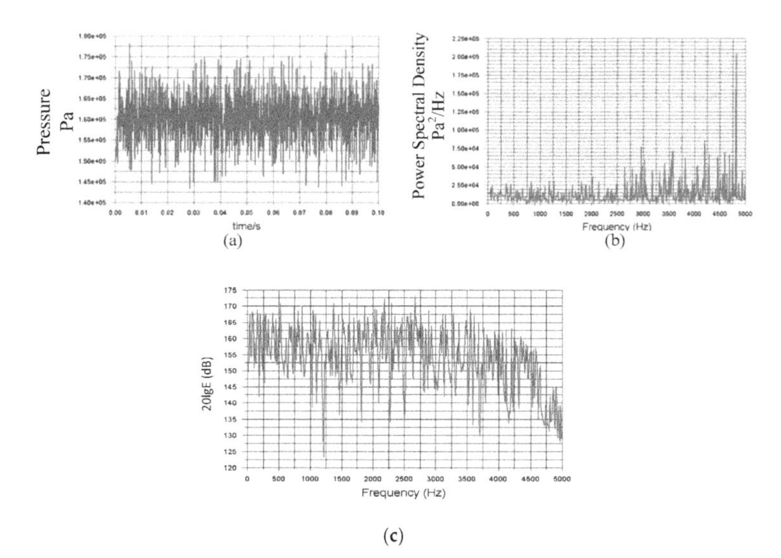

Figure 7. Pressure curves in time and frequency domains when the inlet velocity is 5 m/s. (**a**) Time domain. (**b**) Frequency domain. (**c**) 20logE frequency domain.

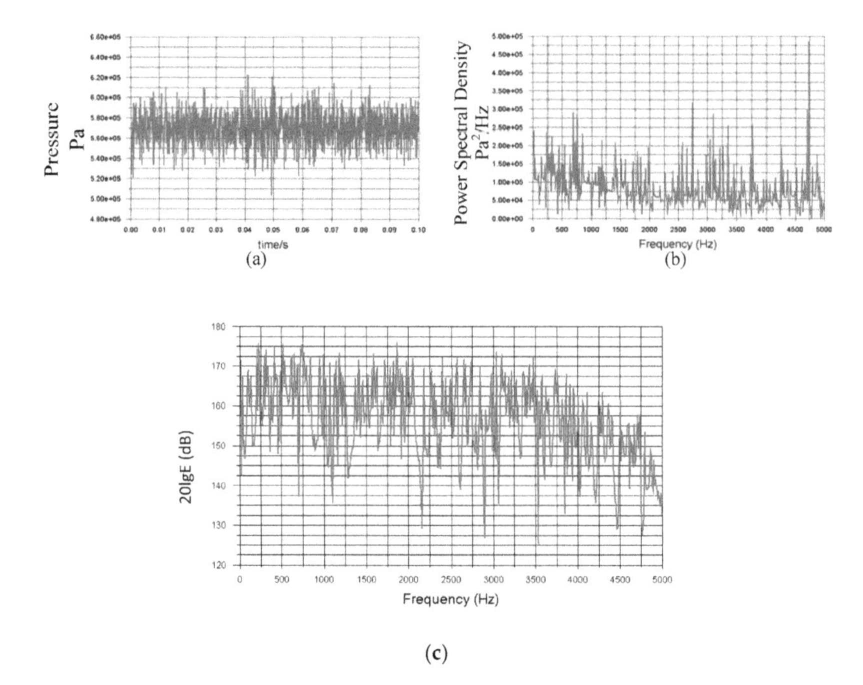

Figure 8. Pressure curves in time and frequency domains when the inlet velocity is 10 m/s. (**a**) Time domain. (**b**) Frequency domain. (**c**) 20logE frequency domain.

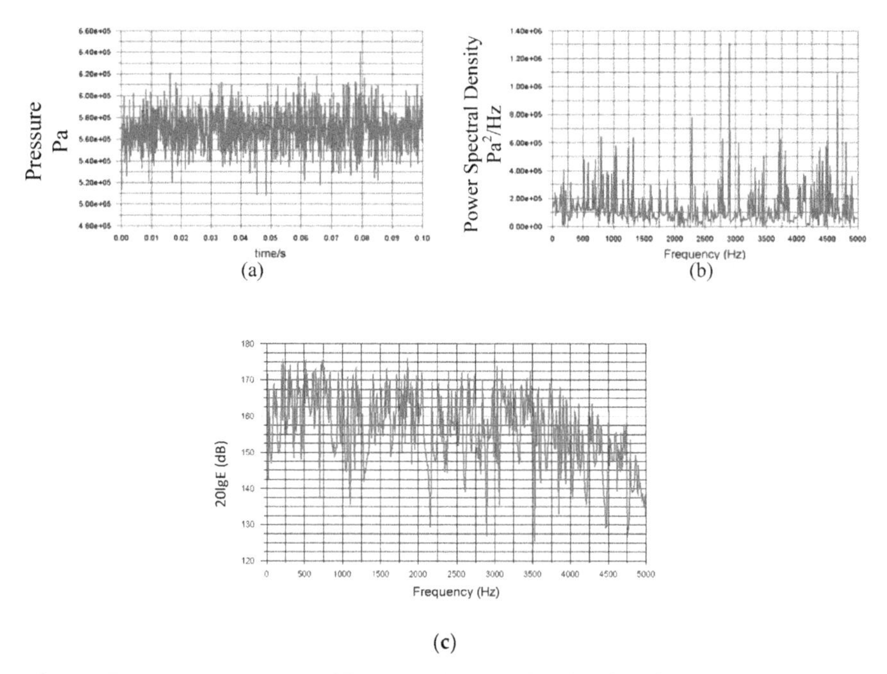

Figure 9. Pressure curves in time and frequency domains when the inlet velocity is 25 m/s. (**a**) Time domain. (**b**) Frequency domain. (**c**) 20logE frequency domain.

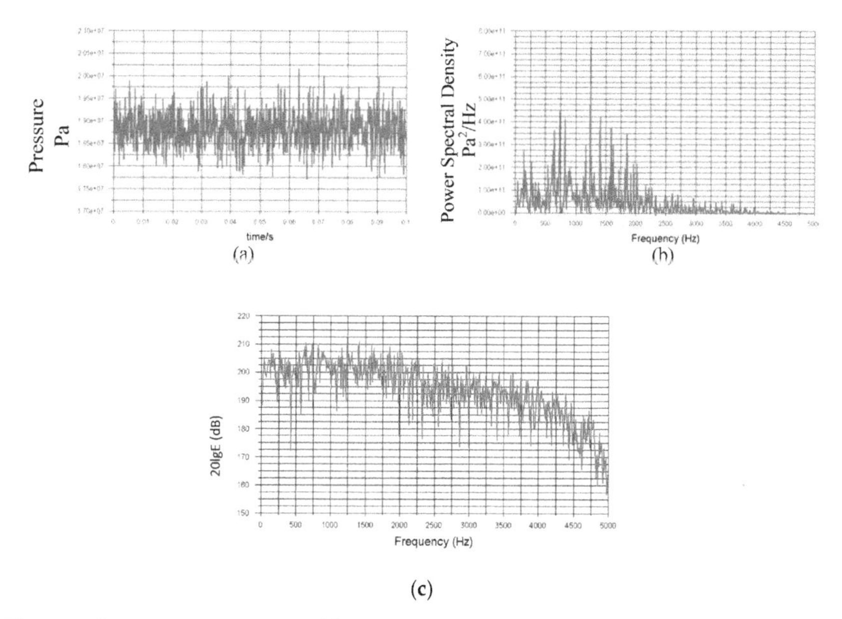

Figure 10. Pressure curves in time and frequency domains when the inlet velocity is 50 m/s. (**a**) Time domain. (**b**) Frequency domain. (**c**) 20logE frequency domain.

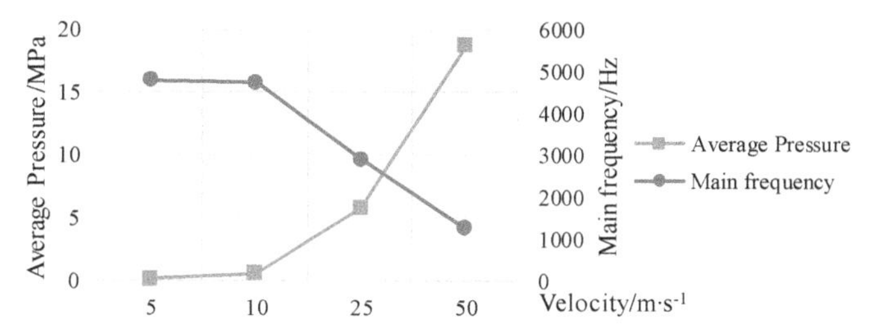

Figure 11. Pressure oscillation and main frequency of the flappers at different velocities.

3.3. *Variation of Main Frequency and Amplitude of Oscillation under Different Displacements*

Under different working conditions, the deflection angle of the flapper of the jet flapper servo valve is different, and the pressure on both sides of the flapper will also be different. Therefore, this section studies the pressure oscillation of the flappers under different displacements while keeping the inlet oil velocity unchanged. When the flapper is at the zero position, the distance between the flapper and the nozzles on both sides is 0.2 mm. The deflection displacement of each flapper is 0.05 mm, and it is in the right direction. Since the distance between the two sides of the flapper is different, the oil flow field on both sides is no longer the same, so in order to compare the pressure difference between the two sides, point 1 and point 2 in Figure 4 are used to observe. Since the flapper at zero point 1 is the same as point 2, only point 1 is selected, as shown in Figure 12, and its pressure oscillates near 0.57 MPa with the main pulsation frequency above 3000 Hz. In Figures 13–15, point 1 is closer to the nozzle than point 2, so its overall pressure relative to point 2 is higher, and as can be seen from the spectrum diagram, the pressure of point 1 is also higher. The frequency distribution of the force pulsation is relatively average, but there is a tendency to shift to low, while the pressure of point 2 is mainly high frequency, mainly over 2500 Hz. By analyzing the pressure oscillation curves, the relationship between the force acting on the flapper and the main frequency and the deflection distance of the flapper is obtained as shown in Figure 16. It can be seen that the smaller the distance between the nozzle and the flapper, the stronger the pressure fluctuation, but the relative pressure fluctuation amplitude is positively correlated with the distance as a whole; the main frequency is also related to the distance: the larger the distance is, the greater the main frequency of the pressure oscillation will be. This is because as the distance between the nozzle and the baffle decreases, the fluid resistance of the variable throttle hole increases, the kinetic energy of oil converts into pressure energy increases, and the more hydraulic pressure is generated on the baffle; furthermore, as the distance increases, the greater the turbulence degree of the oil flowing through the nozzle, the greater the pressure oscillation effect of the fluid. This is why the main frequency increases.

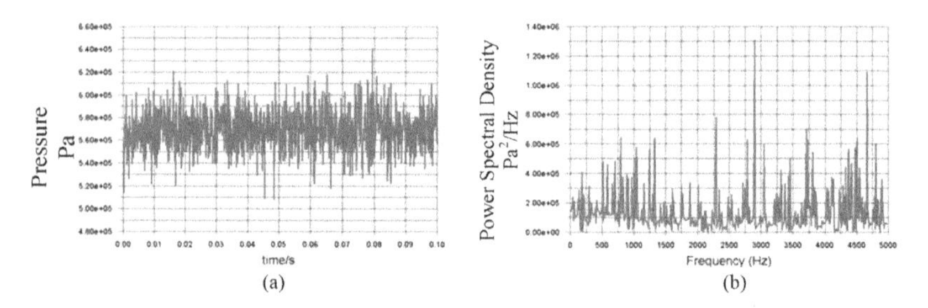

Figure 12. Pressure oscillation curve of the flapper when the displacement of the flapper is $x = 0.00$ mm. (**a**) Point 1 in the time domain. (**b**) Point 1 in the frequency domain.

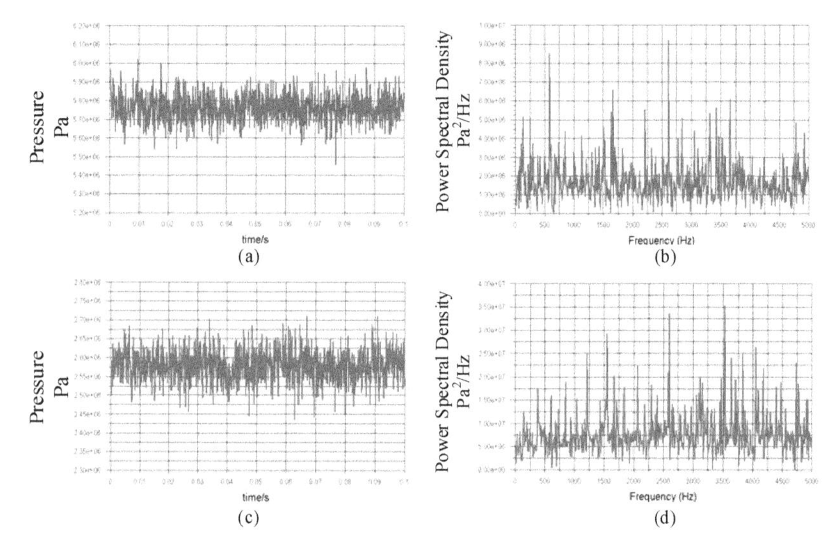

Figure 13. Pressure oscillation curve of the flapper when the displacement of the flapper is $x = 0.05$ mm. (**a**) Point 1 in the time domain. (**b**) Point 1 in the frequency domain. (**c**) Point 2 in the time domain. (**d**) Point 2 in the frequency domain.

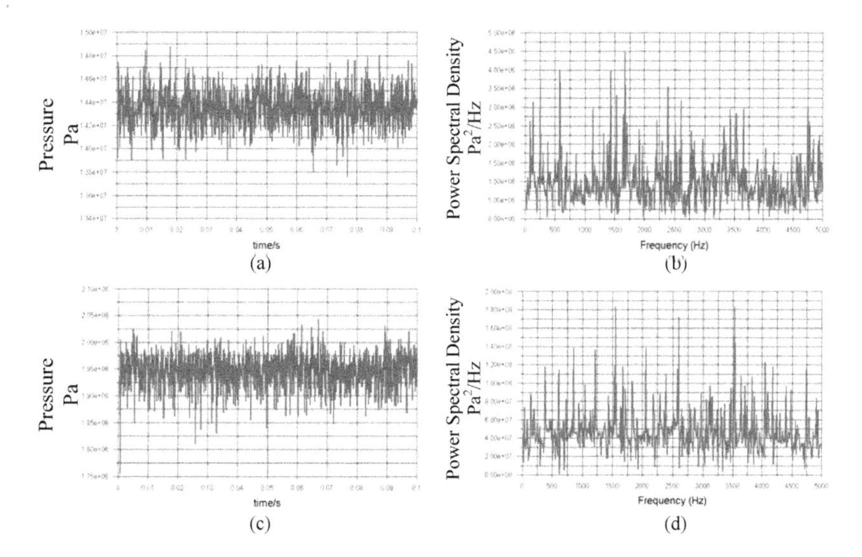

Figure 14. Pressure oscillation curve of the flapper when the displacement of the flapper is $x = 0.10$ mm. (**a**) Point 1 in the time domain. (**b**) Point 1 in the frequency domain. (**c**) Point 2 in the time domain. (**d**) Point 2 in the frequency domain.

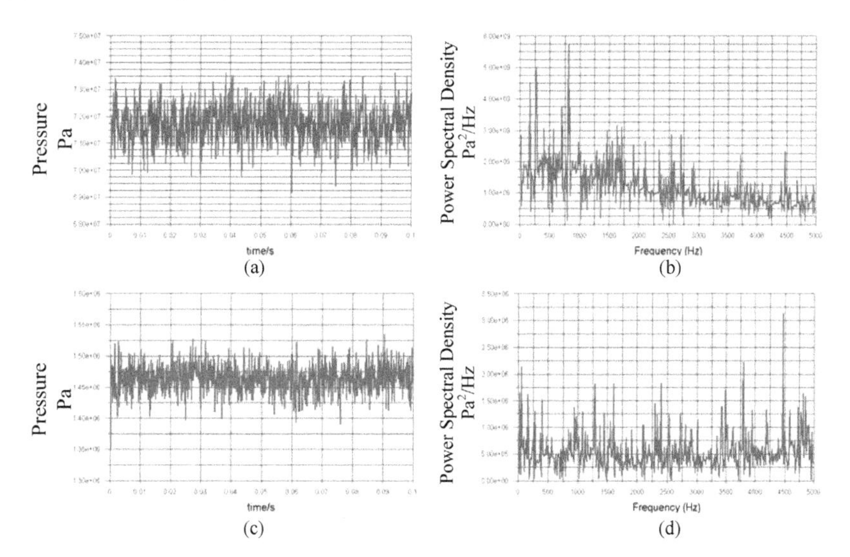

Figure 15. Pressure oscillation curve of the flapper when the displacement of the flapper is $x = 0.15$ mm. (**a**) Point 1 in the time domain. (**b**) Point 1 in the frequency domain. (**c**) Point 2 in the time domain. (**d**) Point 2 in the frequency domain.

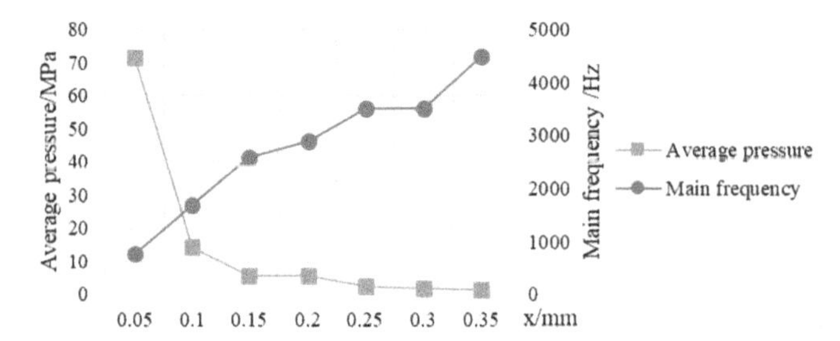

Figure 16. Average pressure and main frequency of the flapper at different actual distances between the nozzle and flapper.

3.4. Oscillation Characteristics Considering Motion Coupling of the Main Valve

In this section, the influence of the main valve on the flow field will be considered to establish a more realistic flow field in the servo valve. Similarly, the two-dimensional model is still used in this chapter. The flow field of the whole valve is shown in Figure 17.

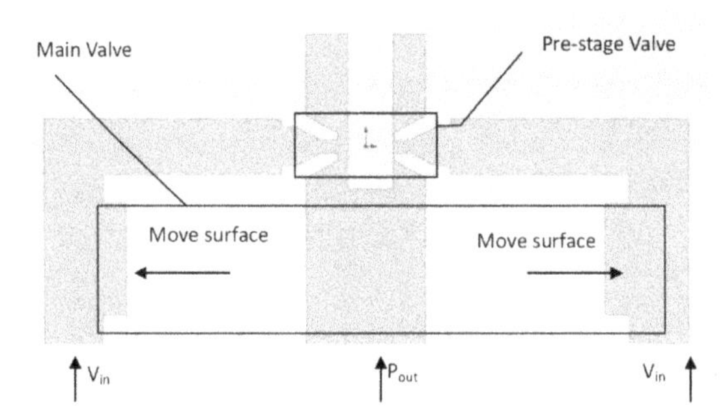

Figure 17. Watershed model of the servo valve considering the effect of the main valve.

3.4.1. UDF Implementation Logic

During the deflection of the flapper, the displacement of the main valve core will occur with the offset of the flapper. The interface between the valve core and the oil is the moving surface, and the moving distance should be linear with the displacement of the flapper. The moving surface is shown in Figure 18. In order to realize the above moving process, the mathematical model of the main spool motion should be written first.

The main valve mainly plays the role of a power amplifier. The main working process is to move to the side with weak pressure under the pressure difference between the two sides of the pilot valve, so as to realize the opening and closing of the working oil port and realize the function of promoting load movement. In the whole movement process, the main valve is an inlet throttle, which is mainly affected by the steady-state hydrodynamic force, transient hydrodynamic force of opening and closing the working oil outlet, and oil resistance and pressure on both sides of the valve core. Therefore, there is a feedback rod, so there is a feedback force acted by the feedback rod. Its formula is as follows:

$$m\frac{d^2x(t)}{dt^2} + C_f\frac{dx(t)}{dt} + F_{Rt} - F_{Rs} + F_f = \mathrm{A}(p_1 - p_2) = \mathrm{A}\Delta p(t) \quad (1)$$

where F_{Rt} is a transient hydrodynamic force, F_{Rs} is a steady hydrodynamic force, and F_f is a feedback force exerted by the feedback rod. Its expressions are shown below and Table 5 shows the meaning and value of each parameter.

$$F_{Rt} = C_{q1} w L_3 \sqrt{2\rho p_v} \frac{dx}{dt} \tag{2}$$

$$F_{Rs} = -2C_{q1} C_v w p_v x \cos\theta \tag{3}$$

$$F_f = (r+b) k_f x \tag{4}$$

Table 5. Parameters for mathematical modeling of servo valves.

Parameter	Value	Parameter	Value
Spool mass m/kg	0.01	Import and Export Center Distance of Load L_3/m	1.34×10^{-2}
Spool displacement x/m		Pressure difference between inlet and outlet of main valve p_v/MPa	5
Coefficient of viscous resistance C_f/kg·s^{-1}	17.4	Oil Flow Velocity Coefficient of Main Valve C_v	0.99
Spool end area	3.14×10^{-6}	Directional angle of jet-flow θ/°	69
Pressure at both ends of spool p_1, p_2		Distance between nozzle hole axis and armature Center r/m	2.0×10^{-3}
Density of hydraulic oil ρ/kg·m^{-3}	889	Axis pitch from core center to nozzle hole b/m	3.0×10^{-3}
The Flow coefficient of the main valve C_{q1}	0.8	Stiffness of Feedback Bar k_f/N·m^{-1}	3.4×10^3
Area gradient of the throttle hole w/m^2	1.256×10^{-2}	Inlet flow Q/m^3·s^{-1}	8.3×10^{-5}

By substituting the above formulas and the values of each parameter into Equation (1), the following results are obtained:

$$\frac{d^2x(t)}{dt^2} + 3010\frac{dx(t)}{dt} + 3,566,600x(t) = 3.14 \times 10^{-4} \Delta p(t) \tag{5}$$

Let $u = dx/dt$, after Laplacian transformation, obtaining:

$$U(s)s + 3010U(s) + 3,566,600\frac{U(s)}{s} = 3.14 \times 10^{-4} \Delta P(s) \tag{6}$$

$$\frac{U(s)}{\Delta P(s)} = \frac{3.14 \times 10^{-4} s}{s^2 + 3010s + 35,666,000} \tag{7}$$

So the velocity curve of the spool in the time domain is:

$$u(t) = 3.14 \times 10^{-4} e^{-1505t} \sin(5779t + 1.32) \Delta p(t) \tag{8}$$

Then, the UDF logic block diagram is written as follows in Figure 18:

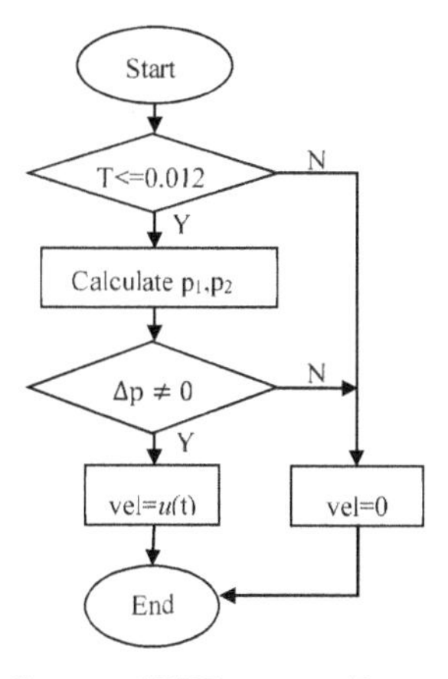

Figure 18. Execution logic block diagram of UDF program (the symbol T means time, and the value 0.012 is the time of the moving grid motion).

3.4.2. Dynamic Grid Design and Simplification for Flapper Deflection

In Section 2.3 we found and validated the reliability of the optimal grid size. Although the flow field of the main valve has been added to this chapter, the final goal of this chapter is to divide the grid model of the whole valve on the basis of the optimal grid size found in the above chapter because the force of the flapper in the nozzle pilot valve area is still the ultimate goal. It has 80,431 grid nodes and 78,741 grid elements. Its element quality, skewness and orthogonal quality are 0.927, 0.032 and 0.994, respectively, which meet the requirements of grid generation. In addition, in order to realize the motion characteristics of the main valve, a moving grid is needed. In this paper, Fluent is used to provide three dynamic grid models: smoothing, layering and regridding. The smoothing model chooses the spring method to smooth the transition; the layering model chooses the height base to set the split factor and collapse factor as default values; and the regridding model chooses the local cell regridding mode, with the maximum surface skewness set to 0.7, and the rest to remain as default.

3.4.3. Principal Frequency and Amplitude Change Rule in the Coupling of Main Valve

(1) Variation of flow field of servo valve with the change of inlet velocity

From Figures 19–22, it can be seen that the pressure trend of the flapper is the same as that of the previous one. Figure 23 shows that, with the increase of velocity, the average pressure of the flapper increases and the main frequency decreases, which is the same with the previous one. Unlike the previous one, the relative pressure amplitude does not increase linearly with the speed, but decreases linearly. Moreover, the comparison shows that the peak frequency is also reduced by adding the motion characteristics of the main valve. This is mainly due to the large inertia of the main valve, resulting in its low response frequency. Due to the existence of the feedback rod, this effect is reflected in the flapper, so that the flapper pressure frequency generated by fluid impact is reduced.

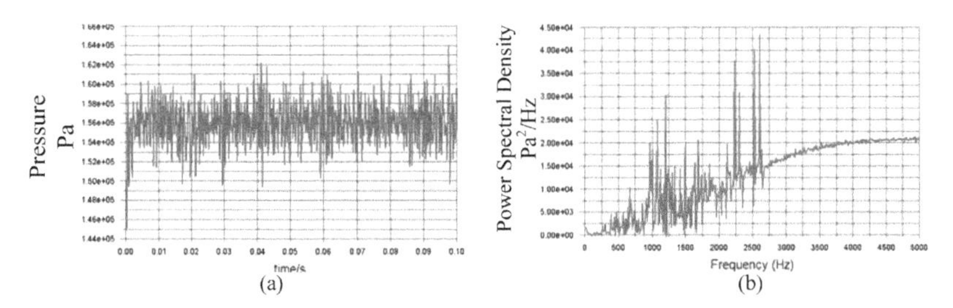

Figure 19. Pressure oscillation curve of the flapper considering the impression of the main valve and velocity of 5 m/s. (**a**) Time domain. (**b**) Frequency domain.

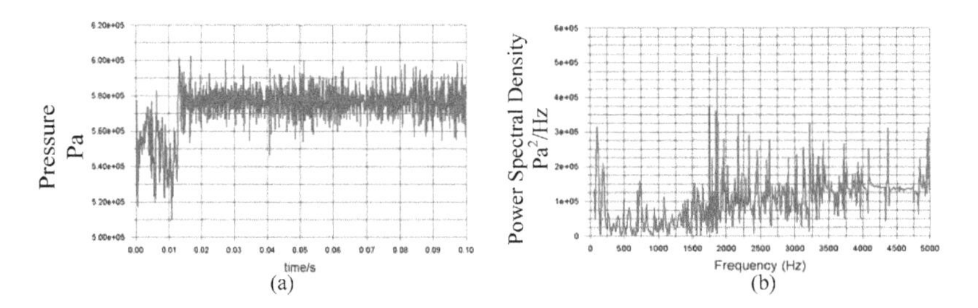

Figure 20. Pressure oscillation curve of the flapper considering the impression of the main valve and velocity of 10 m/s. (**a**) Time domain. (**b**) Frequency domain.

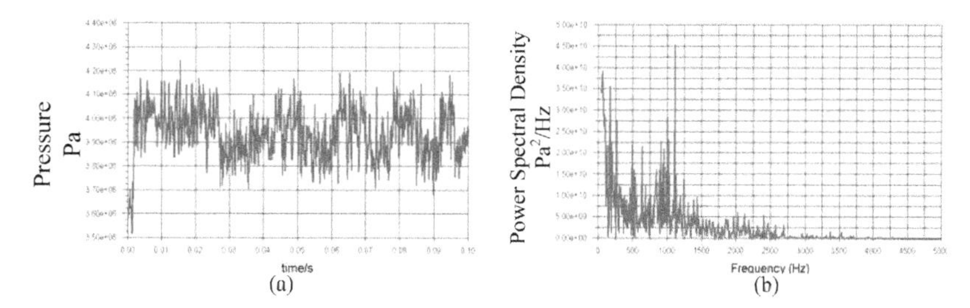

Figure 21. Pressure oscillation curve of the flapper considering the impression of the main valve and velocity of 25 m/s. (**a**) Time domain. (**b**) Frequency domain.

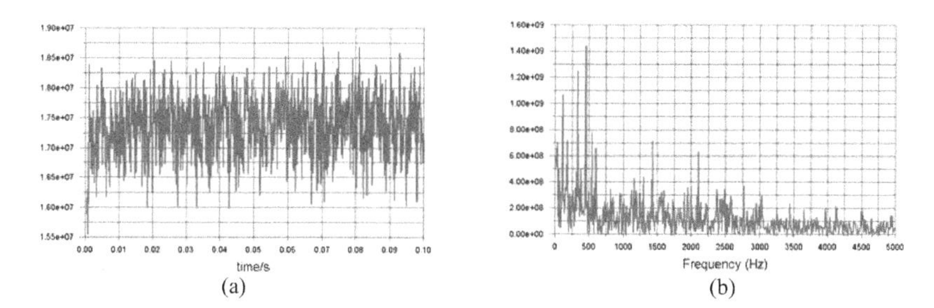

Figure 22. Pressure oscillation curve of the flapper considering the impression of the main valve and velocity of 50 m/s. (**a**) Time domain. (**b**) Frequency domain.

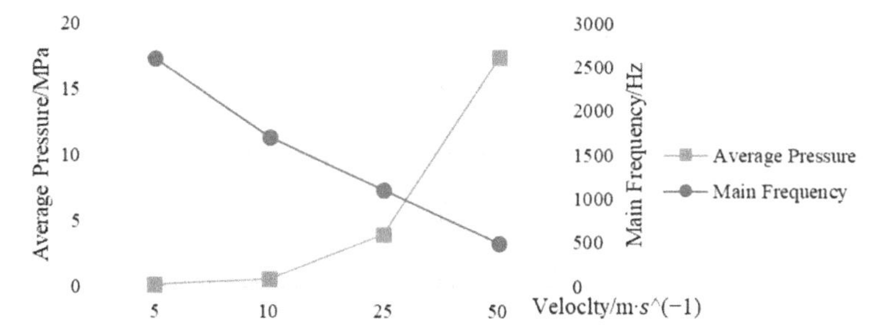

Figure 23. Pressure and main frequency of the servo valve flapper at different speeds.

(2) Variation of flow field of the servo valve with the change of the flapper displacement

Points 1 and 2 are taken as monitoring points. When the flapper deflects, the pressure curve of the pilot valve under the influence of the main valve is obtained, as shown in Figures 24–27. The pressure under displacement is summarized in Figure 28. The analysis shows that when the distance between the flapper and the nozzle increases, the pressure on the flapper decreases, the relative amplitude increases, and the main frequency increases. The overall trend is consistent with the previous one. It can be seen that the motion characteristics of the main valve do not affect the deflection characteristics of the flapper. At the same time, it can be seen that, whether the main valve exists or not, the pressure level of the pilot valve flapper is not affected at different distances, but the frequency of pressure oscillation is significantly reduced.

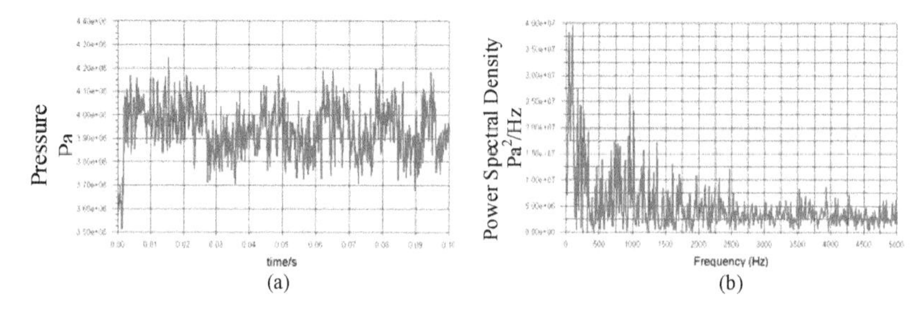

Figure 24. Pressure oscillation curve of the flapper under the influence of the main valve when the displacement of the flapper is $x = 0.00$ mm. (**a**) Point 1 in the time domain. (**b**) Point 1 in the frequency domain.

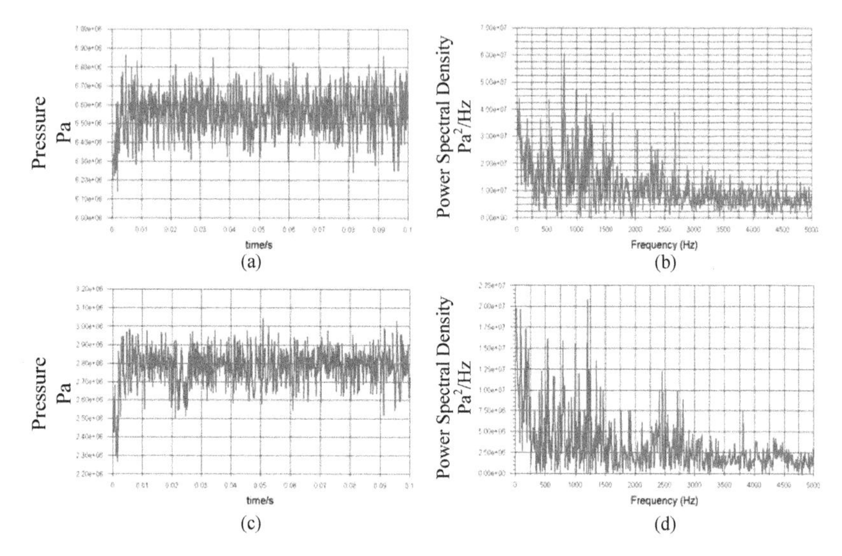

Figure 25. Pressure oscillation curve of the flapper under the influence of the main valve when the displacement of the flapper is $x = 0.05$ mm. (**a**) Point 1 in the time domain. (**b**) Point 1 in the frequency domain. (**c**) Point 2 in the time domain. (**d**) Point 2 in the frequency domain.

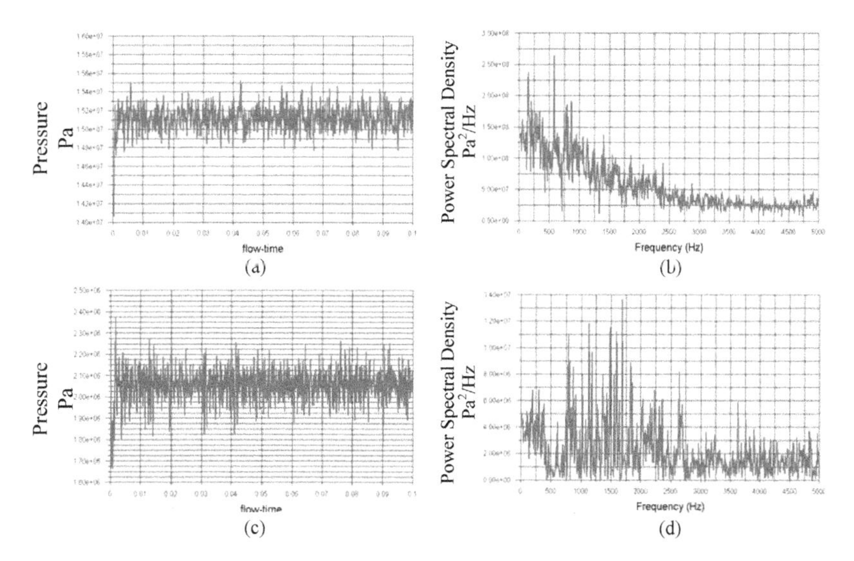

Figure 26. Pressure oscillation curve of the flapper under the influence of the main valve when the displacement of the flapper is $x = 0.10$ mm. (**a**) Point 1 in the time domain. (**b**) Point 1 in the frequency domain. (**c**) Point 2 in the time domain. (**d**) Point 2 in the frequency domain.

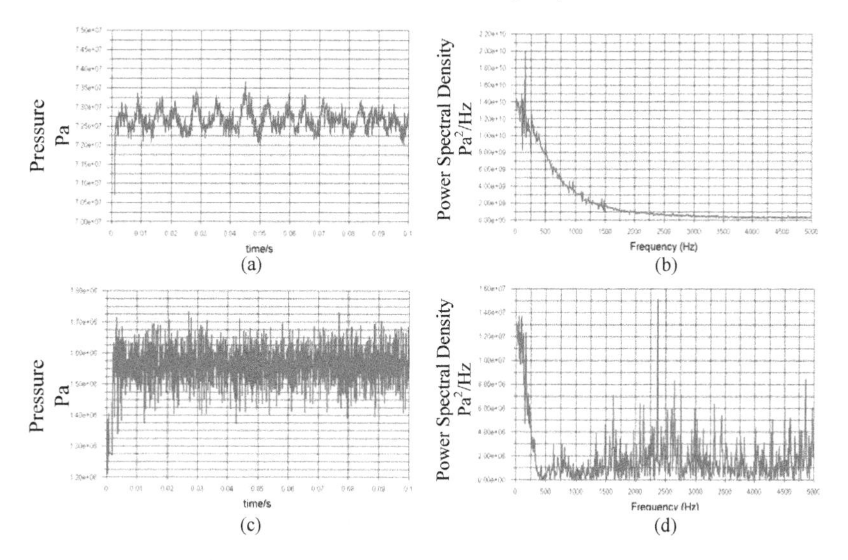

Figure 27. Pressure oscillation curve of the flapper under the influence of the main valve when the displacement of the flapper is $x = 0.15$ mm. (**a**) Point 1 in the time domain. (**b**) Point 1 in the frequency domain. (**c**) Point 2 in the time domain. (**d**) Point 2 in the frequency domain.

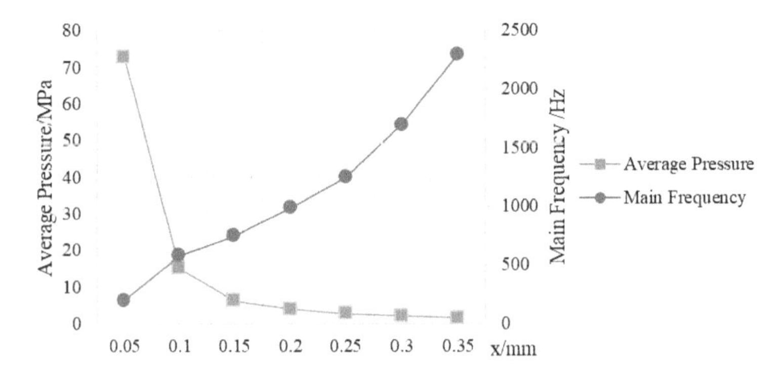

Figure 28. Pressure and main frequency of the flapper when the actual distance between the flapper and nozzle changes.

4. Conclusions

In this paper, the CFD approach with LES turbulent modeling is utilized for the double flapper nozzle servo valve oscillating flow numerical analysis. The vortex street flow phenomenon could be observed when the flow passes the nozzle flapper channel, the vortex alternating in both sides producing the periodical flow oscillation. With detailed discussion, it could be determined that:

(1) Without the influence of the main valve, the greater the inlet velocity is, the greater the average pressure on the flapper is, and the lower the main frequency of pressure oscillation is.

(2) Without the influence of the main valve, the larger the actual distance between the nozzle and the flapper is, the smaller the average pressure on the flapper is, and the larger the main frequency of pressure oscillation is.

(3) When considering the influence of the main valve, the change trend of hydraulic power and the main frequency of the servo valve flapper are rarely changed, but the main frequency of the pressure oscillation is significantly lower than that without considering the main valve.

Author Contributions: Investigation S.L. and L.L; Simulation and Analysis S.L.; Methodology L.L.; Software S.L.; Writing and Editing S.L., K.Z. and L.L.; Validation L.L. and K.Z.

Funding: The authors are grateful to the National Natural Science Foundation of China (no. 51605333 and no. 51805317) for financial support.

Conflicts of Interest: The authors declare no conflict of interest.

Nomenclature

d	Internal diameter	D	External diameter
L_1	Spacing	α	Angle
L	Length	D_1	Diameter
L_2	Thickness	L_3	Import and Export Center Distance of Load
m	Spool mass	p_v	Pressure difference between inlet and outlet of main valve
x	Spool displacement	C_v	Oil Flow Velocity Coefficient of Main Valve
C_f	Coefficient of viscous resistance	θ	Directional angle of jet-flow
A	Spool end area	r	Distance between nozzle hole axis and armature Center
p_1, p_2	Pressure at both ends of spool	b	Axis pitch from core center to nozzle hole
ρ	Density of hydraulic oil	K_f	Stiffness of Feedback Bar
C_{q1}	The Flow Coefficient of Main Valve	Q	Inlet flow
w	Area Gradient of Throttle Hole		

References

1. Jones, J.C. *Developments in Design of Electrohydraulic Control Valves*; Moog Technical Paper; Moog: Melbourne, VIC, Australia, 1997.
2. Moog, J.W.C. Electrohydraulic Servo Mechanism. U.S. Patent 2,625,136, 13 January 1953.
3. Howard, C.T. Flow Control Servo Valve. U.S. Patent 2,790,427, 30 April 1957.
4. Atchley, R.D. Valve. U.S. Patent 3,017,864, 23 January 1962.
5. Vanderlaan, R.D.; Meulendyk, J.W. Direct Drive Valve-Ball Drive Mechanism. U.S. Patent 4,672,992, 16 June 1987.
6. Laux, K. Motor-to-Spool Coupling for Rotary-to-Linear Direct Drive Valve. U.S. Patent 5,263,680, 23 November 1993.
7. Samakwong, T.; Assawinchaichote, W. PID controller design for electro-hydraulic servo valve system with genetic algorithm. *Procedia Comput. Sci.* **2016**, *86*, 91–94. [CrossRef]
8. Brito, A.G.; Leite, F.W.C.; Hemerly, E.M. Identification of a Hammerstein model for an aerospace electrohydraulic servovalve. *IFAC Proc. Vol.* **2013**, *46*, 459–463. [CrossRef]
9. Khodaee, Z.; Zareinejad, M.; Ghidary, S.S. Modeling of a two-stage flapper-nozzle electrohydraulic servo valve exposed to acceleration. In Proceedings of the 2014 Second RSI/ISM International Conference on Robotics and Mechatronics (ICRoM) IEEE, Tehran, Iran, 15–17 October 2014; pp. 268–273.
10. Ye, S.; Zhang, J.; Xu, B.; Zhu, S.; Xiang, J.; Tang, H. Theoretical investigation of the contributions of the excitation forces to the vibration of an axial piston pump. *Mech. Syst. Signal Process.* **2019**, *129*, 201–217. [CrossRef]
11. Li, S.; Aung, N.Z.; Zhang, S.; Cao, J.; Xue, X. Experimental and numerical investigation of cavitation phenomenon in flapper–nozzle pilot stage of an electrohydraulic servo-valve. *Comput. Fluids* **2013**, *88*, 590–598. [CrossRef]
12. Chen, M.; Aung, N.Z.; Li, S.; Zou, C. Effect of oil viscosity on self-excited noise production inside the pilot stage of a two-stage electrohydraulic servovalve. *J. Fluids Eng.* **2019**, *141*, 011106. [CrossRef]
13. Qian, J.Y.; Chen, M.R.; Liu, X.L.; Jin, Z.J. A numerical investigation of the flow of nanofluids through a micro Tesla valve. *J. Zhejiang Univ. Sci. A* **2019**, *20*, 50–60. [CrossRef]
14. Chao, Q.; Zhang, J.; Xu, B.; Huang, H.; Zhai, J. Effects of inclined cylinder ports on gaseous cavitation of high-speed electro-hydrostatic actuator pumps: A numerical study. *Eng. Appl. Comput. Fluid Mech.* **2019**, *13*, 245–253. [CrossRef]
15. Qian, J.Y.; Gao, Z.X.; Liu, B.Z.; Jin, Z.J. Parametric study on fluid dynamics of pilot-control angle globe valve. *J. Fluids Eng.* **2018**, *140*, 111103. [CrossRef]
16. Zhang, J.H.; Wang, D.; Xu, B.; Gan, M.Y.; Pan, M.; Yang, H.Y. Experimental and numerical investigation of flow forces in a seat valve using a damping sleeve with orifices. *J. Zhejiang Univ. Sci. A* **2018**, *19*, 417–430. [CrossRef]
17. Yin, Y.B. *Theory and Application of Advanced Hydraulic Component*; Shanghai Scientific & Technical Publishers: Shanghai, China, 2017.
18. Wang, F.J. *Computational Fluid Dynamics Analysis*; Tsinghua University Press: Beijing, China, 2004.

Article

Bifurcation Characteristic Research on the Load Vertical Vibration of a Hydraulic Automatic Gauge Control System

Yong Zhu [1], Shengnan Tang [1,2,*], Chuan Wang [1,3,*], Wanlu Jiang [4], Xiaoming Yuan [2,4] and Yafei Lei [4,*]

[1] National Research Center of Pumps, Jiangsu University, Zhenjiang 212013, China; zhuyong@ujs.edu.cn
[2] State Key Laboratory of Fluid Power and Mechatronic Systems, Zhejiang University, Hangzhou 310027, China; yuanxiaoming@ysu.edu.cn
[3] School of Hydraulic Energy and Power Engineering, Yangzhou University, Yangzhou 225002, China
[4] Hebei Provincial Key Laboratory of Heavy Machinery Fluid Power Transmission and Control, Yanshan University, Qinhuangdao 066004, China; wljiang@ysu.edu.cn
* Correspondence: 2111811013@stmail.ujs.edu.cn (S.T.); wangchuan@ujs.edu.cn (C.W.); yafeilei@stmail.ysu.edu.cn (Y.L.); Tel.: +86-0511-88799918 (S.T. & C.W.); +86-0335-8061729 (Y.L.)

Received: 28 August 2019; Accepted: 2 October 2019; Published: 10 October 2019

Abstract: As the core control system of a rolling mill, the hydraulic automatic gauge control (HAGC) system is key to ensuring a rolling process with high speed, high precision and high reliability. However, a HAGC system is typically a mechanical-electric-hydraulic coupling system with nonlinear characteristics. The vertical vibration of the load easily occurs during the working process, which seriously affects the stability of the system and the causes are difficult to determine. In this work, the theory and method of nonlinear dynamics were employed. The load vertical vibration model of the HAGC system was established. Then, the multi-scale method was utilized to solve the obtained model, and the singularity theory was further applied to derive the transition set. Moreover, the research object of this article focused on some nonlinear factors such as excitation force, elastic force and damping force. The effects of the above feature parameters on bifurcation behavior were emphatically explored. The bifurcation characteristic of the load vertical vibration of the HAGC system was revealed. The research results indicate that the bifurcation curves in each sub-region, divided by the transition set, possess their own topological structure. The changes of the feature parameters, such as the nonlinear stiffness coefficient, liquid column height, nonlinear damping coefficient, and external excitation have an influence on the vibration amplitude of the HAGC system. By reasonably adjusting the nonlinear stiffness coefficient to effectively avoid the resonance region, the stability of the system will be facilitated. Furthermore, this is conducive to the system's stability as it properly controls the size of the liquid column height of the hydraulic cylinder. The appropriate nonlinear damping coefficient can decrease the unstable area, which is beneficial to the stability of the system. However, large external excitation is not conducive to the stability of the system.

Keywords: flow control; vertical vibration; bifurcation characteristic; nonlinear dynamics; hydraulic automatic gauge control system; rolling mill

---

1. Introduction

In the metallurgical industry, the rolling mill represents a core piece of equipment. However, it has been found that there are always some parameter changes or disturbance factors during the rolling process, which may induce vibration in the rolling mill [1,2]. This may even cause the load roll system to produce vertical vibration with large amplitude, leading to production accidents such as steel-heaping, strip breakage, component damage and so on [3–5]. The hydraulic automatic gauge control (HAGC) system is considered the core control system of a rolling mill. The stability of the HAGC system is crucial to ensuring a rolling process with high speed, high precision and high reliability.

For a long time, many scholars and research institutions have explored the rolling mill's vibration characteristics, and rich results have been demonstrated. Zhong et al. [6] explored the mechanism of interfacial coupling and electromechanical coupling in the vibration of rolling mills. It was found that the various vibration phenomena of a high-speed rolling mill were related to various interactions of the rolling process. Wang et al. [7] conducted a series of studies on the control of plate thickness and plate shape, which laid a theoretical foundation for improving the control accuracy of the HAGC system. Chen et al. [8] studied the nonlinear parametric vibration of a four-roll cold strip mill and explored the influence of the deformation resistance of rolled piece on parametric resonance. The effects of rolling speed and tension on the nonlinear vibration of the rolling mill were investigated by Sun et al. [9], and the influence of external excitation frequency on system stability was analyzed. Liu et al. [10] researched the vertical, nonlinear parametric vibration characteristics of the mill roll system, and discussed the influence of parameters such as damping, stiffness and external excitation on system stability. In consideration of the change of parameters, Liu et al. [11] analyzed the partial bifurcation phenomenon of the load roll system, finding that both internal resonance and main resonance have an amplitude jump phenomenon. Yan et al. [12] studied the coupling vibration mechanism and vibration suppression method of the rolling mill. The characteristics and laws of vibration of the hot rolling mill were studied by the analysis of the measured signals, both in the time and frequency domain [13,14]. Yang et al. [15] researched the nonlinear modeling and stability of the vertical vibration of a cold rolling mill. The singular value theory was used to analyze the stability of the system. Moreover, the effects of parameters such as damping and stiffness on the vibration characteristics of the system were researched. In terms of periodic excitation, Bi et al. [16,17] studied the bifurcation mechanism of oscillations of the dynamic system. Wang and Qian et al. [18,19] investigated the effects of important components such as pump [20–23] and valve [24,25] on the vibration characteristics. Bai et al. [26–28] researched the vibration of the pump under varied conditions. Zhang et al. [29–31] studied the influence of excitation forces on the vibration of the pump and the measure of noise reduction. Niziol et al. [32] investigated the effect of rolling speed on the vibration of a mill roll system, and analyzed the steady state domain of the vibration of the roller system under different rolling speeds. Heidari et al. [33,34] studied the influence of friction damping and lubrication state on the flutter of cold strip mill, and analyzed the effects of some main parameters of lubricant on the critical vibration velocity. In order to study the self-excited vibration and nonlinear paramagnetic vibration of the rolling process, a nonlinear mathematical model for the vibration of the rolling mill was established by Drzymala et al. [35].

The above research results have provided a theoretical guidance for the vibration mechanism analysis and vibration suppression of the rolling mill system. Nevertheless, based on the existing research results, some problems have been discovered. In most of the research results, the influence of the HAGC system is often overlooked when the vertical vibration of the rolling mill is analyzed [36,37]. In essence, the HAGC system is a nonlinear closed-loop system, which has many factors affecting its stability. If it is unstable, it will certainly have an influence on the vibration characteristics of the load roll system [38,39]. In addition, with the development of nonlinear science, it is of important theoretical significance to reveal the essence and mechanism of the system dynamic process by applying the nonlinear dynamics theory, which simultaneously presents broad application prospects [40–42]. This area is worthy of further study when the theory and method of nonlinear dynamics are utilized to explore the vibration mechanism of the HAGC system, and then to obtain the influence law and incentive of nonlinear factors on the vibration characteristics of the system.

In this paper, based on the theory and method of nonlinear dynamics, the influence of some nonlinear parameters of the HAGC system on load vertical vibration will be researched. Moreover, the bifurcation characteristics of load vertical vibration will be emphatically explored to reveal the instability mechanism of the HAGC system. In Section 2, the vertical vibration model of the load of the

HAGC system is established. In Section 3, the bifurcation characteristic of load vertical vibration is analyzed. In Section 4, the effects of some feature parameters on bifurcation behavior are thoroughly explored. In Section 5, some conclusions are provided.

2. Vertical Vibration Model of Load

The external load of the HAGC system consists of several rollers, which present the symmetrical structure [43]. The basic load structure of the commonly used, four-high mill's HAGC system is shown in Figure 1.

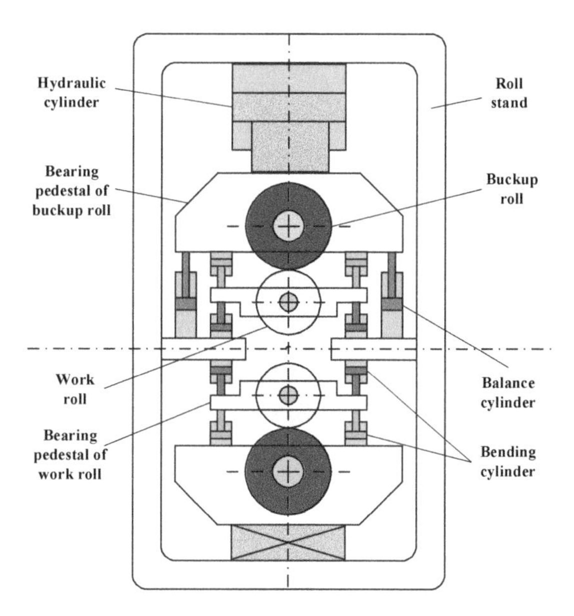

Figure 1. Structural schematic diagram of load.

In addition, the structure and vibration of load possess symmetry [44]. Moreover, some nonlinear factors such as nonlinear stiffness, nonlinear damping and nonlinear excitation are also considered. Then, a nonlinear load vertical vibration model of the HAGC system is built, as displayed in Figure 2.

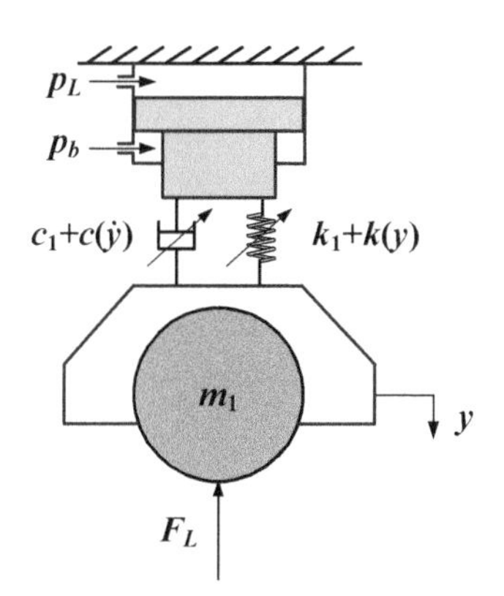

Figure 2. Vertical vibration model of load.

According to Newton's second law, the nonlinear load vertical vibration equation of a HAGC system can be represented as:

$$m_1\ddot{y} + c_1\dot{y} + k_1 y + \alpha F_k(y) + \beta F_c(\dot{y}) = \Delta F \quad (1)$$

$$\Delta F = p_L A_p - p_b A_b - F_L \quad (2)$$

where, m_1 is equivalent mass; c_1 and k_1 are linear damping coefficient and linear stiffness coefficient, respectively; y is vibration displacement. $F_k(y)$ and $F_c(\dot{y})$ are nonlinear elastic force and nonlinear friction of hydraulic cylinder, respectively; α and β are the action coefficient of nonlinear stiffness and nonlinear damping, respectively; A_p and A_b are the effective working area of rodless cavity and rod cavity, respectively; p_L and p_b are the working pressure of rodless cavity and rod cavity, respectively; F_L is external load force. ΔF is external disturbance excitation. ΔF is mainly caused by the factors such as the pulsation of oil source, the fluctuations of rolled metal thickness or tension etc., which can be expressed by $F\cos\omega t$ [45]. F is the amplitude of external excitation.

Among them, the expression of nonlinear elastic force $F_k(y)$ can be expressed by [46]:

$$F_k(y) = \left(\frac{A_p\beta_e}{L_1} + \frac{A_b\beta_e}{L - L_1}\right)y + \left[\frac{A_p\beta_e}{L_1^3} + \frac{A_b\beta_e}{(L - L_1)^3}\right]y^3 \quad (3)$$

where L_1 is the liquid column height of control cavity of hydraulic cylinder.

The expression of nonlinear friction $F_c(\dot{y})$ can be displayed as [47,48]:

$$F_c(\dot{y}) = \begin{cases} F_N(\mu_s - \mu_1\dot{y} + \mu_2\dot{y}^3) & \dot{y} > 0 \\ F & \dot{y} = 0 \\ F_N(-\mu_s - \mu_1\dot{y} + \mu_2\dot{y}^3) & \dot{y} < 0 \end{cases} \quad (4)$$

where

$$\mu_1 = \frac{3(\mu_s - \mu_m)}{2v_m} \quad (5)$$

$$\mu_2 = \frac{\mu_s - \mu_m}{2v_m^3} \quad (6)$$

where μ_m is the maximum dynamic friction factor; μ_s is static friction factor; v_m is vibration velocity; F_N is the positive pressure of piston acts on cylinder wall, which mainly depends on the factors such as the tightness degree of assembly, the hardness of sealing material, and the radial component of load.

3. Analysis of Bifurcation Characteristic

As below, the bifurcation characteristic of load vertical vibration is analyzed in detail. Moreover, the multi-scale method is utilized to solve the nonlinear equation of load vertical vibration.

Firstly, the small parameter factor ε and a series of slow-change time scales T_n are employed [49,50].

$$T_n = \varepsilon^n t \quad n = 0, 1, 2\cdots \quad (7)$$

Then, T_n are regarded as independent variables; there are:

$$\frac{\mathrm{d}}{\mathrm{d}t} = \frac{\partial}{\partial T_0} + \varepsilon\frac{\partial}{\partial T_1} + \cdots + \varepsilon^n\frac{\partial}{\partial T_n} = D_0 + \varepsilon D_1 + \cdots + \varepsilon^n D_n \quad (8)$$

$$\begin{aligned} \frac{\mathrm{d}^2}{\mathrm{d}t^2} &= \frac{\mathrm{d}}{\mathrm{d}t}\left(\frac{\partial}{\partial T_0} + \varepsilon\frac{\partial}{\partial T_1} + \cdots + \varepsilon^n\frac{\partial}{\partial T_n}\right) \\ &= (D_0 + \varepsilon D_1 + \cdots + \varepsilon^n D_n)^2 \\ &= D_0^2 + 2\varepsilon D_0 D_1 + \varepsilon^2(D_1^2 + 2D_0 D_2) + \cdots \end{aligned} \quad (9)$$

where D_n is a symbol of partial differential operator:

$$D_n = \frac{\partial}{\partial T_n} \quad n = 0, 1, 2 \cdots \tag{10}$$

Equation (1) can be further sorted as:

$$\ddot{y} + \omega_0^2 y = \frac{1}{m_1}\left[F\cos(\omega t) - \alpha F_k(y) - \beta F_c(\dot{y}) - c_1\dot{y}\right] \tag{11}$$

where $\omega_0 = \sqrt{k_1/m_1}$.

Then, the right side of Equation (11) is multiplied by ε; there is:

$$\ddot{y} + \omega_0^2 y = \frac{\varepsilon}{m_1}\left[F\cos(\omega t) - \alpha F_k(y) - \beta F_c(\dot{y}) - c_1\dot{y}\right] \tag{12}$$

Suppose that ω is close to ω_0, that is:

$$\omega = \omega_0 + \varepsilon\sigma \tag{13}$$

where ω is perturbation frequency, ω_0 is natural frequency, σ is the frequency tuning factor.

Equation (13) is substituted into Equation (12); there is:

$$\ddot{y} + \omega_0^2 y = \varepsilon\left[\frac{F}{m_1}\cos(\omega_0 + \varepsilon\sigma)t + f(y, \dot{y})\right] \tag{14}$$

Among them:

$$f(y, \dot{y}) = -\frac{1}{m_1}\left[\alpha F_k(y) + \beta F_c(\dot{y}) + c_1\dot{y}\right] \tag{15}$$

Assume that Equation (14) has the solution with the following form when it is under the external excitation:

$$y = y_0(T_0, T_1) + y_1(T_0, T_1) \tag{16}$$

Substituting Equation (16) into Equation (14), Equations (8) and (9) are introduced. Then, the partial differential equations can be obtained:

$$\begin{cases} D_0^2 y_0 + \omega_0^2 y_0 = 0 \\ D_0^2 y_1 + \omega_0^2 y_1 = -2D_0 D_1 y_0 + f(y_0, D_0 y_0) + \frac{F}{m_1}\cos(\omega_0 T_0 + \sigma T_1) \end{cases} \tag{17}$$

The solution for the first equation of Equation (17) is set as:

$$y_0(T_0, T_1) = a(T_1)\cos[\omega_0 T_0 + \psi(T_1)] = A(T_1)e^{\mathrm{i}\omega_0 T_0} + \mathrm{cc} \tag{18}$$

$$A(T_1) = \frac{a(T_1)}{2}e^{\mathrm{i}\psi(T_1)} \tag{19}$$

where $a(T_1)$ and $\psi(T_1)$ are the slow-change functions of vibration amplitude and phase angle, respectively. cc indicates the conjugated plural.

Substituting Equation (18) into the second equation of Equation (17), there is:

$$\begin{gathered} D_0^2 y_1 + \omega_0^2 y_1 = \mathrm{i}\omega_0\left(-2D_1 - \frac{c_1}{m_1}\right)A\mathrm{e}^{\mathrm{i}\omega_0 T_0} + \frac{F}{2m_1}\mathrm{e}^{\mathrm{i}(\omega_0 T_0 + \sigma T_1)} \\ -\frac{\alpha\beta_e}{m_1}\left[\left(\frac{A_p}{L_1} + \frac{A_b}{L - L_1}\right)A\mathrm{e}^{\mathrm{i}\omega_0 T_0} + \left(\frac{A_p}{L_1^3} + \frac{A_b}{(L - L_1)^3}\right)\left(A^3\mathrm{e}^{\mathrm{i}3\omega_0 T_0} + 3A^2\overline{A}\mathrm{e}^{\mathrm{i}\omega_0 T_0}\right)\right] \\ -\frac{\beta F_N}{m_1}\left[\mu_s \mathrm{sgn}(\dot{y}) - \mathrm{i}\omega_0\mu_1 A\mathrm{e}^{\mathrm{i}\omega_0 T_0} - \mathrm{i}\omega_0^3\mu_2\left(A^3\mathrm{e}^{\mathrm{i}3\omega_0 T_0} - 3A^2\overline{A}\mathrm{e}^{\mathrm{i}\omega_0 T_0}\right)\right] + \mathrm{cc} \end{gathered} \tag{20}$$

The secular term is further eliminated. In Equation (20), the coefficient of $e^{i\omega_0 T_0}$ is set to zero; there is:

$$\begin{aligned} i\omega_0\left(2D_1 + \tfrac{c_1}{m_1}\right)A - \tfrac{F}{2m_1}e^{i\sigma T_1} + \tfrac{\alpha\beta_e}{m_1}\left[\left(\tfrac{A_p}{L_1} + \tfrac{A_b}{L-L_1}\right)A + \left(\tfrac{A_p}{L_1^3} + \tfrac{A_b}{(L-L_1)^3}\right)(3A^2\overline{A})\right] \\ + \tfrac{\beta F_N}{m_1}\left[-i\omega_0\mu_1 A + i3\omega_0^3\mu_2 A^2\overline{A}\right] = 0 \end{aligned} \tag{21}$$

Then, Equation (19) is substituted into Equation (21). Moreover, the real and imaginary parts are separated; there are:

$$\begin{cases} D_1 a = -\frac{c_1}{2m_1}a + \frac{\beta F_N}{m_1}\left(\frac{\mu_1}{2}a - \frac{9a^3}{8}\omega_0^2\mu_2\right) + \frac{F}{2m_1\omega_0}\sin(\sigma T_1 - \psi) \\ D_1\psi = \frac{\alpha\beta_e}{2m_1\omega_0}\left[\left(\frac{A_p}{L_1} + \frac{A_b}{L-L_1}\right) + \frac{3a^2}{4}\left(\frac{A_p}{L_1^3} + \frac{A_b}{(L-L_1)^3}\right)\right] - \frac{F}{2m_1\omega_0 a}\cos(\sigma T_1 - \psi) \end{cases} \tag{22}$$

Suppose that $\varphi = \sigma T_1 - \psi$, then Equation (22) can be changed into:

$$\begin{cases} D_1 a = -\frac{c_1}{2m_1}a + \frac{\beta F_N}{m_1}\left(\frac{\mu_1}{2}a - \frac{9a^3}{8}\omega_0^2\mu_2\right) + \frac{F}{2m_1\omega_0}\sin\varphi \\ D_1\varphi = \sigma - \frac{\alpha\beta_e}{2m_1\omega_0}\left[\left(\frac{A_p}{L_1} + \frac{A_b}{L-L_1}\right) + \frac{3a^2}{4}\left(\frac{A_p}{L_1^3} + \frac{A_b}{(L-L_1)^3}\right)\right] + \frac{F}{2m_1\omega_0 a}\cos\varphi \end{cases} \tag{23}$$

In Equation (23), when $D_1 a = 0$ and $D_1\varphi = 0$, the system has a stable vibration amplitude and frequency; there are:

$$\begin{cases} \frac{c_1}{m_1}a - \frac{\beta F_N a}{m_1}\left(\mu_1 - \frac{9a^2}{4}\omega_0^2\mu_2\right) = \frac{F}{m_1\omega_0}\sin\varphi \\ 2\sigma a - \frac{\alpha\beta_e a}{m_1\omega_0}\left[\left(\frac{A_p}{L_1} + \frac{A_b}{L-L_1}\right) + \frac{3a^2}{4}\left(\frac{A_p}{L_1^3} + \frac{A_b}{(L-L_1)^3}\right)\right] = -\frac{F}{m_1\omega_0}\cos\varphi \end{cases} \tag{24}$$

In Equation (24), the square sum is performed to eliminate the parameter φ; there are:

$$\begin{aligned} \left\{\frac{c_1}{m_1}a - \frac{\beta F_N a}{m_1}\left(\mu_1 - \frac{9a^2}{4}\omega_0^2\mu_2\right)\right\}^2 + \\ \left\{2\sigma a - \frac{\alpha\beta_e a}{m_1\omega_0}\left[\left(\frac{A_p}{L_1} + \frac{A_b}{L-L_1}\right) + \frac{3a^2}{4}\left(\frac{A_p}{L_1^3} + \frac{A_b}{(L-L_1)^3}\right)\right]\right\}^2 = \left(\frac{F}{m_1\omega_0}\right)^2 \end{aligned} \tag{25}$$

Equation (25) is unfolded and organized; there is:

$$a^6 + \lambda_1 a^4 + \lambda_2 a^2 + \mu = 0 \tag{26}$$

Among them:

$$\begin{aligned} \lambda_1 = & -\left\{\frac{\alpha\beta_e}{3\omega_0}\left[\frac{A_p}{L_1^3} + \frac{A_b}{(L-L_1)^3}\right]\left[m_1\sigma - \frac{\alpha\beta_e}{2\omega_0}\left(\frac{A_p}{L_1} + \frac{A_b}{L-L_1}\right)\right] - \frac{\beta F_N\mu_2\omega_0^2}{2}\left(c_1 - \beta F_N\mu_1\right)\right\} \\ & \div\left\{\frac{\alpha^2\beta_e^2}{16\omega_0^2}\left[\frac{A_p}{L_1^3} + \frac{A_b}{(L-L_1)^3}\right]^2 + \frac{9\beta^2F_N^2\mu_2^2\omega_0^4}{16}\right\} \end{aligned} \tag{27}$$

$$\begin{aligned} \lambda_2 = & \left\{\left[2\sigma - \frac{\alpha\beta_e}{m_1\omega_0}\left(\frac{A_p}{L_1} + \frac{A_b}{L-L_1}\right)\right]^2 + \left(\frac{c1}{m1} - \frac{\beta F_N\mu_1}{m1}\right)^2\right\} \\ & \div\left\{\frac{9\alpha^2\beta_e^2}{16m_1^2\omega_0^2}\left[\frac{A_p}{L_1^3} + \frac{A_b}{(L-L_1)^3}\right]^2 + \frac{81\beta^2F_N^2\mu_2^2\omega_0^4}{16m_1^2}\right\} \end{aligned} \tag{28}$$

$$\mu = -\left(\frac{F}{m_1}\right)^2 \div \left\{\frac{9\alpha^2\beta_e^2}{16m_1^2}\left[\frac{A_p}{L_1^3} + \frac{A_b}{(L-L_1)^3}\right]^2 + \frac{81\beta^2F_N^2\mu_2^2\omega_0^6}{16m_1^2}\right\} \tag{29}$$

Equation (26) can be further written as:

$$a^7 + \lambda_1 a^5 + \lambda_2 a^3 + \mu a = 0 \tag{30}$$

According to the singularity theory [51], Equation (30) is the universal unfolding of the paradigm $a^7 + \mu a = 0$, λ_1 and λ_2 are the open fold parameters, and μ is the external disturbance quantity. If the values of the open fold parameters are different, the system will have different bifurcation patterns.

Assume:

$$G(a,\mu,\lambda) = a^7 + \lambda_1 a^5 + \lambda_2 a^3 + \mu a \tag{31}$$

Then, the derivations about a and μ in Equation (31) are respectively performed, and the following can be obtained:

$$\dot{G}_a(a,\mu,\lambda) = 7a^6 + 5\lambda_1 a^4 + 3\lambda_2 a^2 + \mu \tag{32}$$

$$\dot{G}_\mu(a,\mu,\lambda) = a \tag{33}$$

$$\ddot{G}_a(a,\mu,\lambda) = 42a^5 + 20\lambda_1 a^3 + 6\lambda_2 a \tag{34}$$

In the singularity theory, the transition set is a very important concept. This is the set of folding parameters corresponding to the non-persistent bifurcation diagram of the universal unfolding $G(a,\mu,\lambda)$. The sets of bifurcation point, lag point, and double limit point correspond to the three types of the non-persistence bifurcation diagram [52,53].

(1) Bifurcation point set

$B = \{\lambda \in \mathbf{R}^n |$ exist $(a,\mu) \in \mathbf{R} \times \mathbf{R}$, make $G = \dot{G}_a = \dot{G}_\mu = 0$ at $(a,\mu,\lambda)\}$;

(2) Lag point set

$H = \{\lambda \in \mathbf{R}^n |$ exist $(a,\mu) \in \mathbf{R} \times \mathbf{R}$, make $G = \dot{G}_a = \ddot{G}_a = 0$ at $(a,\mu,\lambda)\}$;

(3) Double limit point set

$D = \{\lambda \in \mathbf{R}^n |$ exist $(a_i,\mu)(i = 1,2) \in \mathbf{R} \times \mathbf{R}, a_1 \neq a_2$, make $G = \dot{G}_a = 0$ at $(a_i,\mu,\lambda)\}$;

(4) Transition set: $\sum = B \cup H \cup D$

The transition set can divide the neighborhood of the origin of the real space $\mathbf{R}^n$ into several subregions, and the bifurcation diagram of the universal unfolding $G(a,\mu,\lambda)$ in each subregion is persistent. In the same subregion, the bifurcation diagrams corresponding to different folding parameters λ are equivalent. Then, all the persistent bifurcation maps of $G(a,\mu,\lambda)$ can be enumerated according to these subregions.

Next, singularity theory is applied to solve the transition set of bifurcation Equation (30), and the bifurcation characteristics of the system are explored. The subscripts R and I represent the new transition set generated by the conventional transition set and nonlinear action, respectively [54].

Bifurcation point set:

$$B_R = \varnothing \quad B_I = \varnothing \tag{35}$$

Lag point set:

$$H_R = \left\{ \frac{3\lambda_2^2}{\lambda_1^2} - \lambda_2 = 0 \right\} \quad H_I = \varnothing \tag{36}$$

Double limit point set:

$$D_R = \varnothing \quad D_I = \left\{ 6a^4 + 4\lambda_1 a^2 + 2\lambda_2 = 0 \right\} \tag{37}$$

Transition set:

$$\sum = B_R \cup H_R \cup D_R \cup B_I \cup H_I \cup D_I \tag{38}$$

At this point, the transition set of the system under the open fold parameters λ_1 and λ_2 is shown in Figure 3. The topological structures of the bifurcation curves in different subregions divided by the transition set are displayed in Figure 4.

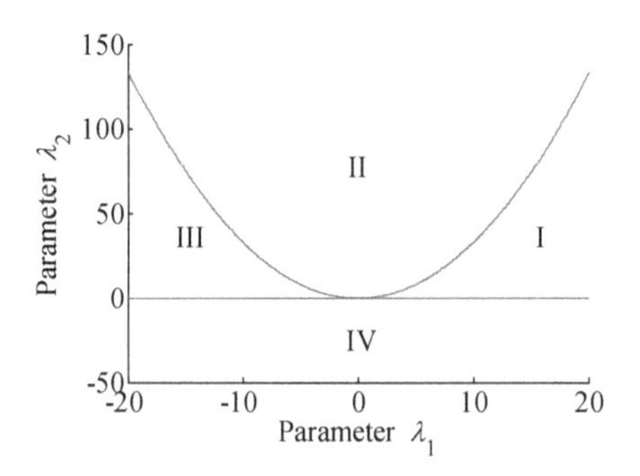

Figure 3. Transition set.

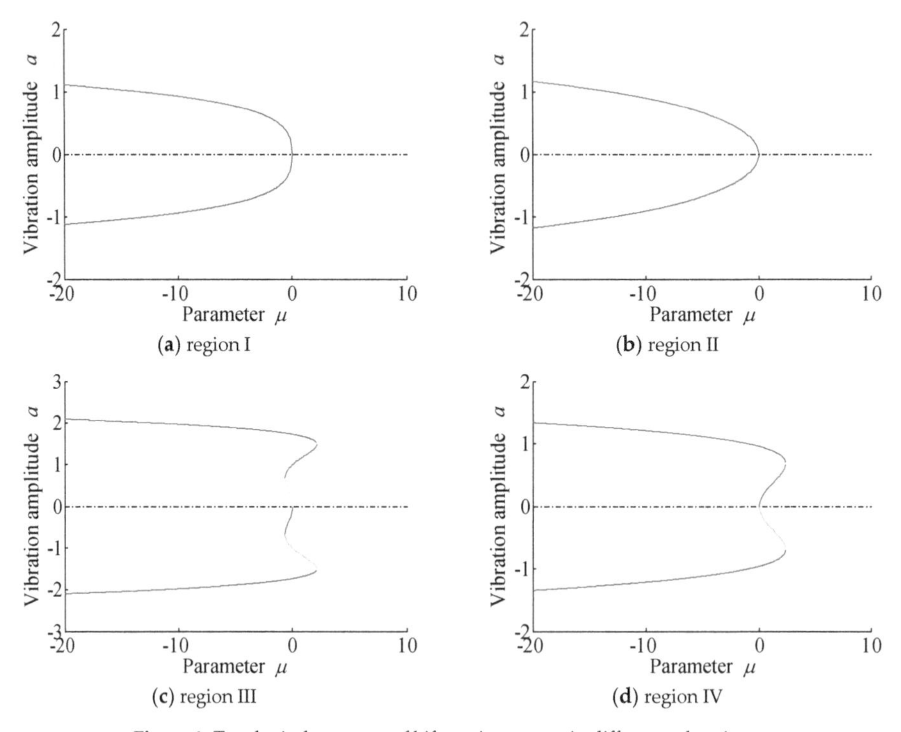

Figure 4. Topological structure of bifurcation curves in different subregions.

According to Figures 3 and 4, the system plane is divided into four subregions (I, II, III, IV) by the transition set under the open fold parameters λ_1 and λ_2. The bifurcation curves in each subregion have their own topological structure. Furthermore, with the change of λ_1 and λ_2, the topological structure changes at the transition set. It indicates that the system has different vibration behaviors in diverse subregions and will exhibit different bifurcation behaviors under various parameter combinations. Therefore, by analyzing the bifurcation characteristics of the system, the parameter region that causes the system to be unstable can be determined. Meanwhile, from the obtained topological structures of the bifurcation curve in different subregions, the bifurcation state of the system can be changed by the disturbance parameter μ. In a certain bifurcation form, the change of μ will lead to the change of vibration amplitude, which in turn changes the vibration behavior of the system.

4. Research on Bifurcation Behavior

The load of the HAGC system will be affected by different forces during the working process. The influence factors are diverse and complex. The research object of this article will focus on some nonlinear factors such as the excitation force, elastic force and damping force. The selected nonlinear forces are important influence factors. The effects of the above feature parameters on bifurcation behavior are explored.

The actual physical parameters of the 650 4/6-roll cold rolling mill from the "National Engineering Research Center for Equipment and Technology of Cold Strip Rolling" are employed in the following research. Some unmeasured parameters are empirical estimates. The parameters are shown in Table 1. The photos of the 650 4/6-roll cold rolling mill are displayed in Figure 5.

Table 1. Parameters of the numerical experiment.

Physical Quantity	Value	Unit	Physical Quantity	Value	Unit
equivalent mass m_1	8656	kg	bulk modulus β_e	780×10^6	Pa
equivalent linear stiffness coefficient k_1	1.6724×10^9	N/m	dynamic friction factor μ_m	0.01	—
equivalent linear damping coefficient c_1	1.2923×10^5	N·s/m	static friction factor μ_s	0.02	—
total stroke L	260	mm	vibration velocity v_m	0.01	m/s
effective working area A_p	19.635×10^{-2}	m^2	positive pressure F_N	0.04×10^6	N
effective working area A_b	3.0159×10^{-2}	m^2	oil density ρ	872	kg/m^3

(a)

(b)

Figure 5. Photos of the 650 4/6-roll cold rolling mill.

The influence of nonlinear elastic force can be reflected by the nonlinear stiffness coefficient α. Hence, the effect of α on bifurcation characteristics is firstly researched. According to Equation (11) derived in the previous section, the bifurcation diagram when α changes is revealed in Figure 6. The jump phenomenon of the vibration amplitude will gradually be enhanced with the increase in α. The degree of jump phenomenon is especially severe in the resonance region. However, when it is far away from the resonance region, the degree of jump will be reduced. The primary reason for the above result is that the change of the nonlinear stiffness coefficient α affects the natural frequency ω_0 of the system. As α increases, the natural frequency of the system increases. The change of natural frequency results in the resonance phenomenon when natural frequency couples with external excitation frequency ω. However, the resonance phenomenon can cause the increase in system instability. Therefore, if α is reasonably adjusted to effectively avoid the resonance region, the stability of the system will be facilitated.

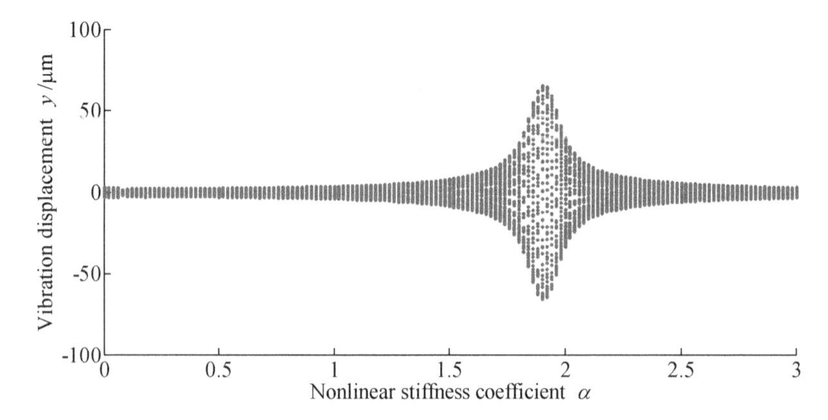

Figure 6. Bifurcation diagram with the variation of α.

Additionally, the nonlinear elastic force can be influenced by the liquid column height L_1 of control cavity. So, the effect of L_1 on bifurcation characteristics was further studied. The bifurcation diagram with the variation of L_1 is displayed in Figure 7. When L_1 is close to the two ends of hydraulic cylinder, the jump phenomenon of the vibration amplitude is more serious. When L_1 is away from the ends of the hydraulic cylinder, the degree of amplitude jump is relatively reduced. The bifurcation phenomenon near the middle position (130 mm) is more complex. The foremost reason is that the stiffness of the hydraulic spring is related to the piston position of the hydraulic cylinder. When the piston is in the middle position, the liquid compressibility is most affected. At this time, the hydraulic spring stiffness is small and the natural frequency of the system is low. Therefore, it shows poor stability in the system. Hence, properly controlling the size of L_1 is conducive to the stability of the system.

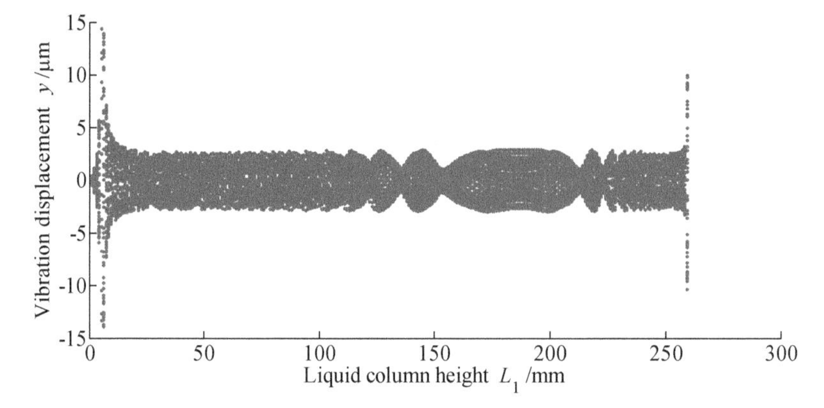

Figure 7. Bifurcation diagram with the variation of L_1.

The effect of nonlinear damping force can be reflected by the nonlinear damping coefficient β. Hence, the influence of β on bifurcation characteristics was analyzed. The bifurcation diagram with the change of β is illustrated in Figure 8. As can be observed, the jump phenomenon of the vibration amplitude will gradually decrease with the increase in β, and the vibration amplitude is effectively suppressed. However, when the value of β exceeds a certain threshold, the suppression effect for vibration amplitude is no longer obvious, and the jump phenomenon of vibration amplitude still exists. The main reason is that the appropriate β can narrow the frequency band of resonance and decrease the unstable area, which is beneficial to the stability of the system.

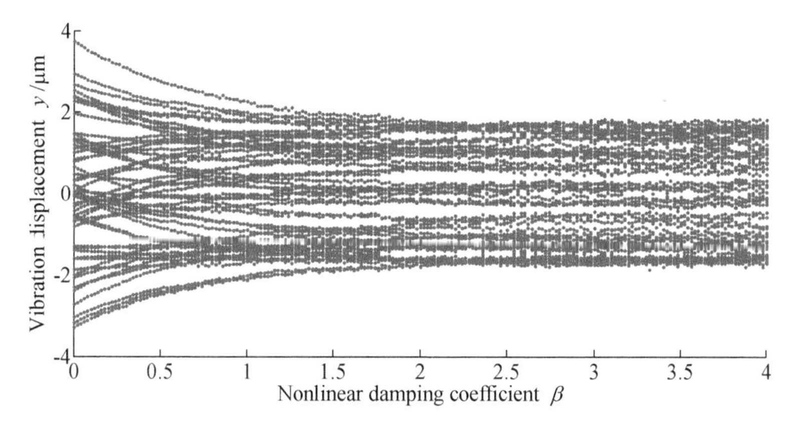

Figure 8. Bifurcation diagram with the variation of β.

Since the influence of nonlinear excitation force can be reflected by the external excitation amplitude F, the effect of F on bifurcation characteristics was also investigated. The bifurcation diagram with the variation of F is demonstrated in Figure 9. It will be observed that the jump phenomenon of vibration amplitude will gradually strengthen with the increase in F. Furthermore, the degree of jump will gradually increase. The main reason is that the increase in F can widen the frequency band of resonance and augment the instability of the system.

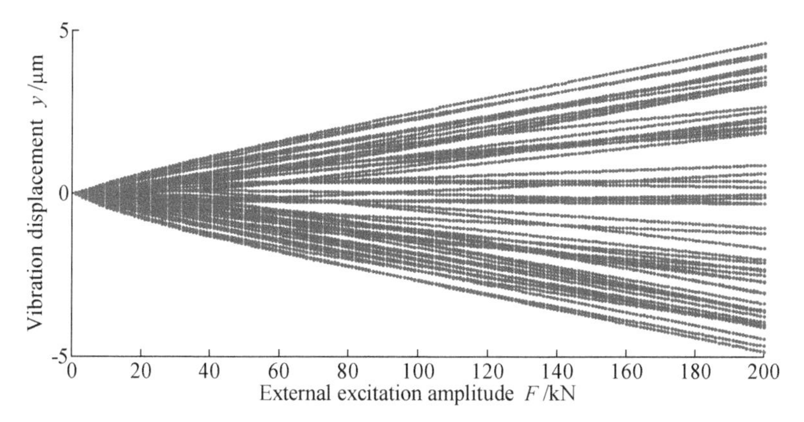

Figure 9. Bifurcation diagram with the variation of F.

5. Conclusions

On the basis of the theory of nonlinear dynamics, the bifurcation characteristics of load vertical vibration of the HAGC system were researched. The effects of some important parameters on bifurcation characteristics were emphatically explored. Through in-depth research, some conclusions are drawn:

(1) The bifurcation curves in each subregion have their own topological structure. With the change of the open fold parameters, the topological structure changes at the transition set. Moreover, the system has different vibration behaviors in diverse subregions, and will exhibit different bifurcation behaviors under various parameter combinations. Therefore, by analyzing the bifurcation characteristics of the system, the parameter region that causes the system to be unstable can be determined.

(2) With the increase in the nonlinear stiffness coefficient α, the jump phenomenon of the vibration amplitude will gradually be enhanced. Especially, the degree of jump phenomenon is severe in the

resonance region. However, the degree of jump will be reduced when it is far away from the resonance region. Therefore, if α is reasonably adjusted to effectively avoid the resonance region, the stability of the system will be facilitated.

(3) The jump phenomenon of the vibration amplitude is more serious when the liquid column height L_1 is close to the two ends of hydraulic cylinder. The degree of amplitude jump is relatively reduced when L_1 is located in the middle section of the hydraulic cylinder. However, the bifurcation phenomenon near the middle position is more complex. Hence, properly controlling the size of L_1 is conducive to the stability of the system.

(4) With the increase in the nonlinear damping coefficient β, the jump phenomenon of the vibration amplitude will gradually decrease, and the vibration amplitude is effectively suppressed. However, the suppression effect for vibration amplitude is no longer obvious when the value of β exceeds a certain threshold, and the jump phenomenon of the vibration amplitude still exists. The appropriate β can narrow the frequency band of resonance and decrease the unstable area, which is beneficial to the stability of the system.

(5) With the increase in the external excitation amplitude F, the jump phenomenon of the vibration amplitude will gradually strengthen. Moreover, the degree of jump will gradually increase, which is not conducive to the stability of the system.

The acquired results provide a theoretical basis for vibration traceability and suppression of the HAGC system. This research can provide an important basis for the further study on nonlinear dynamic behaviors of the HAGC system.

Author Contributions: Conceptualization, Y.Z. and W.J.; Methodology, S.T.; Investigation, Y.Z. and S.T.; Writing—Original Draft Preparation, Y.Z.; Writing—Review & Editing, X.Y. and Y.L.; Supervision, C.W.

Funding: This research was funded by National Natural Science Foundation of China (No. 51805214, 51875498), China Postdoctoral Science Foundation (No. 2019M651722), Natural Science Foundation of Hebei Province (No. E2018203339, E2017203129), Nature Science Foundation for Excellent Young Scholars of Jiangsu Province (No. BK20190101), Open Foundation of National Research Center of Pumps, Jiangsu University (No. NRCP201604) and Open Foundation of the State Key Laboratory of Fluid Power and Mechatronic Systems (No. GZKF-201820).

Conflicts of Interest: The authors declare no conflict of interest.

References

1. Zheng, Y.J.; Li, Y.G.; Shen, G.X.; Xie, M.L. Spatial vibration characteristics of six-high cold strip rolling mills. *Ironmak. Steelmak.* **2017**, *45*, 1–6. [CrossRef]
2. Shao, Y.; Deng, X.; Yuan, Y.; Mechefske, C.K.; Chen, Z. Characteristic recognition of chatter mark vibration in a rolling mill based on the non-dimensional parameters of the vibration signal. *J. Mech. Sci. Technol.* **2014**, *28*, 2075–2080. [CrossRef]
3. Wu, S.; Shao, Y.; Wang, L.; Yuan, Y.; Mechefske, C.K. Relationship between chatter marks and rolling force fluctuation for twenty-high roll mill. *Eng. Fail. Anal.* **2015**, *55*, 87–99. [CrossRef]
4. Zhu, Y.; Jiang, W.L.; Kong, X.D.; Wang, H.N. Analytical solution for nonlinear vertical vibration model of mill roll system based on improved complexification averaging method. *J. Vibroeng.* **2016**, *18*, 5521–5536.
5. Xue, Z.H.; Cao, X.; Wang, T.Z. Vibration test and analysis on the centrifugal pump. *J. Drain. Irrig. Mach. Eng.* **2018**, *36*, 472–477.
6. Zhong, J.; Tang, H.P. Vibration problems of high speed rolling mill-study of dynamics of complex electromechanically coupled system. *J. Vib. Meas. Diagn.* **2002**, *22*, 1–8.
7. Wang, Y.Q.; Sun, F.; Liu, J.; Sun, M.H.; Xie, Y.H. Application of smith predictor based on single neural network in cold rolling shape control. *Chin. J. Mech. Eng.* **2009**, *22*, 282–286. [CrossRef]
8. Chen, Y.H.; Shi, T.L.; Yang, S.Z. Study on parametrically excited nonlinear vibrations on 4-H cold rolling mills. *Chin. J. Mech. Eng.* **2003**, *39*, 56–60. [CrossRef]
9. Sun, J.L.; Peng, Y.; Liu, H.M. Vertical vibration of moving strip in rolling process based on beam theory. *Chin. J. Mech. Eng.* **2009**, *22*, 680–687. [CrossRef]

10. Liu, B.; Jiang, J.H.; Liu, F.; Northeastern University at Qinhuangdao; Institute of Information Technology and Engineering; Yanshan University. Nonlinear vibration characteristic of strip mill under the coupling effect of roll-rolled piece. *J. Vibroengineering* **2016**, *18*, 5492–5505. [CrossRef]
11. Liu, H.R.; Shi, P.M.; Chen, H.; Hou, D.X. Study on nonlinear parametrically exited coupling vibrations of roller system on 4-H rolling mills. *China Mech. Eng.* **2011**, *22*, 1397–1401.
12. Yan, X.Q. Machinery-electric-hydraulic coupong vibration control of hot continuous rolling mills. *Chin. J. Mech. Eng.* **2011**, *47*, 61–65. [CrossRef]
13. Tang, S.N.; Zhu, Y.; Li, W.; Cai, J.X. Status and prospect of research in preprocessing methods for measured signals in mechanical systems. *J. Drain. Irrig. Mach. Eng.* **2019**, *37*, 822–828.
14. Zhu, Y.; Tang, S.N.; Quan, L.X.; Jiang, W.L.; Zhou, L. Extraction method for signal effective component based on extreme-point symmetric mode decomposition and Kullback-Leibler divergence. *J. Braz. Soc. Mech. Sci. Eng.* **2019**, *41*, 100. [CrossRef]
15. Yang, X.; Li, J.Y.; Tong, C.N. Nonlinear vibration modeling and stability analysis of vertical roller system in cold rolling mill. *J. Vib. Meas. Diagn.* **2013**, *33*, 302–306.
16. Bi, Q.S.; Ma, R.; Zhang, Z.D. Bifurcation mechanism of the bursting oscillations in periodically excited dynamical system with two time scales. *Nonlinear Dyn.* **2015**, *79*, 101–110. [CrossRef]
17. Bi, Q.S.; Li, S.L.; Kurths, J.; Zhang, Z.D. The mechanism of bursting oscillations with different codimensional bifurcations and nonlinear structures. *Nonlinear Dyn.* **2016**, *85*, 993–1005. [CrossRef]
18. Wang, C.; He, X.; Shi, W.; Wang, X.; Wang, X.; Qiu, N. Numerical study on pressure fluctuation of a multistage centrifugal pump based on whole flow field. *AIP Adv.* **2019**, *9*, 035118. [CrossRef]
19. Wang, C.; Hu, B.; Zhu, Y.; Wang, X.; Luo, C.; Cheng, L. Numerical study on the gas-water two-phase flow in the self-priming process of self-priming centrifugal pump. *Processes* **2019**, *7*, 330. [CrossRef]
20. Wang, C.; He, X.; Zhang, D.; Hu, B.; Shi, W. Numerical and experimental study of the self-priming process of a multistage self-priming centrifugal pump. *Int. J. Energy Res.* **2019**, *43*, 4074–4092. [CrossRef]
21. Wang, C.; Shi, W.; Wang, X.; Jiang, X.; Yang, Y.; Li, W.; Zhou, L. Optimal design of multistage centrifugal pump based on the combined energy loss model and computational fluid dynamics. *Appl. Energy* **2017**, *187*, 10–26. [CrossRef]
22. He, X.; Jiao, W.; Wang, C.; Cao, W. Influence of surface roughness on the pump performance based on Computational Fluid Dynamics. *IEEE Access* **2019**, *7*, 105331–105341. [CrossRef]
23. Wang, C.; Chen, X.X.; Qiu, N.; Zhu, Y.; Shi, W.D. Numerical and experimental study on the pressure fluctuation, vibration, and noise of multistage pump with radial diffuser. *J. Braz. Soc. Mech. Sci. Eng.* **2018**, *40*, 481. [CrossRef]
24. Qian, J.Y.; Chen, M.R.; Liu, X.L.; Jin, Z.J. A numerical investigation of the flow of nanofluids through a micro Tesla valve. *J. Zhejiang Univ. Sci. A* **2019**, *20*, 50–60. [CrossRef]
25. Qian, J.Y.; Gao, Z.X.; Liu, B.Z.; Jin, Z.J. Parametric study on fluid dynamics of pilot-control angle globe valve. *Asme J. Fluids Eng.* **2018**, *140*, 111103. [CrossRef]
26. Bai, L.; Zhou, L.; Jiang, X.P.; Pang, Q.L.; Ye, D.X. Vibration in a multistage centrifugal pump under varied conditions. *Shock Vib.* **2019**, *2019*, 2057031. [CrossRef]
27. Bai, L.; Zhou, L.; Han, C.; Zhu, Y.; Shi, W.D. Numerical study of pressure fluctuation and unsteady flow in a centrifugal pump. *Processes* **2019**, *7*, 354. [CrossRef]
28. Wang, L.; Liu, H.L.; Wang, K.; Zhou, L.; Jiang, X.P.; Li, Y. Numerical simulation of the sound field of a five-stage centrifugal pump with different turbulence models. *Water* **2019**, *11*, 1777. [CrossRef]
29. Ye, S.G.; Zhang, J.H.; Xu, B.; Zhu, S.Q. Theoretical investigation of the contributions of the excitation forces to the vibration of an axial piston pump. *Mech. Syst. Signal Process.* **2019**, *129*, 201–217. [CrossRef]
30. Zhang, J.H.; Xia, S.; Ye, S.; Xu, B.; Song, W.; Zhu, S.; Xiang, J. Experimental investigation on the noise reduction of an axial piston pump using free-layer damping material treatment. *Appl. Acoust.* **2018**, *139*, 1–7. [CrossRef]
31. Pei, J.; Zhang, F.; Appiah, D.; Hu, B.; Asomani, S.N. Performance prediction based on effects of wrapping angle of a side channel pump. *Energies* **2019**, *12*, 139. [CrossRef]
32. Niziol, J.; Światoniowski, A. Numerical analysis of the vertical vibrations of rolling mills and their negative effect on the sheet quality. *J. Mater. Process. Technol.* **2005**, *162–163*, 546–550. [CrossRef]
33. Heidari, A.; Forouzan, M.R.; Akbarzadeh, S. Development of a rolling chatter model considering unsteady lubrication. *ISIJ Int.* **2014**, *54*, 165–170. [CrossRef]

34. Heidari, A.; Forouzan, M.R.; Akbarzadeh, S. Effect of friction on tandem cold rolling mills chattering. *ISIJ Int.* **2014**, *54*, 2349–2356. [CrossRef]
35. Drzymala, Z.; Świątoniowski, A.; Bar, A. Non-linear vibrations in cold rolling mills. *Mécanique Ind.* **2003**, *4*, 151–158. [CrossRef]
36. Xu, H.; Cui, L.L.; Shang, D.G. A study of nonlinear coupling dynamic characteristics of the cold rolling mill system under different rolling parameters. *Adv. Mech. Eng.* **2017**, *9*, 1–15. [CrossRef]
37. Wang, Q.L.; Li, X.; Hu, Y.J.; Sun, J.; Zhang, D.H. Numerical analysis of intermediate roll shifting-induced rigidity characteristics of UCM cold rolling mill. *Steel Res. Int.* **2018**, *89*, 1700454. [CrossRef]
38. Chen, T.U.; Liu, X.; Gong, Y. Research on fatigue failure of agc servo hydraulic cylinder. *Mach. Tool Hydraul.* **2017**, *45*, 158–161.
39. Mosayebi, M.; Zarrinkolah, F.; Farmanesh, K. Calculation of stiffness parameters and vibration analysis of a cold rolling mill stand. *Int. J. Adv. Manuf. Technol.* **2017**, *91*, 1–11. [CrossRef]
40. Zhu, Y.; Jiang, W.L.; Kong, X.D.; Zheng, Z. Study on nonlinear dynamics characteristics of electro-hydraulic servo system. *Nonlinear Dyn.* **2015**, *80*, 723–737. [CrossRef]
41. Khalid, M.S.U.; Imran, A.; Dong, H.; Ahsan, N.; Wu, B. Bifurcations and route to chaos for flow over an oscillating airfoil. *J. Fluids Struct.* **2018**, *80C*, 262–274. [CrossRef]
42. Yu, Y.; Zhang, C.; Han, X.J.; Bi, Q.S. Dynamical behavior analysis and bifurcation mechanism of a new 3-D nonlinear periodic switching system. *Nonlinear Dyn.* **2013**, *73*, 1873–1881. [CrossRef]
43. Zhu, Y.; Qian, P.F.; Tang, S.N.; Jiang, W.L.; Li, W.; Zhao, J.H. Amplitude-frequency characteristics analysis for vertical vibration of hydraulic AGC system under nonlinear action. *AIP Adv.* **2019**, *9*, 035019. [CrossRef]
44. Liu, F.; Liu, B.; Shi, P.M.; Hou, D.X. Vibration behavior of roll system under nonlinear constraints of the hydraulic cylinder. *J. Mech. Eng.* **2014**, *50*, 59–65. [CrossRef]
45. Ling, Q.H.; Yan, X.Q.; Zhang, Q.D.; Zhang, Y.F. Research on vibration characteristics of the hot rolling mill by dual power source driven. *J. Vib. Meas. Diagn.* **2014**, *34*, 534–538.
46. Liu, B.; Li, P.; Liu, F.; Liu, H.R.; Jiang, J.H. Vibration behavior and control of roll system under nonlinear stiffness of a hydraulic cylinder. *China Mech. Eng.* **2016**, *27*, 3190–3196.
47. Xuan, B.T.; Hafizah, N.; Yanada, H. Modeling of dynamic friction behaviors of hydraulic cylinders. *Mechatronics* **2012**, *22*, 65–75.
48. Pennestrì, E.; Rossi, V.; Salvini, P.; Valentini, P.P. Review and comparison of dry friction force models. *Nonlinear Dyn.* **2016**, *83*, 1785–1801.
49. Salahshoor, E.; Ebrahimi, S.; Maasoomi, M. Nonlinear vibration analysis of mechanical systems with multiple joint clearances using the method of multiple scales. *Mech. Mach. Theory* **2016**, *105*, 495–509. [CrossRef]
50. Zhang, Z.D.; Liu, B.B.; Bi, Q.S. Non-smooth bifurcations on the bursting oscillations in a dynamic system with two timescales. *Nonlinear Dyn.* **2015**, *79*, 195–203. [CrossRef]
51. Cirillo, G.I.; Habib, G.; Kerschen, G.; Sepulchre, R. Analysis and design of nonlinear resonances via singularity theory. *J. Sound Vib.* **2017**, *392*, 295–306. [CrossRef]
52. Zhang, C.; Han, X.J.; Bi, Q.S. On symmetry-breaking bifurcation in the periodic parameter-switching Lorenz oscillator. *Sci. China-Technol. Sci.* **2013**, *56*, 2310–2316. [CrossRef]
53. Zhang, R.; Wang, Y.; Zhang, Z.D.; Bi, Q.S. Nonlinear behaviors as well as the bifurcation mechanism in switched dynamical systems. *Nonlinear Dyn.* **2015**, *79*, 465–471. [CrossRef]
54. Wu, Z.Q.; Yu, P.; Wang, K.Q. Bifurcation analysis on a self-excited hysteretic system. *Int. J. Bifurc. Chaos* **2004**, *14*, 2825–2842. [CrossRef]

Article

A Fault Feature Extraction Method for the Fluid Pressure Signal of Hydraulic Pumps Based on Autogram

Zhi Zheng [1], Xianze Li [1,*] and Yong Zhu [2,*]

[1] College of Mechanical Engineering, North China University of Science and Technology, Tangshan 063210, China; zhengzhi@ncst.edu.cn
[2] National Research Center of Pumps, Jiangsu University, Zhenjiang 212013, China
* Correspondence: lixianze129@163.com (X.L.); zhuyong@ujs.edu.cn (Y.Z.); Tel.: +86-0315-8805440 (X.L.)

Received: 23 July 2019; Accepted: 11 September 2019; Published: 3 October 2019

Abstract: Center spring wear faults in hydraulic pumps can cause fluid pressure fluctuations at the outlet, and the fault feature information on fluctuations is often contaminated by different types of fluid flow interferences. Aiming to resolve the above problems, a fluid pressure signal method for hydraulic pumps based on Autogram was applied to extract the fault feature information. Firstly, maximal overlap discrete wavelet packet transform (MODWPT) was adopted to decompose the contaminated fault pressure signal of center spring wear. Secondly, based on the squared envelope of each node, three kinds of kurtosis of unbiased autocorrelation (AC) were computed in order to describe the fault feature information comprehensively. These are known as standard Autogram, upper Autogram and lower Autogram. Then a node corresponding to the biggest kurtosis value was selected as a data source for further spectrum analysis. Lastly, the data source was processed by threshold values, and then the fault could be diagnosed based on the fluid pressure signal.

Keywords: hydraulic pump; feature extraction; fluid pressure; Autogram; kurtosis

1. Introduction

Fluid plays an important role in the power transmission of the hydraulic system or component [1,2]. Because of hydraulic impact and mechanical faults, fluid flow pressure is very complex and nonstationary in hydraulic systems or components. Meanwhile, there are many background noises due to high and low pressure conversion, fluid pressure impact, cavitation phenomenon, fluid pulsation, and so on. The above aforementioned noises can cause problems in the fault feature extraction of the hydraulic systems or components based on the fluid pressure signal. Much running condition feature information is contained in the fluid pressure signal. Therefore, pressure fluctuation can reflect the running condition of the hydraulic system or component [3–8]. Fault diagnosis for the hydraulic systems or components based on the fluid pressure signal has been studied by many scholars at home and abroad [9–16]. Zhang put forward a method of flow measurement based on the new sensor, where the flow rate could be measured in hydraulic system by applying the mathematical model, and then flow detection of a 7 piston-pump was realized based on the sensor signal [9]. Goharrizi utilized the Hilbert- Huang transform to decompose the pressure signal of hydraulic actuators, and the first intrinsic mode function was used as a data source, and then the root mean square extracted from it could be adopted to detect internal leakage and its severity effectively. This was done without requiring prior knowledge about the model of the actuator or leakage [10]. Vásquez proposed an active model-based algorithm of fault detection and isolation. With the help of frequency-domain estimators, continuous-time models in a user-defined frequency band were identified. Then, a method for fault detection and isolation was adopted to diagnose early faults in hydraulic actuators based on the fluid pressure signal [11]. Aiming to resolve the serious influences of pressure fluctuation and other noises in the pressure signal, Tang applied a method of wavelet theory to decompose the signal, obtaining the wavelet energy of fault feature information can be got. Then, the inner leakage fault of the hydraulic cylinder could be diagnosed [12]. In order to diagnose the faults in reciprocating pumps, a fluid pressure signal in pump cylinder was analyzed. Frequency energy was extracted for the feature vectors, and then the improved neural network was used to diagnose the pump fault successfully [13]. Guo proposed a pre-filter combined with threshold self-learning wavelet algorithm. The denoising threshold could be self-learnt in the steady flow state, and its noise suppression effect was better than that of the traditional wavelet algorithms based on fluid pressure signals of hydraulic pipeline [14]. You proposed a fusion method using the hybrid particle swarm optimization algorithm and wavelet packet energy entropy. Neural network weights and threshold were optimized by the algorithm, and wavelet packet energy entropy was used for the eigenvector, and then the fault of hydraulic system could be diagnosed effectively [15]. In contrast to the traditional way of detecting the ship fault in the time domain, Li presented a novel method in the frequency domain. The method decomposed the hydraulic pressure signal using the wavelet-transform technique, and reconstructed it at the low-frequency region; thus, the ship fault could be diagnosed effectively [16].

The hydraulic pump is an important power component, and it has been applied in the fields of robotics [17], engineering machinery [18], underwater machinery [19,20], and wind power machinery [21]. These fields often involve working conditions with high temperature, high pressure, high speed, high humidity, and heavy load, leading to a high failure rate and significant casualties and economic losses [22–33].

In 2018, the Italian scholars Ali Moshrefzadeh and Alessandro Fasana proposed a new method named Autogram based on unbiased autocorrelation (AC) [34]. It was applied on rolling element bearings. Fault feature information of the inner race, outer race and rolling element could be thus extracted effectively. The Autogram method possesses some advantages. Firstly, heavy Gaussian and non-Gaussian background noise have a very bad influence on fault feature extraction, and the procedure of unbiased AC has overcome this disadvantage. Secondly, because of down-sampling operation, the length of the time history halves at each level of decomposition, which can limit the ability to investigate the traditional wavelet transform coefficients. Furthermore, the transform may be interfered with selection of a signal starting point. To resolve these problems, maximal overlap (undecimated) discrete wavelet packet transform (MODWPT) can be used to remove the down-sampling step in discrete traditional wavelet packet transform (DWPT) [35]. Thirdly, there is no need to obtain the prior morphological feature knowledge of a signal. Because Autogram was proposed in 2018, few scholars have studied the method at a global scale.

In this paper, we firstly introduce Autogram into the fault feature extraction for the fluid pressure signal of a hydraulic pump successfully. Based on three kinds of kurtosis and threshold values, we find that only standard Autogram can select the optimal frequency band, and rich feature information of center spring wear fault can be extracted effectively without processing of the threshold value. Application of Autogram is extended to hydraulic pump from rolling element bearings, and the acquired results can provide a theoretical basis for the fault feature extraction of the hydraulic pump. It can also provide an important basis for the further study of multiple and single faults diagnosis of the hydraulic pump and other rotating machinery.

The organization of this paper is as follows: In Section 2, the algorithm of Autogram is introduced; In Section 3, the flowchart of Autogram is described. In Section 4, some examples of simulation experiment validation are presented. In Section 5, the experimental results are demonstrated by applying the Autogram to the fault signal of center spring wear of hydraulic pump. In Section 6, the conclusions of this investigation are summarized.

2. Algorithm of Autogram

In order to be readable, the nomenclature is given as follows:

Nomenclature	
MODWPT	maximal overlap (undecimated) discrete wavelet packet transform
DWPT	discrete traditional wavelet packet transform
AC	unbiased Autocorrelation
FK	fast kurtogram
kurtosis	*kurtosis* is obtained based on Equation (2)
$kurtosis_u$	*kurtosis* is obtained based on Equation (3)
$kurtosis_l$	*kurtosis* is obtained based on Equation (4)
standard Autogram	colormap presentations of the result based on the *kurtosis* is denoted standard Autogram
upper Autogram	colormap presentations of the result based on the $kurtosis_u$ is denoted upper Autogram
lower Autogram	colormap presentations of the result based on the $kurtosis_l$ is denoted lower Autogram
no threshold spectrum	spectrum based on a signal without threshold processing
upper threshold spectrum	spectrum based on upper parts of a signal which are larger than threshold value $\overline{x}_T(i)$
lower threshold spectrum	spectrum based on lower parts of a signal which are smaller than threshold value $\overline{x}_T(i)$
SNR	signal-to-noise ratio
$R_{xx}(\tau)$	unbiased AC analysis of the (periodic) instantaneous autocovariance
x	squared envelope of the signal filtered by MODWPT
τ	delay factor
f_s	sampling frequency
n	node length
$\overline{x}_T(i)$	threshold value
k	length of the windowed signal to be averaged
$x'(t)$	simulated signal
$x_1(t)$	impulsive signal with periodic exponential attenuation
$x_2(t)$	Gauss white noise with standard deviation of 0.5

The fast kurtogram (FK) is adopted to select the signal with the most impulsive frequency band; and it has been a significant method for fault diagnosis of rotating machinery for many years [36]. However, in some harsh backgrounds with low signal-to-noise ratio (SNR), strong non-Gaussian noise, or randomly distributed impulses, its extraction ability is much reduced.

With the aim of resolving the above problems, Italian scholars proposed a method named Autogram to enhance the feature extraction ability in heavy Gaussian and non-Gaussian background noise in 2018. It is an effective tool for processing the impulsive fault signal, and no prior knowledge of the signal is needed [34].

The algorithm is described as follows:

(1) Decomposition of maximal overlap (undecimated) discrete wavelet packet transform (MODWPT)

According to a dyadic tree structure, a fault signal is divided in frequency bands by means of the wavelet transform. The MODWPT removes the down-sampling step of the discrete wavelet packet transform (DWPT), and then it is used as a filter; the signals are consequently produced at each level of decomposition. The signals in each decomposition level correspond to a frequency band and central frequency, known as the node.

(2) Calculating unbiased AC of the squared envelope for each node

Unbiased AC analysis of the (periodic) instantaneous autocovariance of the signal $R_{xx}(t_i, 0)$ is calculated and shown in Equation (1).

$$\hat{R}_{xx}(\tau) = \frac{1}{n-q}\sum_{i=1}^{n-q} x(t_i)x(t_i+\tau) \tag{1}$$

where x is squared envelope of the signal filtered by MODWPT at Step 1, $\tau = q/f_s$ is the delay factor, f_s is sampling frequency, and $q = 0, \ldots, n-1$.

The advantage of AC is that it filters out the uncorrelated components within the fault feature information, i.e. noise and random impulsive contents. Furthermore, fault feature information can be made more obvious. It is even more effective, because it is done for each node separately rather than on the complete original signal, so that SNR for each demodulated band signal is increased.

It can be seen form Equation (1) that the point number of node will decrease with an increment of τ, and therefore AC will have an estimation error, thus the first half of the AC is used in the paper. With the help of the MODWPT, AC can make the diagnostic process more accurate.

(3) Kurtosis of the AC

This step is to select the node, and let the node be data source for further fault feature extraction. The proposed method is different from FK, because the kurtosis of Autogram is computed based on the AC of the node of each level. Subsequently, the kurtosis values of all nodes, similar to FK, are presented in a colormap for which the color scale is proportional to the kurtosis value, and the vertical and horizontal axis represent the level of the MODWPT decomposition and frequency, respectively.

Kurtosis aims to quantify the impulsivity of the AC of each node. Three kinds of equations are illustrated in Equations (2)–(4):

$$\text{Kutrosis}(x) = \frac{\sum_{i=1}^{n/2} [\hat{R}_{xx}(i) - \min(\hat{R}_{xx}(\tau))]^4}{[\sum_{i=1}^{n/2} [\hat{R}_{xx}(i) - \min(\hat{R}_{xx}(\tau))]^2]^2} \tag{2}$$

$$\text{Kutrosis}_u(x) = \frac{\sum_{i=1}^{n/2} \left|\hat{R}_{xx}(i) - \overline{x}_T(i)\right|_+^4}{[\sum_{i=1}^{n/2} \left|\hat{R}_{xx}(i) - \overline{x}_T(i)\right|_+^2]^2} \tag{3}$$

$$\text{Kutrosis}_l(x) = \frac{\sum_{i=1}^{n/2} \left|\hat{R}_{xx}(i) - \overline{x}_T(i)\right|_-^4}{[\sum_{i=1}^{n/2} \left|\hat{R}_{xx}(i) - \overline{x}_T(i)\right|_-^2]^2} \tag{4}$$

where n is node length, operator of $|\blacksquare|_+$ or $|\blacksquare|_-$ illustrate that only positive or negative value is adopted respectively, and the other values are set to 0. $\overline{x}_T(i)$ is the threshold value, and it can be obtained based on the moving mean value of AC.

$$\overline{x}_T(i) = \frac{1}{k} \sum_{j=i}^{i+k-1} \hat{R}_{xx}(i) \tag{5}$$

where k is length of the windowed signal to be averaged.

Colormap presentations, based on the Equations (2)–(4), are denominated standard Autogram, upper Autogram, and lower Autogram, respectively.

Ultimately, the node associated with the largest kurtosis value is considered for further investigation.

(4) Spectrum analysis based on the threshold value

Based on the data source obtained in Step (3), the Fourier transform of the squared envelope based on the no threshold value (original node), the smaller than threshold value, and the larger than threshold value can be obtained. Thus, their spectrums are known as the no threshold spectrum, the lower threshold spectrum, and upper threshold spectrum, respectively.

3. Flowchart of the proposed method

Fluid pressure signal of center spring wear fault is sampled. The MODWPT is adopted to decompose the signal, and some nodes at each level of decomposition can be obtained. The AC of each node is computed, and the node that corresponding to the biggest AC value is selected as the data source for further investigation. Then, the spectrum of data source can be acquired, and the fault feature information can be extracted. The flowchart of the Autogram is shown in Figure 1.

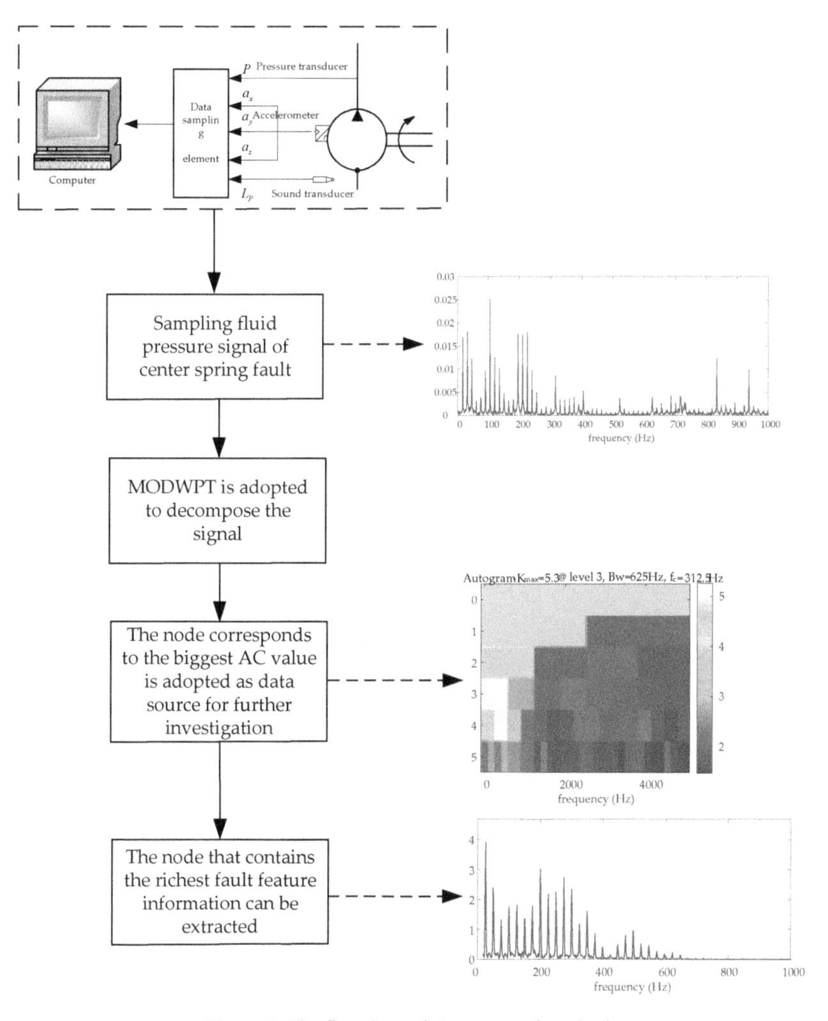

Figure 1. The flowchart of the proposed method.

4. Numerical experiment

4.1. Simulated signal

If a center spring wear fault has occurred, this will cause serious hydraulic shock in the hydraulic pump, which can be reflected as impulsive feature in morphology of the fault signal. Thus, $x_1(t)$ (an impulsive signal with periodic exponential attenuation) is used to simulate the fault signal. There are also some background noises in the fault pump; thus, $x_2(t)$ of Gauss white noise with standard deviation of 0.5 is adopted to simulate the background noises.

The purpose of the simulation experiment is used to validate the effectiveness of Autogram to extract $x_1(t)$ of the simulated impulsive fault in theory.

The simulated signal is expressed in Equation(6):

$$x'(t) = x_1(t) + x_2(t) \tag{6}$$

where $x_1(t)$ is used to simulate an impulsive fault, and it is an exponential attenuation signal with periodic frequency of 30 Hz. It is $3e^{-400t}\sin(300\pi t)$ in a period. $x_2(t)$ is Gauss white noise with a standard deviation of 0.5, and it is used to simulate a background noise. The sampling frequency is 2048 Hz.

The simulated signal in frequency domain is shown in Figure 2.

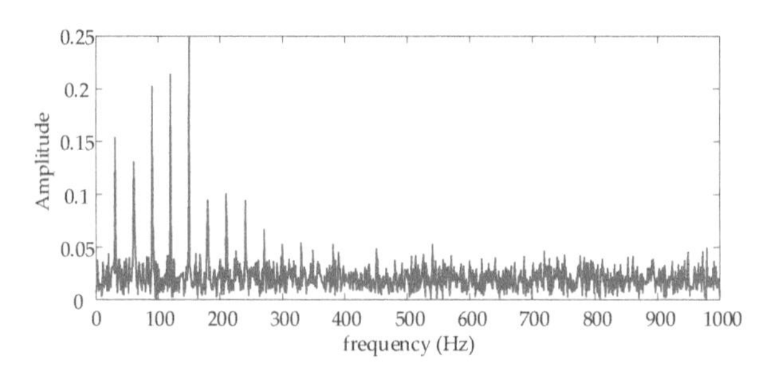

Figure 2. Spectrum of the simulated signal.

4.2. Analysis of the simulated signal based on standard Autogram

Standard Autogram is adopted to extract the simulated fault feature information, and the colormap presentation obtained based on the Equation (2) is displayed in Figure 3.

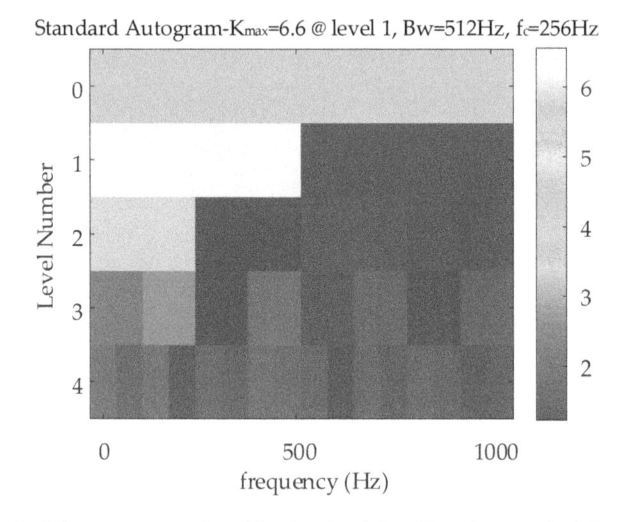

Figure 3. Colormap presentation of the simulated signal based on standard Autogram.

The maximum value *kurtosis* 6.6 is assigned to the node (1, 1), with a center frequency of 256 Hz and a bandwidth of 512 Hz. Thus, the node (1, 1) is used as data source for future investigation, and the no threshold spectrum, upper threshold spectrum, and lower threshold spectrum are shown in Figure 4.

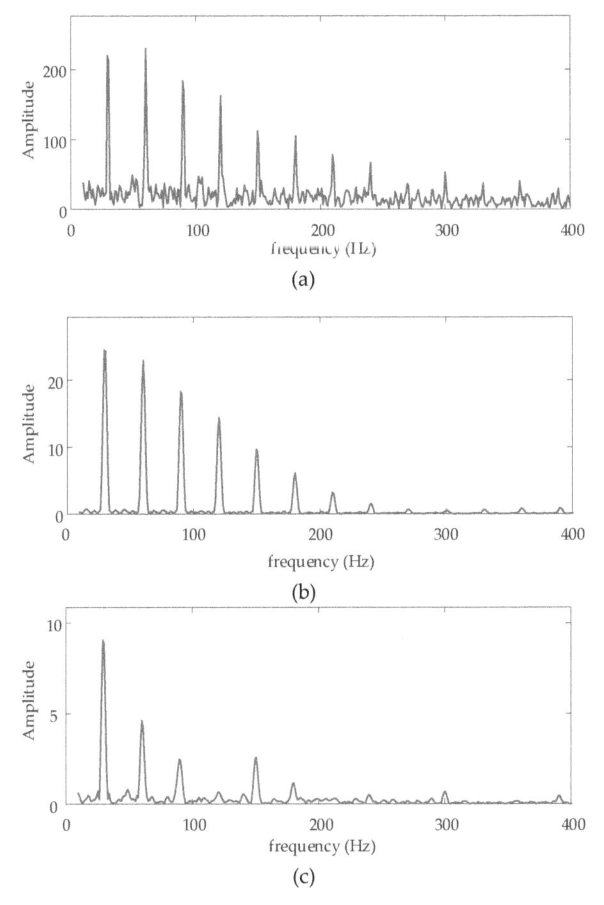

Figure 4. Spectrums of the simulated signal data source node (1, 1) based on standard Autogram. (**a**) No threshold spectrum; (**b**) Upper threshold spectrum; (**c**) Lower threshold spectrum.

Among the three figures in Figure 4, simulated fault feature information at fault feature frequency 30 Hz and all of its harmonics are only extracted only in Figure 4a. The amplitude values based on no threshold processing in Figure 4a are the largest, and all of them are larger than those in the simulated signal spectrum in Figure 2. It can be seen that standard Autogram extracts the information effectively, and the extraction ability of Autogram based on no threshold processing is the strongest.

4.3. *Analysis of thesimulated signal based on Upper Autogram*

The upper Autogram is applied to extract the simulated fault feature information and the colormap presentation can be obtained based on Equation (3), as shown in Figure 5.

The maximum value $kurtosis_u$ is 5.3, and it is assigned to the node (3, 4), with center frequency of 448 Hz and a bandwidth of 128 Hz. Thus, the node (3, 4) is used as data source for further investigation, and the no threshold spectrum, upper threshold spectrum, and lower threshold spectrums are illustrated in Figure 6.

It is very clear that there is no simulated fault feature information at fault feature frequency of 30 Hz with its harmonics in Figure 6, and there are many background noises. Thus, the feature information cannot be effectively extracted by upper Autogram based on lower threshold processing.

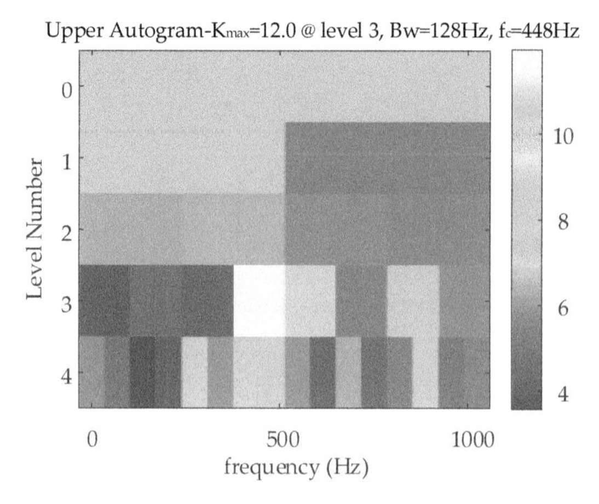

Figure 5. Colormap presentation of the simulated signal based on upper Autogram.

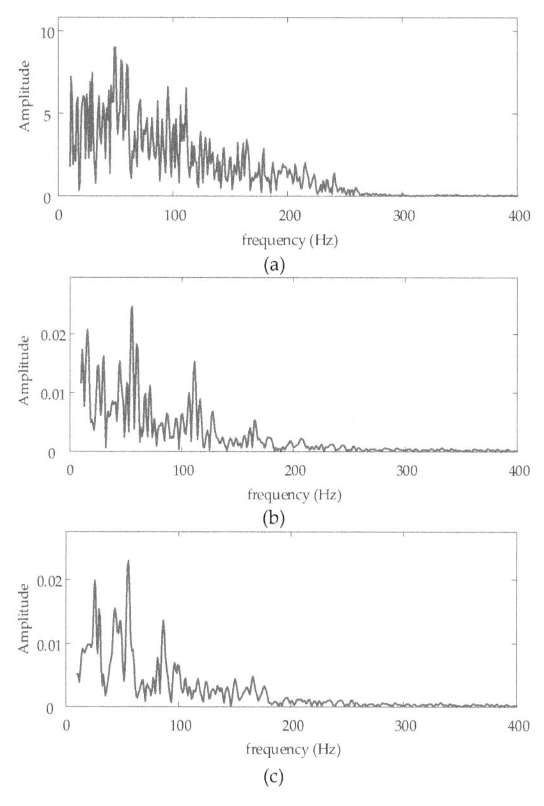

Figure 6. Spectrums of the simulated signal data source node (1, 1) based on upper Autogram. (**a**) No threshold spectrum; (**b**) Upper threshold spectrum; (**c**) Lower threshold spectrum.

4.4. Analysis of the simulated signal based on Lower Autogram

The simulated fault feature information is extracted by lower Autogram, and the colormap presentation based on the Equation (4) is shown in Figure 7.

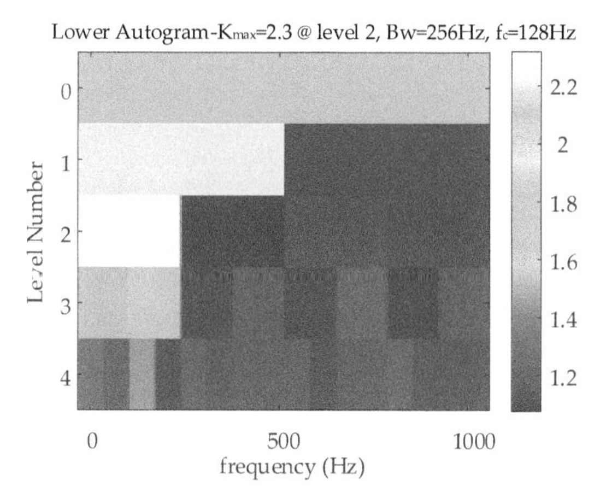

Figure 7. Colormap presentation of the simulated signal based on Lower Autogram.

The maximum value $kurtosis_l$ is 2.3 and corresponds to node (2, 1), the node has a center frequency of 128 Hz and a bandwidth of 256 Hz. Node (2, 1) is adopted as a data source for further investigation, and the no threshold spectrum, upper threshold spectrum, and lower threshold spectrums are demonstrated in Figure 8.

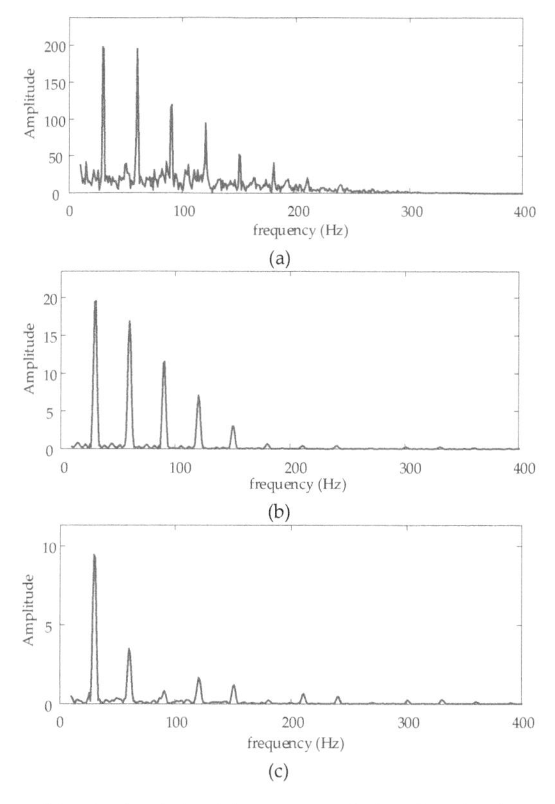

Figure 8. Spectrums of the simulated signal data source node (2, 1) based on lower Autogram. (**a**) No threshold spectrum; (**b**) Upper threshold spectrum; (**c**) Lower threshold spectrum.

It can be seen in Figure 8 that the simulated fault feature information at fault feature frequency 30 Hz with some harmonics can be extracted. The amplitude values obtained based on no threshold processing in Figure 8a are the larger than in Figure 8b,c, and the amplitude values in Figure 8 are all larger than those in Figure 7. Thus, lower Autogram is effective in extracting the information, and its extraction ability based on no threshold processing is the strongest.

Compared with standard Autogram based on no threshold processing in Figure 4, the fault feature information with many harmonics cannot be extracted by lower Autogram, and their amplitude values are small.

From the results of standard Autogram (Figure 4), upper Autogram (Figure 6), and lower Autogram (Figure 8), it can be concluded that standard Autogram has the strongest extraction ability, with background noises more greatly reduced.

4.5. Analysis of center spring wear fault signal based on FK

In order to demonstrate the effectiveness and advantages of the proposed method, the signal is also analyzed by FK, and the colormap presentation is displayed in Figure 9.

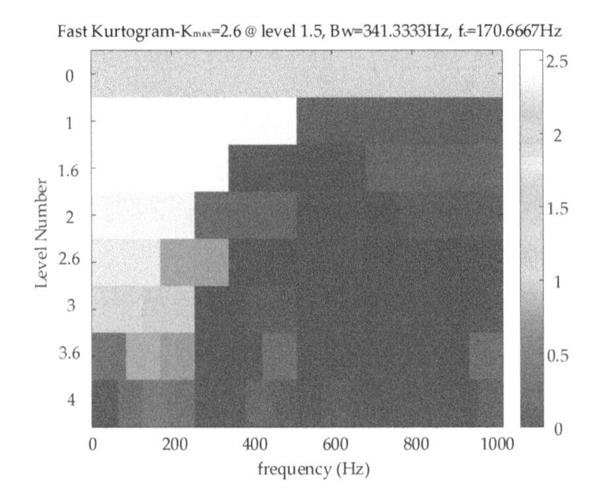

Figure 9. Spectrums of the simulated signal based on FK.

The maximum value of kurtosis based on FK is 2.6, corresponding to a sub-signal at level 1.6; the sub-signal has a center frequency of 170.6667 Hz and a bandwidth of 341.3333 Hz. The sub-signal is used as a data source for further investigation, and its spectrums are shown Figure 10.

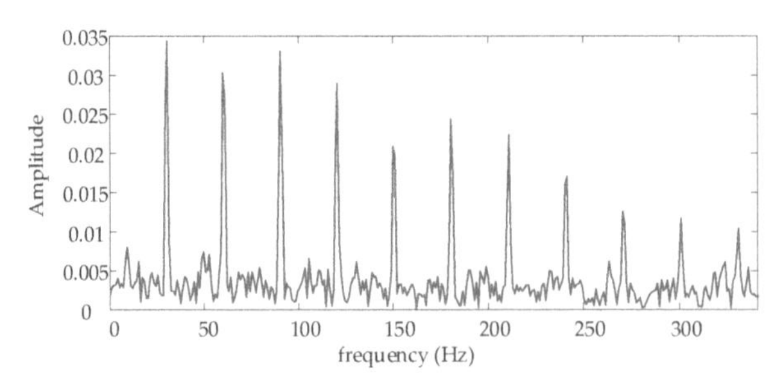

Figure 10. Spectrums of the simulated signal data source sub-signal based on FK.

Based on above, we know that standard Autogram has the strongest extraction ability among the three methods. Compared with standard Autogram, while FK can extract 30 Hz and its harmonics, the amplitude values are too small.

It can be concluded from above that the standard Autogram based on no threshold processing has the strongest extraction ability. It can extract simulated fault feature information, and it performs best among all methods of Autogram and FK.

5. Application to hydraulic pump fault signal

5.1. Experimental scheme

In order to validate the effectiveness of the Autogram, an experiment was executed on a swashplate axial plunger pump with a center spring wear fault. The rotational speed of the pump was set as at 1470 r/min. The pressure transducer was set at the pump outlet, and the fluid pressure of the outlet was set as 10 MPa. The fluid pressure signal of the outlet was sampled at 10 kHz. The center spring wear fault feature frequency was 24.5 Hz [37].

The swashplate axial plunger pump experiment systems are shown in Figure 11.

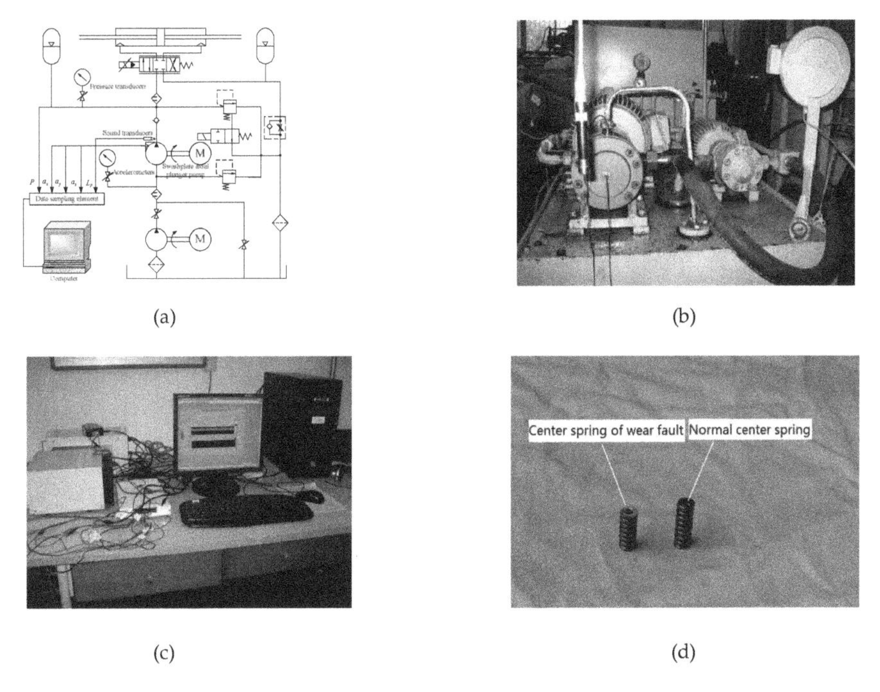

Figure 11. Swashplate axial plunger pump experiment systems. (**a**) Schematic diagram; (**b**) Swashplate axial plunger pump; (**c**) Data acquisition equipment; (**d**) Invalidation spring and the normal spring.

5.2. Center spring wear fault signal of fluid pressure

The center spring wear fault signal of fluid pressure in the frequency domain is shown in Figure 12.

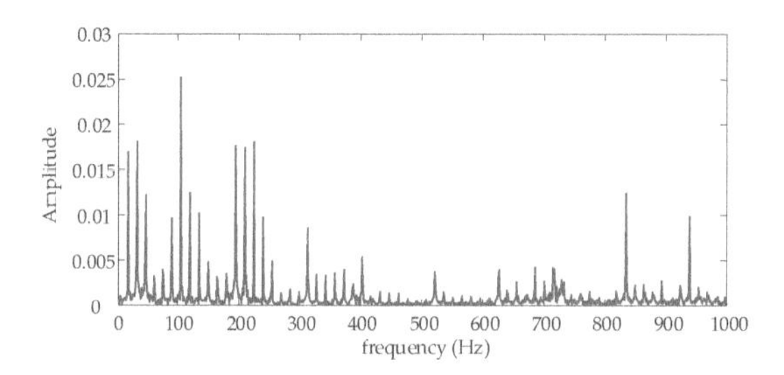

Figure 12. Spectrum of the center spring wear fault signal.

It can be seen from Figure 12 that the fault feature information on center spring wear at fault feature frequency 24.5 with some harmonics is not obvious, and the amplitude values are very small. Thus, the fault feature information is contaminated by a lot of background noise.

5.3. *Analysis of center spring wear fault signal of fluid pressure based on the standard Autogram*

The fault feature information on center spring wear is extracted by standard Autogram, and the colormap presentation based on the Equation (2) is shown in Figure 13.

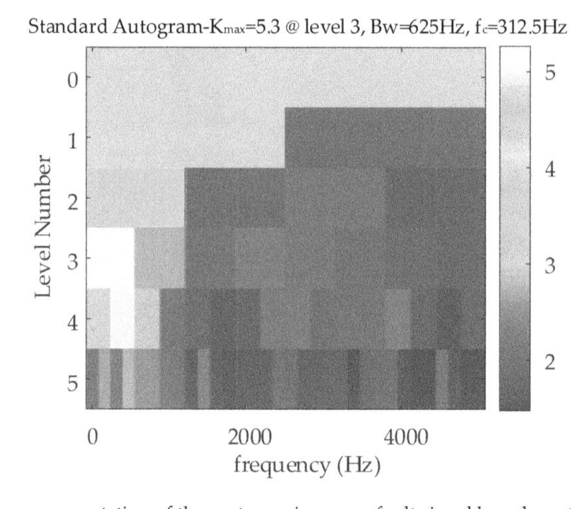

Figure 13. Colormap presentation of the center spring wear fault signal based on standard Autogram.

The maximum *kurtosis* of 5.3 is assigned to the node (3, 1), with center frequency of 312.5 Hz and a bandwidth of 625 Hz. Thus, node (3, 1) is adopted as a data source for further investigation, and the no threshold spectrum, upper threshold spectrum, and lower threshold spectrum are displayed in Figure 14.

Figure 14 shows the fault feature information on center spring wear at fault feature frequency 24.5 Hz with most of its harmonics extracted. The amplitude values obtained based on no threshold processing in Figure 14a are much larger than those in Figure 14b,c, and they are larger than those of original center spring wear signal spectrum in Figure 12. It can be seen that Autogram is effective for extracting the information, and standard Autogram based on no threshold processing has the strongest extraction ability.

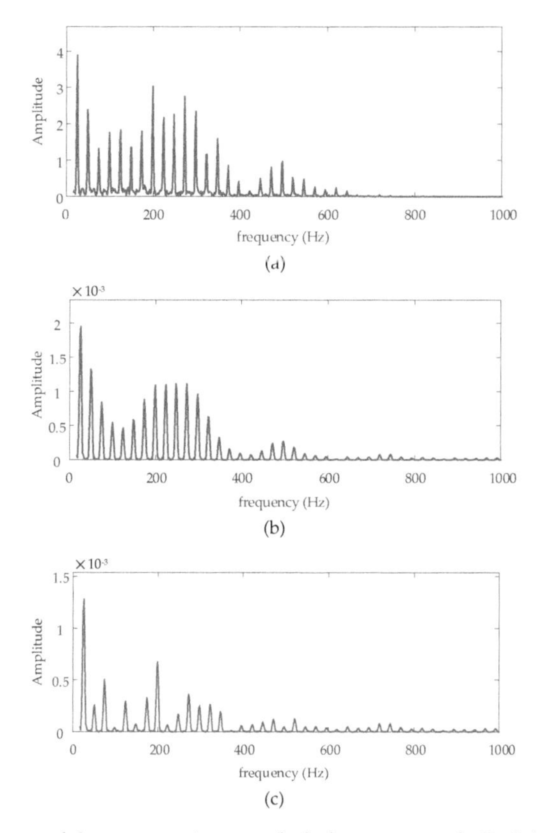

Figure 14. Spectrums of the center spring wear fault data source node (3, 1) based on standard Autogram. (**a**) No threshold spectrum; (**b**) Upper threshold spectrum; (**c**) Lower threshold spectrum.

5.4. Analysis of center spring wear fault signal of fluid pressure based on Upper Autogram.

Upper Autogram is applied to extract the fault feature information, and the colormap presentation based on Equation (3) is shown in Figure 15.

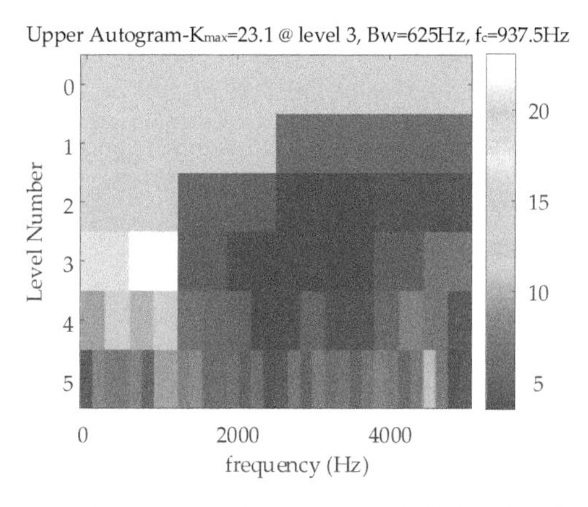

Figure 15. Colormap presentation of the center spring wear fault signal based on Upper Autogram.

In Figure 15, the maximum $kurtosis_u$ is 23.1, and it corresponds to the node (3, 2), with a center frequency of 937.5 Hz and a bandwidth of 625 Hz. Node (3, 2) is used as data source for further investigation, and the no threshold spectrum, upper threshold spectrum, and lower threshold spectrum are shown in Figure 16.

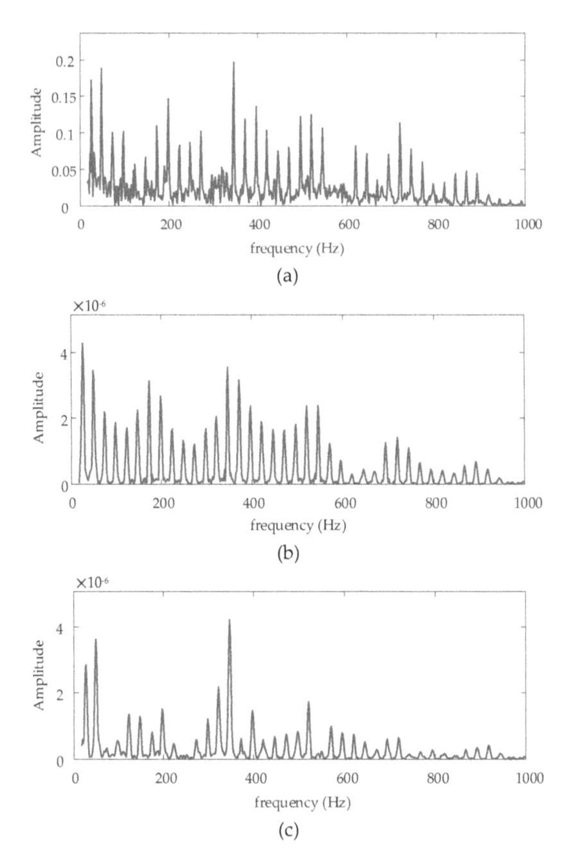

Figure 16. Spectrums of the center spring wear fault data source node (3, 2) based on upper Autogram. (**a**) No threshold spectrum; (**b**) Upper threshold spectrum; (**c**) Lower threshold spectrum.

Figure 16c illustrates that the fault feature information at fault feature frequency 24.5 Hz and its harmonics are extracted. However, in Figure 16a,b, most harmonics are not extracted effectively, and there are many background noises. Thus, upper Autogram based on lower threshold processing has the strongest extraction ability.

Compared with standard Autogram based on no threshold processing in Figure 14a and original center spring wear signal in Figure 12, standard Autogram based on lower threshold processing is not so effective and has the weakest extraction ability.

5.5. Analysis of center spring wear fault signalof fluid pressurebased on Lower Autogram.

Extraction of the fault feature information is executed by lower Autogram, and the colormap presentation can be obtained based on Equation (4). It is illustrated in Figure 17.

In Figure 17, the maximum $kurtosis_l$ is 1.6, and it corresponds to node (3, 4), with a center frequency of 2187.5 Hz and a bandwidth of 625 Hz. Node (3, 4) is used as data source, and its no threshold spectrum, upper threshold spectrum, and lower threshold spectrum are demonstrated in Figure 18.

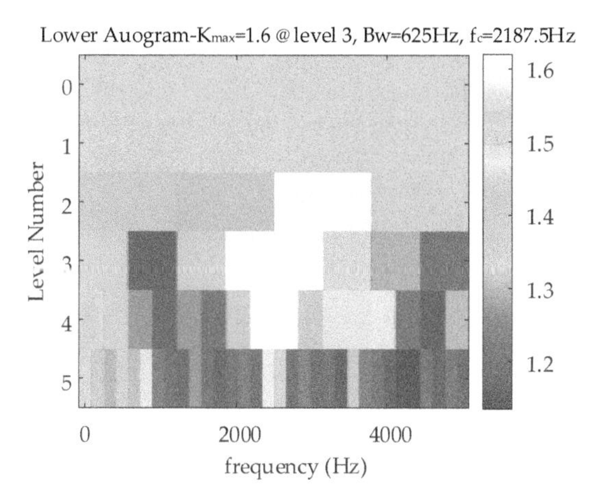

Figure 17. Colormap presentation of the center spring wear fault signal based on lower Autogram.

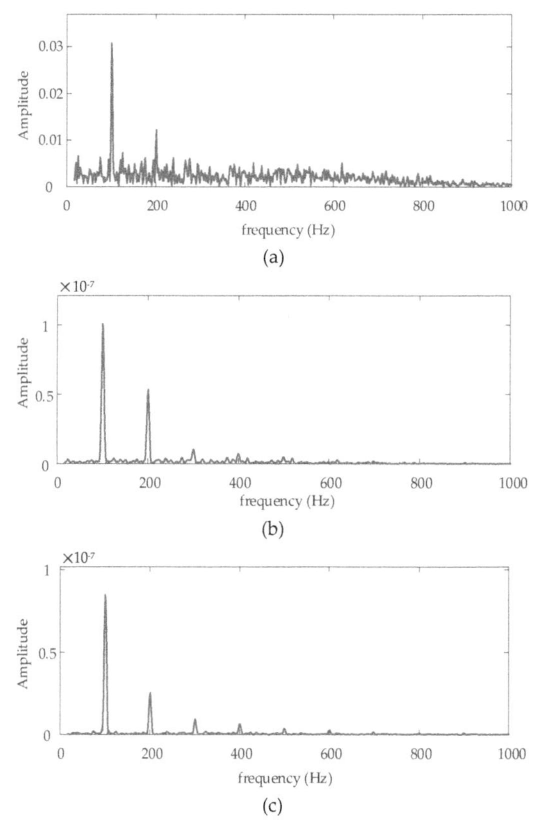

Figure 18. Spectrums of the center spring wear fault data source node (3, 4) based on lower Autogram. (**a**) No threshold spectrum; (**b**) Upper threshold spectrum; (**c**) Lower threshold spectrum.

The fault feature information at fault feature frequency 24.5 Hz and most of its harmonics are not extracted in the three figures in Figure 18, and their amplitude values are smaller compared with those of original center spring wear signal in Figure 12. Thus, lower Autogram based on three kinds of threshold processing has very weak extraction ability.

Comparing with the standard Autogram result in Figure 14 and the upper Autogram result in Figure 16, lower Autogram has the weakest extraction ability, and it is influenced by background noises. Thus, standard Autogram has the strongest extraction ability and can extract the most fault feature information from the background noises.

5.6. Analysis of center spring wear fault signalof fluid pressure based on FK

In order to illustrate the validation and advantages of the Autogram, the signal is also analyzed by FK. Colormap presentation obtained based on FK is illustrated in Figure 19.

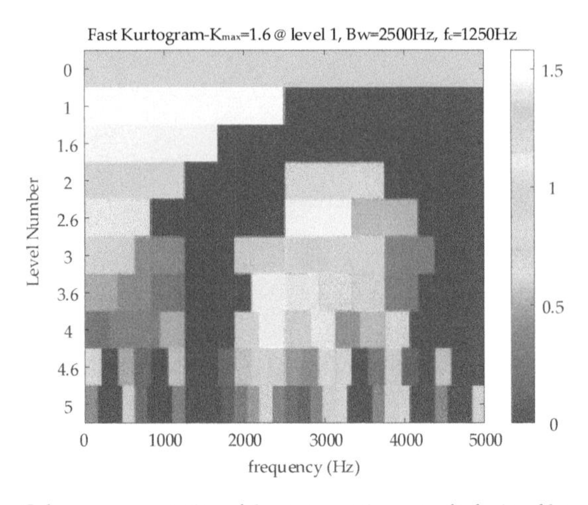

Figure 19. Colormap presentation of the center spring wear fault signal based on FK.

In Figure 19, the maximum kurtosis value is 1.6, and it corresponds to the node (1, 1), with a center frequency of 1230 Hz and a bandwidth of 2500 Hz. Thus, the node is used as a data source for further investigation. Its spectrum is shown in Figure 20.

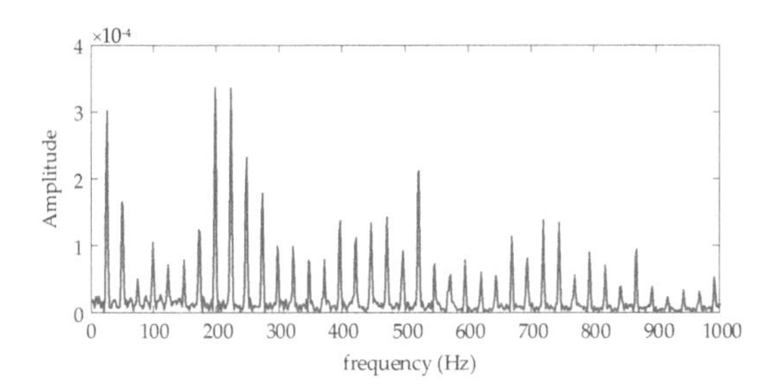

Figure 20. Spectrums of the center spring wear fault data source node (1, 1) based on FK.

Figure 20 signifies that the fault feature information at fault feature frequency 24.5 Hz and most of harmonics are extracted. However, compared with standard Autogram based on no threshold processing in Figure 14a and the original center spring wear in Figure 12, the amplitude values are very small (Figure 20).

The conclusion can be drawn from above that standard Autogram based on no threshold processing can extract the fault feature information on center spring wear successfully, and it performs better than FK in fault feature extraction ability.

6. Conclusions

Pressure fluctuation of hydraulic pump can be caused by the center spring wear fault, and the fault feature information of fluctuation is heavily influenced by fluid flow contamination. Aiming to resolve this problem, Autogram is applied to extract the fault feature information from the fluid pressure signal. The research results indicate that standard Autogram can extract more fault feature information on center spring wear than upper Autogram and lower Autogram, especially standard Autogram based on no threshold processing. Autogram also performs better than FK in extracting the fault feature information.

Author Contributions: Conceptualization, Z.Z.; Methodology, Z.Z.; Investigation, Z.Z. and X.L.; Writing-Original Draft Preparation, Z.Z.; Writing-Review and Editing, Y.Z. and Z.Z.

Funding: This research was funded by the Startup Foundation for the Doctors of North China University of Science and Technology (0088/28412499), the National Natural Science Foundation of China (51805214), and the China Postdoctoral Science Foundation (2019M651722).

Conflicts of Interest: The authors declare no conflict of interest.

References

1. Qian, J.Y.; Chen, M.R.; Liu, X.; Jin, Z. A numerical investigation of the flow of nanofluids through a micro Tesla valve. *J. Zhejiang. Univ. Sci. A* **2019**, *20*, 50–60. [CrossRef]
2. Qian, J.Y.; Gao, Z.X.; Liu, B.Z.; Jin, Z.J. Parametric study on fluid dynamics of pilot-control angle globe valve. *ASME J. Fluids Eng.* **2018**, *140*, 111103. [CrossRef]
3. Zhang, J.; Xia, S.; Ye, S.; Xu, B.; Song, W.; Zhu, S.; Tang, H.; Xiang, J. Experimental investigation on the noise reduction of an axial piston pump using free-layer damping material treatment. *Appl. Acoust.* **2018**, *139*, 1–7. [CrossRef]
4. Ye, S.G.; Zhang, J.H.; Xu, B.; Zhu, S.Q.; Xiang, J.W.; Hesheng, T. Theoretical investigation of the contributions of the excitation forces to the vibration of an axial piston pump. *Mech. Syst. Signal. Pr.* **2019**, *129*, 201–217. [CrossRef]
5. Wang, C.; Chen, X.X.; Qiu, N.; Zhu, Y.; Shi, W.D. Numerical and experimental study on the pressure fluctuation, vibration, and noise of multistage pump with radial diffuser. *J. Braz. Soc. Mech. Sci.* **2018**, *40*. [CrossRef]
6. Chen, T.; Chen, L.; Xu, X.; Cai, Y.F.; Jiang, H.B.; Sun, X.Q. Passive fault-tolerant path following control of autonomous distributed drive electric vehicle considering steering system fault. *Mech. Syst. Signal Proc.* **2019**, *123*, 298–315. [CrossRef]
7. Mao, Y.X.; Liu, G.H.; Zhao, W.X.; Ji, J.H. Vibration prediction in fault-tolerant flux-switching permanent-magnet machine under healthy and faulty conditions. *IET Electr. Power. Appl.* **2017**, *11*, 19–28. [CrossRef]
8. Zhu, Y.; Tang, S.; Quan, L.; Jiang, W.; Zhou, L. Extraction method for signal effective component based on extreme-point symmetric mode decomposition and Kullback-Leibler divergence. *J. Braz. Soc. Mech. Sci.* **2019**, *41*, 100. [CrossRef]
9. Zhang, H.; Zhang, Y.; Liu, D.; Ji, Y.; Jiang, J.; Sun, Y. Research on MEMS sensor in hydraulic system flow detection. *Proc. SPIE* **2011**, *7997*. [CrossRef]
10. Goharrizi, A.Y.; Sepehri, N. Internal leakage detection in hydraulic actuators using empirical mode decomposition and hilbert spectrum. *IEEE Trans. Instrum. Meas.* **2012**, *61*, 368–378. [CrossRef]
11. Vásquez, S.; Kinnaert, M.; Pintelon, R. Active fault diagnosis on a hydraulic pitch system based on frequency-domain identification. *IEEE Trans. Control Syst. Technol.* **2017**, *27*, 1–16. [CrossRef]
12. Tang, H.B.; Wu, Y.X.; Ma, C.X. Inner leakage fault diagnosis of hydraulic cylinder using wavelet energy. *Adv. Mater. Res.* **2011**, *139–141*, 2517–2521. [CrossRef]
13. Xu, J.J.; Yang, S.Y.; Yuan, J. Fault diagnosis of wavelet packet neural network on pump valves of reciprocating pumps based on pressure signal. *J. Dalian Marit. Univ.* **2007**, *33*, 22–25.
14. Guo, X.L.; Yang, K.L.; Guo, Y.X. Hydraulic pressure signal denoising using threshold self-learning wavelet algorithm. *J. Hydrodyn.* **2008**, *20*, 433–439. [CrossRef]
15. You, Z.P.; Ye, X.P.; Zhang, W.H. Hydraulic system fault diagnosis method based on HPSO and WP-EE. *Appl. Mech. Mater.* **2014**, *577*, 438–442. [CrossRef]

16. Li, S.; Zhang, C.H.; Shi, M. Neural network prediction model for ship hydraulic pressure signal under wind wave background. *J. Shanghai Jiaotong Univ.* **2015**, *20*, 224–227. [CrossRef]
17. Desbiens, A.B.; Bigué, J.P.L.; Véronneau, C.; Masson, P.; Iagnemma, K.; Plante, J.S. On the potential of hydrogen-powered hydraulic pumps for soft robotics. *Soft Robot.* **2017**, *4*, 367–378. [CrossRef] [PubMed]
18. Lee, M.C.; Chung, S.H.; Cho, J.H.; Chung, S.T.; Kwon, Y.S.; Kim, J.H.; Joun, M.S. Three-dimensional finite element analysis of powder compaction process for forming cylinder block of hydraulic pump. *Powder Metall.* **2013**, *51*, 89–94. [CrossRef]
19. Yin, F.; Nie, S.L.; Ji, H.; Huang, Y.Q. Non-probabilistic reliability analysis and design optimization for valve-port plate pair of seawater hydraulic pump for underwater apparatus. *Ocean. Eng.* **2018**, *163*, 337–347. [CrossRef]
20. Wu, D.; Liu, Y.; Li, D.; Zhao, X.; Li, C. Effect of materials on the noise of a water hydraulic pump used in submersible. *Ocean. Eng.* **2017**, *131*, 107–113. [CrossRef]
21. Xie, F.; Rui, X.; Gang, S.; Cuntang, W. Flow characteristics of accelerating pump in hydraulic-type wind power generation system under different wind speeds. *Int. J. Adv. Manuf. Technol.* **2017**, *92*, 189–196.
22. Wang, Y.; Li, H.G.; Ye, P. Fault feature extraction of hydraulic pump based on CNC de-noising and HHT. *J. Fail. Anal. Prev.* **2015**, *15*, 139–151. [CrossRef]
23. Li, H.; Sun, J.; Ma, H.; Tian, Z.; Li, Y. A novel method based upon modified composite spectrum and relative entropy for degradation feature extraction of hydraulic pump. *Mech. Syst. Signal. Process.* **2019**, *114*, 399–412. [CrossRef]
24. Leandro, W.; Richard, M.C.; Lais, M.L.; Luiz, F.F.I.; Elvys, I.M.C.; Luan, d.C.C. Didactic system of supervision and data acquisition to evaluate the performance of hydraulic pumps. *IEEE Lat. Am. Trans.* **2018**, *16*, 1113–1120.
25. Zhang, Z.C.; Chen, H.X.; Ma, Z.; Wei, Q.; He, J.-W.; Liu, H.; Liu, C. Application of the hybrid RANS/LES method on the hydraulic dynamic performance of centrifugal pumps. *J. Hydrodyn.* **2019**, *31*, 637–640. [CrossRef]
26. Sun, J.; Li, H.R.; Xu, B.H. Prognostic for hydraulic pump based upon DCT-composite spectrum and the modified echo state network. *Springerplus* **2016**, *5*, 1293. [CrossRef] [PubMed]
27. Sun, J.; Li, H.R.; Wang, W.G.; Xu, B.-H. Degradation feature extraction of hydraulic pump based on morphological undecimated decomposition fusion and DCT high order singular entropy. *J. Vib. Shock.* **2015**, 34. [CrossRef]
28. Tian, Y.; Lu, C.; Wang, Z.L. Approach for hydraulic pump fault diagnosis based on WPT-SVD and SVM. *Appl. Mech. Mater.* **2015**, *764–765*, 191–197. [CrossRef]
29. Lu, C.; Wang, S.P.; Wang, X.J. A multi-source information fusion fault diagnosis for aviation hydraulic pump based on the new evidence similarity distance. *Aerosp. Sci. Technol.* **2017**, *71*, 392–401. [CrossRef]
30. Zhang, T.X.; He, D. A reliability-based robust design method for the sealing of slipper-swash plate friction pair in hydraulic piston pump. *IEEE Trans. Rel.* **2018**, *67*, 1–11. [CrossRef]
31. Sun, H.; Yuan, S.Q.; Luo, Y. Cyclic Spectral Analysis of Vibration Signals for Centrifugal Pump Fault Characterization. *IEEE Sens. J.* **2018**, *18*, 2925–2933. [CrossRef]
32. Zhu, Y.; Qian, P.; Tang, S.; Jiang, W.; Li, W.; Zhao, J. Amplitude-frequency characteristics analysis for vertical vibration of hydraulic AGC system under nonlinear action. *AIP Adv.* **2019**, *9*, 035019. [CrossRef]
33. Zhou, H.W.; Liu, G.H.; Zhao, W.X.; Yu, X.D.; Gao, M.H. Dynamic Performance Improvement of Five-Phase Permanent-Magnet Motor with Short-Circuit Fault. *IEEE Trans. Ind. Electron.* **2018**, *65*, 145–155. [CrossRef]
34. Moshrefzadeh, A.; Fasana, A. The Autogram: An effective approach for selecting the optimal demodulation band in rolling element bearings diagnosis. *Mech. Syst. Signal. Process.* **2018**, *105*, 294–318. [CrossRef]
35. Walden, A.T. Wavelet analysis of discrete time series. *Eur. Cong. Math.* **2001**, *202*, 627–641.
36. Jerome, A. Fast computation of the kurtogram for the detection of transient faults. *Mech. Syst. Signal. Process.* **2007**, *21*, 108–124.
37. Jiang, W.; Zheng, Z.; Zhu, Y.; Li, Y. Demodulation for hydraulic pump fault signals based on local mean decomposition and improved adaptive multiscale morphology analysis. *Mech. Syst. Signal. Process.* **2015**, *58–59*, 179–205. [CrossRef]

Article

Numerical Analysis of Two-Phase Flow in the Cavitation Process of a Waterjet Propulsion Pump System

Weixuan Jiao [1], Li Cheng [1,*], Jing Xu [2] and Chuan Wang [1,*]

[1] College of Hydraulic Science and Engineering, Yangzhou University, Yangzhou 225009, China; DX120170049@yzu.edu.cn
[2] Ningbo Jushen Pumps Industry Co., Ltd., Ningbo 315100, China; xujing1990mail@126.com
* Correspondence: chengli@yzu.edu.cn (L.C.); wangchuan198710@126.com (C.W.); Tel.: +86-0514-8792-1191 (L.C.)

Received: 2 September 2019; Accepted: 19 September 2019; Published: 2 October 2019

Abstract: The waterjet propulsion system has been widely used in the military and civil fields because of its advantages of in terms of high efficiency and energy savings. In order to study the three-dimensional cavitation flow in the waterjet propulsion pump, the cavitation process of the waterjet propulsion pump was simulated numerically using the Zwart–Gerber–Belamri cavitation model and the RNG (Renormalization Group) k-ε model. The simulation results of cavitation on the waterjet propulsion pump and pump system show that, in the initial stage of cavitation, vapors first collect on the leading edge of the suction surface of the blade near the rim of the impeller. As the total pressure at the impeller inlet decreases, the cavitation region expands toward the trailing edge and the vapor fraction volume gradually increases. In order to simulate the cavitation state of the waterjet propulsion pump under the actual working conditions, a numerical simulation of the entire waterjet propulsion pump system with inlet passage was carried out. After assembling the inlet passage, the flow pattern at the impeller inlet becomes uneven, leading to irregular changes in the cavitation region of the impeller. The potential danger regions of cavitation are the lip of inlet passage and the upper and lower connecting curved section of the inlet passage. The performance of waterjet propulsion pump system changes greatly when the net positive suction head available (*NPSHa*) value of the pump reaches the critical value.

Keywords: waterjet propulsion system; pump; two-phase flow; cavitation; numerical simulation

---

1. Introduction

Waterjet propulsion is largely used in the military and civil fields because of its simple transmission mechanism, low noise, and good maneuverability [1–3]. While compared to propeller propulsion waterjet propulsion can utilize the inflow stamping to increase the anti-cavitation capability at high speeds, cavitation problems may arise under some special conditions such as "low ship speed and high rotating speed" [4,5]. This special working condition usually occurs in the process of ship acceleration or turning. At this time, the net positive suction head of the waterjet propulsion system is obviously lower than the *NPSHr* (Net Positive Suction Head Required) of the pump at this speed, and the propulsion pump is prone to cavitation. Cavitation will cause an increase in shaft power, a reduction in flow rate, head, and efficiency, and a decrease in thrust. At the same time, it will also cause noise and vibration in the pump [6–10].

With the development of technology and the problems encountered in practical applications, there is an increasing demand for pump performance indicators in engineering, and cavitation has gradually become an important factor limiting the further improvement of pump performance [11,12].

Cavitation refers to the process of forming vapor bubbles in the low pressure region of the liquid flow field [13]. Previous studies have shown that the cavitation process occurring in hydraulic machinery is harmful, and its damage is mainly manifested in three aspects. First, cavitation can cause damage to the surface of hydraulic machinery materials [14,15]. Secondly, cavitation will lead to a significant reduction in hydraulic performance of hydraulic machinery [16,17]. Thirdly, cavitation not only affects the steady state fluid flow, but also affects the unsteady flow characteristics or dynamic response of the flow [18,19]. For a long time, scholars at home and abroad have started detailed and in-depth research on the mechanism of hydraulic machinery cavitation, and accumulated a large amount of knowledge and experience through the use of numerical simulations and experiments [20–23]. As early as 1917, Rayleigh [24] conducted a theoretical analysis of the cavitation. Rayleigh solved the flow problem of a spherical cavity infinite flow field with a vacuum or internal pressure constant and gave the Rayleigh bubble athletics equation. This laid the foundation for the modern numerical simulation of cavitation dynamics. Bal et al. [25] described a method for simulating the uniform motion of a two-dimensional or three-dimensional cavitating hydrofoil under a free surface. This method is suitable for 2D and 3D hydrofoils under fully wetted or cavitating flow conditions. Brennen et al. [26] simulated the unsteady cavitation flow in the two-dimensional potential flow and successfully predicted the occurrence of two-dimensional cavitation. Singhal et al. [27] and Schmidt et al. [28] performed full-flow numerical calculation and analysis of the cavitation of the axial flow pump under design conditions, and predicted the development of cavitation flow and cavitation occurrence region in the flow passage. Although numerical simulation technology can predict the occurrence and development of cavitation, the experiment is still one of the most efficient way to study cavitation. Gopalan [29] used the PIV (Particle Image Velocimetry) and high-speed photography to study the flow structure in the closure region of sheet cavitation. The results showed that the collapse of cavitation in the enclosed region is the main cause of the whirlpool disaster. Escaler et al. [30] evaluated the cavitation detection in actual hydraulic turbines based on analysis of structural vibrations, acoustic emissions, and hydrodynamic pressures measured in the machine. They validated the proposed technique by experimenting with real prototypes of different types of cavitation. Through the performance test of centrifugal pumps, Johnson et al. [31] described the hydraulic characteristics of centrifugal pumps in cavitation state more accurately and systematically.

As effective research methods for vibration, noise, and cavitation, numerical simulation and model tests have been adopted by many experts and scholars [32–35]. Experts and scholars have made great progress in the study of cavitation, but the cavitation problem under special conditions in the waterjet propulsion system is still an important factor that restricts the full play of its hydraulic performance. Moreover, the present research on cavitation of waterjet propulsion system mainly focuses on the cavitation characteristics of waterjet propulsion pump. There are few studies on cavitation of waterjet propulsion system with inlet passage. In particular, the influence of inlet passage on the cavitation performance of propulsion pump system and pump is worth further study. Hence, it is of great practical value and academic significance to study the cavitation process of the waterjet propulsion system with and without the inlet passage. In this paper, the cavitation characteristics of the whole flow field of the waterjet propulsion pump system were analyzed by means of a two-phase flow numerical simulation.

2. Methodology

2.1. Turbulence Model and Cavitation Model

During the development of cavitation flow, the fluid was treated as a vapor liquid mixture. The gas-liquid two-phase mixture model (Mixture model) was used for two-phase flow calculations in this paper. The mixture model assumes that the fluid is homogeneous, and the two-phase fluid components are assumed to share the same velocity and pressure. The continuity equation of vapor/liquid, momentum equation, and vapor volume fraction mass transport equation for the two-phase flow are generally expressed as follows:

$$\frac{\partial \rho_m}{\partial t} + \frac{\partial(\rho_m u_j)}{\partial x_j} = 0 \tag{1}$$

$$\frac{\partial \rho_m u_i}{\partial t} + \frac{\partial(\rho_m u_i u_j)}{\partial x_j} = \rho f_i - \frac{\partial p}{\partial x_i} + \frac{\partial}{\partial x_j}\left[(\mu_m + \mu_t)\left\{\frac{\partial u_i}{\partial x_j} + \frac{\partial u_j}{\partial x_i} - \frac{2}{3}\frac{\partial u_i}{\partial x_j}\delta_{ij}\right\}\right] \tag{2}$$

$$\frac{\partial(\alpha_v \rho_V)}{\partial t} + \frac{\partial(\alpha_v \rho_V u_j)}{\partial x_j} = R_E - R_C \tag{3}$$

where p is the mixture pressure, Pa; ρ_V is vapor volume density, kg/m^3; u_i is the velocity in i direction, m/s; u_j is the velocity in j direction, m/s; f_i is the body force in the i direction; μ is the laminar viscosity; t is the time, s; μ_t is the turbulent viscosity; α_v is the volume fraction of vapor; and R_E and R_C respectively represent the source terms for evaporation and condensation, kg/(m·s). The mixture density ρ_m is defined as follows:

$$\rho_m = \rho_V \alpha_v + \rho_l(1 - \alpha_v) \tag{4}$$

where ρ_l is the liquid volume density.

The RNG k-ε (Renormalization Group k-ε) model can simulate the region of cavitation clearly and has better adaptability to the simulation of cavitation flow [36–38]. Therefore, under the assumption of mixture homogeneous flow model, the RNG k-ε model was selected to calculate the cavitation characteristics of the waterjet propulsion system.

The cavitation model is a mathematical model that describes the mutual transformation between the liquid volume and the vapor volume. The transport equation model is the most commonly used cavitation model, which mainly includes three types: the Zwart–Gerber–Belamri model [39], the Kunz model [40], and the Schnerr–Sauer model [41]. In this paper, the Zwart model in ANSYS-CFX software was selected to simulate and analyze the cavitation process.

$$R_e = F_{vap}\frac{3\alpha_{ruc}(1-\alpha_v)\rho_V}{R_B}\sqrt{\frac{2}{3}\frac{(P_V - P)}{\rho_l}} \quad \mathrm{P} < \mathrm{P_v} \tag{5}$$

$$R_c = F_{cond}\frac{3\alpha_v \rho_V}{R_B}\sqrt{\frac{2}{3}\frac{(P - P_V)}{\rho_l}} \quad \mathrm{P} > \mathrm{P_v} \tag{6}$$

where P_V is the vapor pressure; P is the flow field pressure; F_{vap} and F_{cond} are empirical coefficients for the vaporization and condensation processes, respectively; α_{ruc} is the non-condensable gas fraction in the liquid; and R_B is the typical bubble size in the water. According to numerous literature discussions [42–44], α_{ruc} takes the value of 5×10^{-4}; R_B takes the value of 1×10^{-6} m; and F_{vap} and F_{cond} take the values 50 and 0.01, respectively.

2.2. Geometric Model and Mesh Generation

As can be seen from Figure 1 that the propulsion pump is the core component of waterjet propulsion pump system, which consists of an impeller with six blades and a guide vane with seven vanes. Therefore, in order to ensure that the pump system has better anti-cavitation ability, it is necessary to require the pump itself to have a good anti-cavitation performance. As shown in Figure 1, the waterjet propulsion pump system is the main research object. The calculation domain of the waterjet propulsion pump system includes import extension, inlet passage, impeller, guide vane and nozzle. The inlet flow of the inlet passage is a non-uniform flow, so the inlet section needs to be extended by a distance to guarantee the accuracy and convergence of the calculation, as shown in the import extension in Figure 1a. The basic geometrical parameters of the waterjet propulsion pump system are shown in Table 1. The rotational speed of the waterjet propulsion pump is 700 r/min. D_0 represents the inlet diameter of the impeller. Section P1 and Section P2 are respectively the inlet and outlet sections for calculating the head of the pump section.

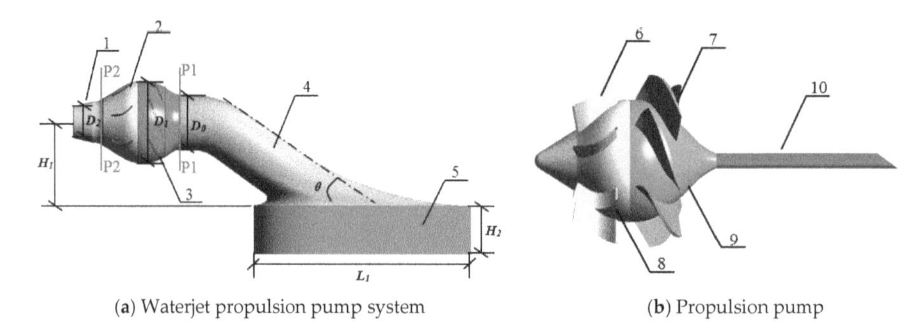

(a) Waterjet propulsion pump system (b) Propulsion pump

Figure 1. Three-dimensional schematic diagram of the waterjet propulsion pump system. 1. Nozzle; 2. Guide vane; 3. Impeller; 4. Inlet passage; 5. Import extension; 6. Guide vane blades; 7. Impeller blades; 8. Guide vane hub; 9. Impeller hub; 10. Shaft; P1. Impeller inlet section; P2. Guide vane outlet section.

Table 1. Basic geometrical parameters of the waterjet propulsion pump system.

Geometric Parameter	Value	Geometric Parameter	Value
Inlet diameter of the impeller D_0	1 D_0	Outlet diameter of the impeller D_1	1.19 D_0
Outlet diameter of the nozzle D_2	0.55 D_0	Height of the inlet passage H_1	1.2 D_0
Height of the import extension H_2	0.9 D_0	Length of the import extension L_1	4 D_0
Number of the impeller blades Z_2	6	Number of the guide vane blades Z_1	7
Dip angle of the inlet passage θ	28°	Rotational speed n (r/min)	700

The entire computational domain was divided into three parts to generate mesh, namely, the nozzle part, the propulsion pump part and the inlet passage part. Structured mesh with hexahedral cells can improve the computational efficiency of CFD (Computational Fluid Dynamics), so the meshes of the computational domain in this paper are divided into structured mesh by ICEM CFD software (ANSYS Inc., Pittsburgh, PA, USA). ICEM CFD is the integrated computer engineering and manufacturing code for computational fluid dynamics, which is a professional preprocessing software. Since both the impeller and guide vane are periodic meshes, the generated single channel meshes are rotated and duplicated to generate the computational domain of impeller and guide vane. Since different turbulence models have different requirements for grid Y plus values, the RNG k-ε model requires $y+$ values between 30 and 100. The grid size of the boundary layer was controlled to ensure that Y plus meets the requirements of turbulence model in this paper. The grid of the computational domain is shown in Figure 2.

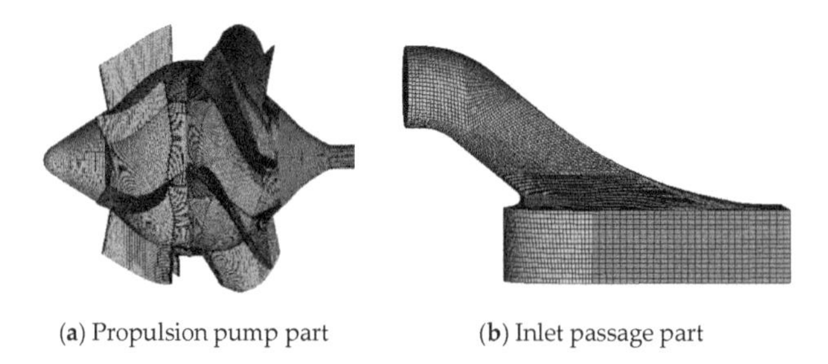

(a) Propulsion pump part (b) Inlet passage part

Figure 2. Grid of the computational domain by using ICEM CFD software.

The number of grids has a great influence on the calculation accuracy and the solution speed. In theory, the denser the calculation domain grid is, the higher the calculation accuracy is. However, in the actual calculation process, too many grids will greatly increase the calculation period and waste computing resources. In order to find the appropriate mesh size, a mesh sensitivity analysis was carried out. For this calculation model, the impeller is the core of the whole calculation domain. The mesh sensitivity of the propulsion pump section was verified by changing the size of the impeller mesh. As shown in Figure 3, when the grid number of the impeller reaches 1.6×10^6, the head of the pump section changes less. Equation (6) was used to calculate the head of the pump section. The calculated cross sections are sections P1 and P2 in Figure 1. Finally, the number of meshes of the impeller is 1.72 million, and the total number of calculation domain grids of the waterjet propulsion pump system is 5.23 million.

$$H = \frac{P_{2-2} - P_{1-1}}{\rho g} \tag{7}$$

where $P_{1\text{-}1}$ and $P_{2\text{-}2}$ are the pressure of the sections P1 and P2, respectively, Pa; ρ is water density, kg/m^3; and g is gravitational acceleration, m/s^2.

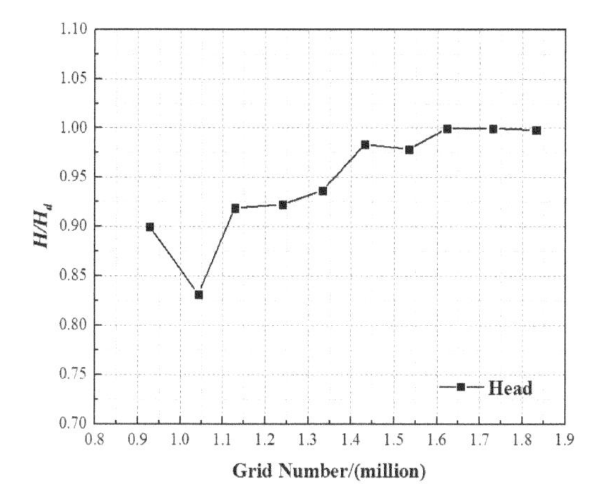

Figure 3. Mesh sensitivity analysis.

2.3. *Setting of the Boundary Condition*

The unsteady cavitating turbulent flow was simulated using the high-performance computational fluid dynamics software (ANSYS CFX 14.5, ANSYS Inc., Pittsburgh, PA, USA) based on the Reynolds-averaged Navier–Stokes (RANS) equation with the RNG k- ε turbulence model. The total pressure condition was applied as outlet boundary condition, and mass flow rate was adopted as the outlet boundary condition. The process of cavitation was controlled by changing the total pressure of the inlet. Assuming that there are no bubbles in the inlet fluid, the volume fractions of the liquid phase and the vapor phase at the inlet were set to 1 and 0, respectively. No slip condition was applied at solid boundaries. The effect of temperature was not considered in the calculation [45]. The mixture model was chosen as the multiphase flow model, the Zwart model was chosen as the cavitation model, and the Rayleigh–Plesset equation was used to control bubble motion. Before the calculation of the cavitation model was embedded, the pump segment under the condition of no cavitation was calculated, and the calculation result was used as the initial value of the cavitation simulation to improve the convergence speed and calculation accuracy. When applying the cavitation model, two physical parameters were given: the vaporization pressure of the liquid at normal temperature (25 °C) P_V = 3574 Pa, and the surface tension of the cavitation bubble, 0.074 N/m.

3. Experiment

3.1. Establishment of the Test Bed

The test bed of waterjet propulsion system is shown in Figure 4. The test bed consists of two closed loops. The main function of the first loop is to ensure the water circulation of the test bed and provide the bottom speed. The main function of the second loop is to test the performance of the waterjet propulsion pump. The size and shape of impeller and guide vane in the test device are consistent with the numerical model. The head, flow rate, torque, and speed were tested to obtain the hydraulic performance of the waterjet propulsion pump. The pump is driven by a DC electromotor (Changchuan, Nanjing, Jiangsu, China) at speeds varying from 700 rev/min to 2400 rev/min and equipped with an auxiliary axial pump to regulate the flow rate. The test rig has two electromagnetic flowmeters (Shanghai Guanghua Instrument Co., Ltd., Shanghai, China) with an absolute accuracy of ±0.5%. These flowmeters are used to test the flow rate of the main circulation pipeline system and second circulation pipeline system. The head of the waterjet propulsion pump is measured by the differential pressure transmitter (Yokogawa Sichuan Instrument Co., Ltd., Chongqing, China) with an absolute accuracy of ±0.2%.

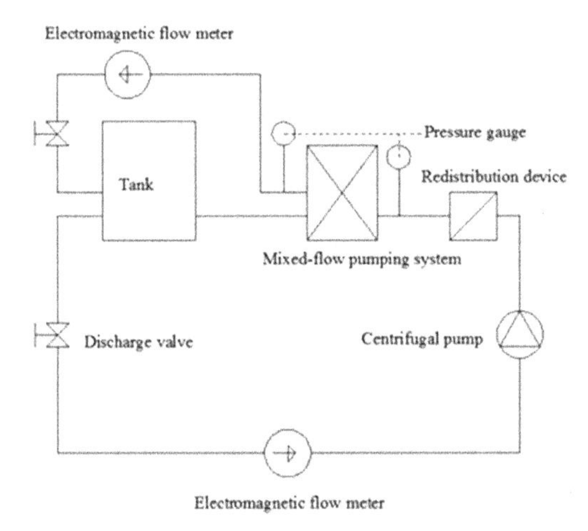

Figure 4. Test rig.

3.2. Test Verification

The external characteristics of the pump system was calculated for the numerical model without cavitation. Figure 5 shows the comparison of test results and numerical simulation results. As shown in Figure 5, under the design flow condition, the difference between the numerical prediction result of the head and the experimental result is the smallest. At the design flow rate, the head difference between the test data and the numerical data is 1.2%. The comparison between numerical simulation results and experimental results shows the numerical results are relatively reliable.

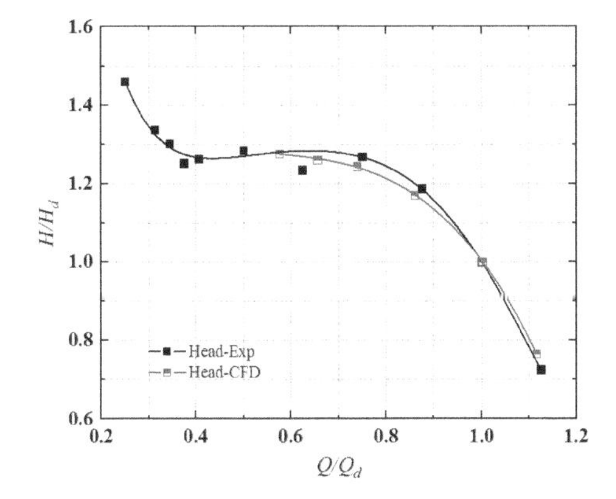

Figure 5. Comparison of calculation and experiment result. * Q_d represents the design flow rate; and H_d represents the head of the waterjet propulsion pump system under the designed flow rate condition.

4. Results and Discussion

The most important factors affecting cavitation are pressure and velocity, so the cavitation number σ is used as a parameter to characterize the possibility of cavitation, σ, defined as:

$$\sigma = \frac{P_s - P_V}{\frac{1}{2}\rho U^2} \tag{8}$$

where P_s is the reference static pressure, which is expressed as the pump inlet pressure in this study; P_V is the vapor pressure; and U is the reference velocity, which is expressed as the inlet tip speed.

The net positive suction head (*NPSH*) is the difference between the total head of the liquid at the pump inlet and the pressure head when the liquid is vaporized. The net positive suction head-available (*NPSHa*) refers to the excess energy of the liquid at the pump inlet that exceeds the vaporization pressure at that temperature.

$$NPSH_a = \frac{P_s}{\rho g} + \frac{v_s^2}{2g} - \frac{P_{cav}}{\rho g} = \frac{P_a}{\rho g} - H_x - \sum h_s - \frac{P_{cav}}{\rho g} \tag{9}$$

where $\frac{P_s}{\rho g}$ is the pressure head of pump inlet section; $\frac{v_s^2}{2g}$ is the velocity head of pump inlet section; $\frac{P_{cav}}{\rho g}$ is the vaporization pressure value; $\frac{P_a}{\rho g}$ is the atmospheric pressure; H_x is the actual water suction head of the pump; and h_s is the hydraulic loss from the intake surface to the pump inlet.

Since the calculation model is based on the center of the impeller, the installation height is 0. The *NPSHa* of the propulsion pump is calculated by the following formula:

$$NPSH_a = (P_{local} - P_V)/\rho g \tag{10}$$

where P_{local} is total pressure of inlet section.

4.1. Cavitation Characteristics of Impeller in Waterjet Propulsion Pump

The propulsion pump is the core component of the waterjet propulsion pump system. Therefore, to ensure that the pump system has better anti-cavitation capability, it is necessary to require the pump itself to have excellent anti-cavitation performance. The analysis and research on the cavitation characteristics of the propulsion pump can provide reference for the study of the cavitation performance of the whole waterjet pump system. This paper predicts the cavitation characteristics of the impeller by analyzing the cavitation form, cavitation occurrence region and development trend of the impeller, and the influence of cavitation on head and efficiency of the propulsion pump. As shown in Figure 6, the propulsion pump inlet is extended for a distance to guarantee the accuracy and convergence of the calculation, and the outlet section is connected to the nozzle. The inlet of the extension section is chosen as the inlet of the overall calculation domain, and the nozzle outlet is the outlet of the overall calculation domain. The setting of the boundary conditions is not changed.

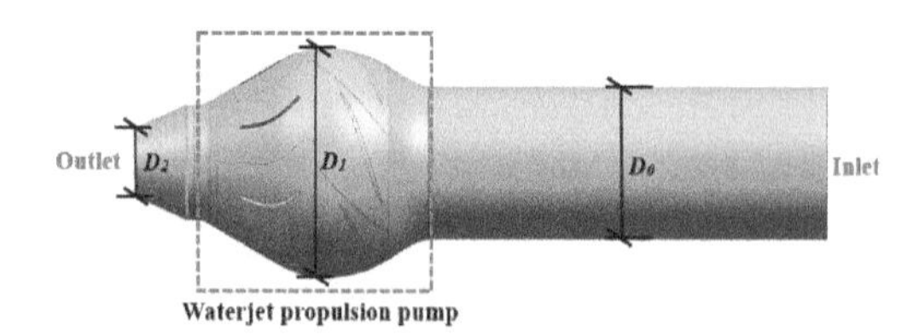

Figure 6. Three-dimensional schematic diagram of the waterjet propulsion pump.

Based on the calculation results of the performance curve, the design flow rate Q_{bep} corresponding to the highest efficiency point was chosen for the simulation calculation of the cavitation flow. During the process of gradually reducing the total pressure of the inlet section from 0.5 atm to 0.1 atm, the pressure distribution of the working surface and suction surface of impeller blade and the change of the area of the cavitation region were simulated. The pressure distribution on the suction surface of the impeller under different *NPSHa* is shown in Figure 7. The blue region represents the lowest air content region, and the red region represents the highest air content region. The air content represents the number of vacuoles per unit volume on the blade surface. In the figure, the blue region represents the low pressure region and the red region represents the high pressure region. It can be seen from the figure that when the value of *NPSHa* is greater than 1.87 m, the low pressure region is mainly concentrated at the leading edge of the blade. When the *NPSHa* value decreases, the area of the low pressure region along the edge of the airfoil gradually increases, indicating that the area and possibilities of the cavitation occurrence become larger.

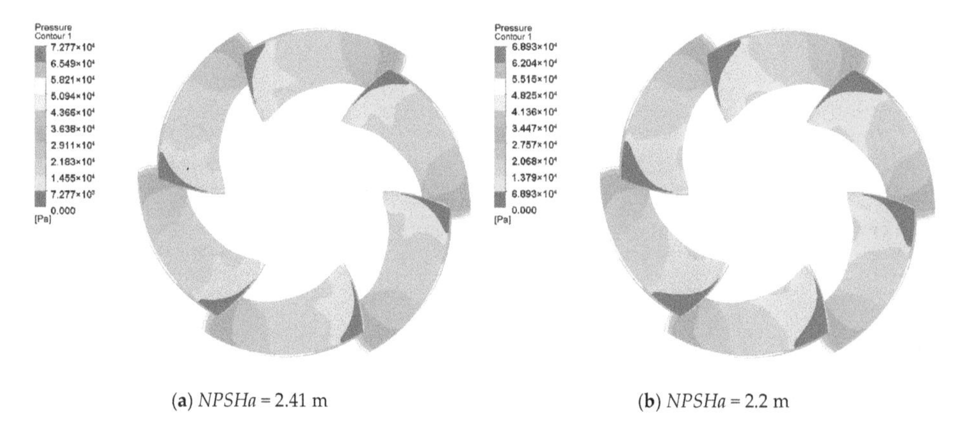

(**a**) *NPSHa* = 2.41 m (**b**) *NPSHa* = 2.2 m

Figure 7. *Cont.*

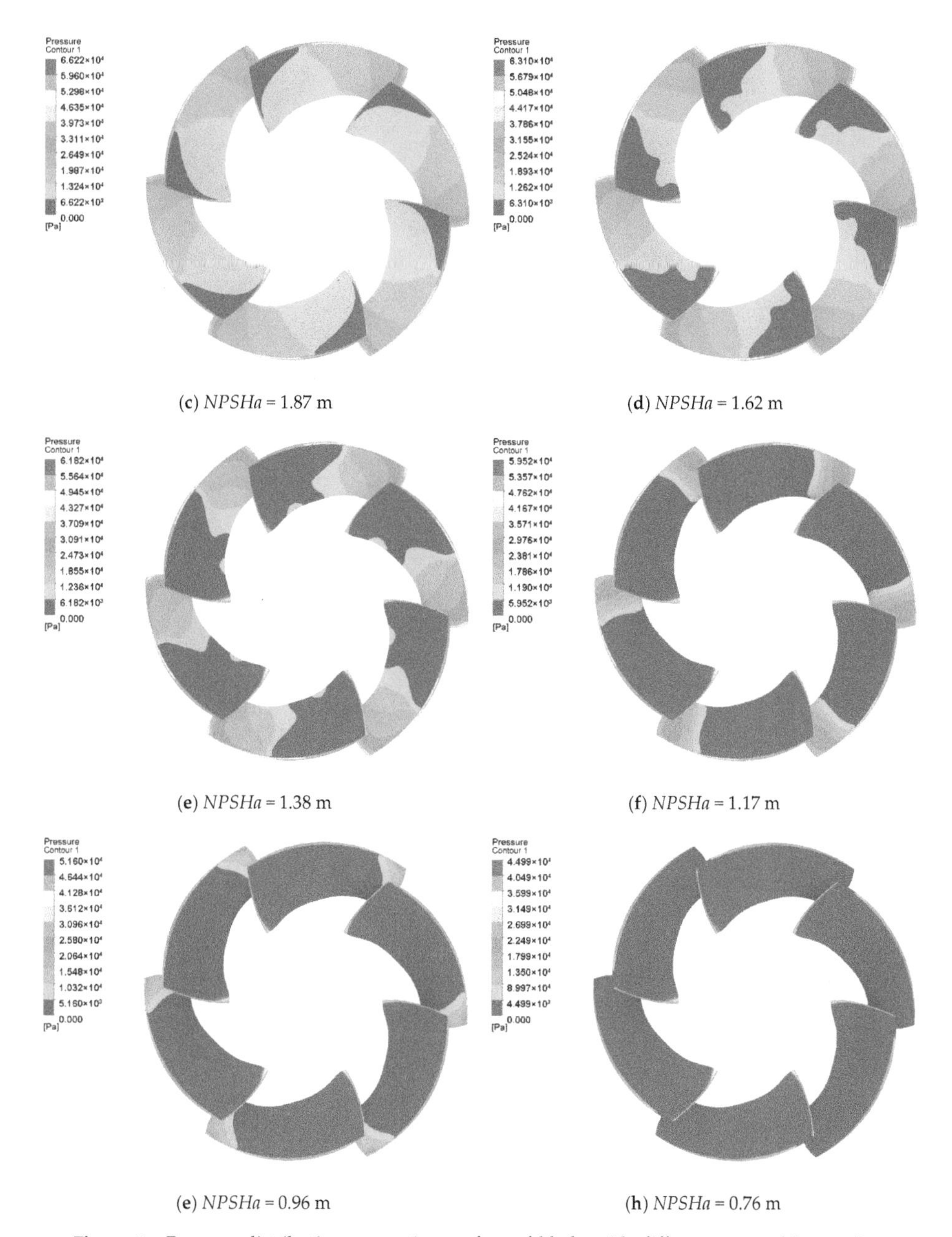

Figure 7. Pressure distribution on suction surface of blade with different net positive suction head-available (*NPSHa*) values.

Pressure distribution on pressure surface of blade with different *NPSHa* values as shown in Figure 8. When the *NPSHa* value is less than 0.96 m, the region on the pressure surface of the blade which is lower than the vaporization pressure begins to appear and the region becomes larger gradually. As the *NPSHa* value decreases, the low pressure region on the pressure surface of the blade gradually develops from the hub near the blade leading edge to the rim.

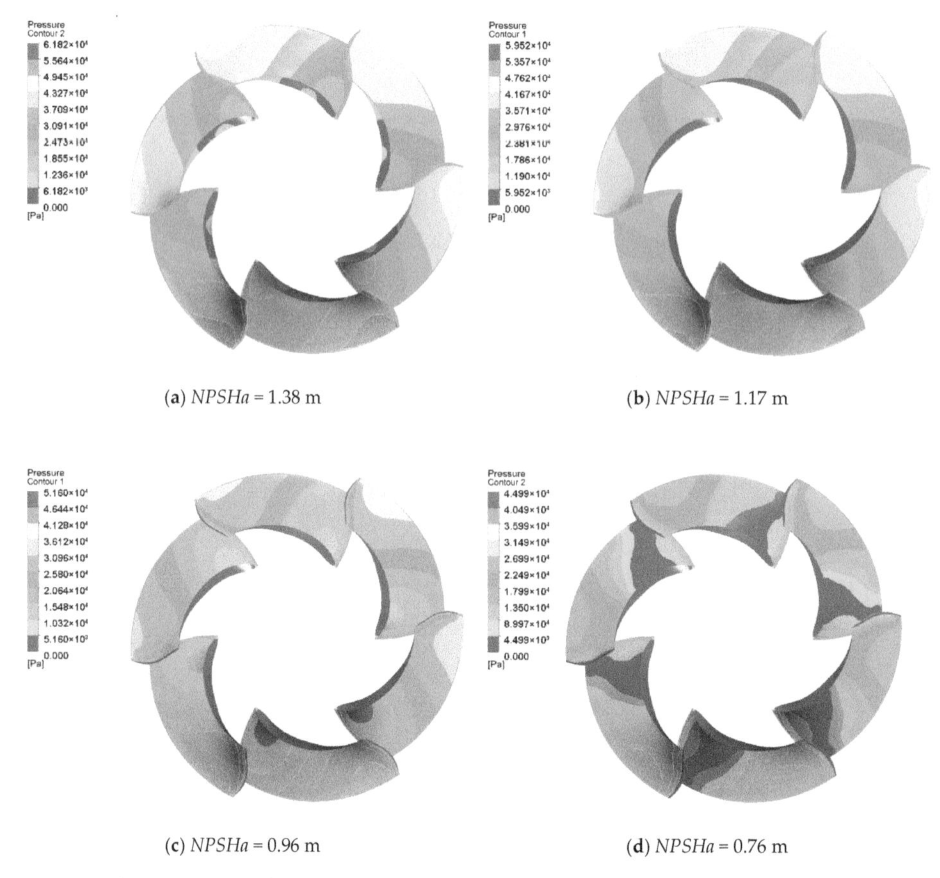

(**a**) *NPSHa* = 1.38 m (**b**) *NPSHa* = 1.17 m

(**c**) *NPSHa* = 0.96 m (**d**) *NPSHa* = 0.76 m

Figure 8. Pressure distribution on pressure surface of blade with different *NPSHa* values.

Figure 9 shows the vapor volume fraction distribution of suction surface of propulsion pump blade in different cavitation stages. The overall distribution trend of vapor volume fraction is consistent with the pressure distribution curve, and cavitation occurs in the low pressure region. The red region indicates the existence of the largest vapor volume fraction in this region, while the blue region indicates the existence of the smallest vapor volume fraction in this region. It can be seen from the figure that the cavitation of the impeller in the propulsion pump first appears around the leading edge of the blade suction surface. The flow pattern at the inlet of the impeller has a great effect on the cavitation performance. When the *NPSHa* value is equal to 1.87 m, the cavitation mainly collects on the leading edge of the blade suction surface near the rim of the impeller. The vapor volume fraction near the rim is the largest. With the decrease of the total pressure, the vapor region gradually spreads along the blade trailing edge and hub direction. When the *NPSHa* value is equal to 1.38 m, cavitation begins to appear in the middle of the surface of the blade, and the cavitation area accounts for about half of the area of the suction surface of the blade. In the process of reducing the *NPSHa* value from 2.41 to 0.76, the maximum vapor volume fraction appeared at the leading edge of the blade near the rim, and as the *NPSHa* value decreased, the maximum vapor volume fraction area gradually expanded toward the blade trailing edge along the water flow direction. As the total pressure of the inlet further decreases, the vapor region gradually spreads to the entire blade. At the same time, the maximum vapor volume fraction area is gradually moving towards the outlet side of the blade. From Figure 9h, it can be seen that when the *NPSHa* value is reduced to 0.76 m, the suction surface of impeller blade is basically covered by bubbles, and cavitation has been fully developed. As shown in Figure 10, when suction surface cavitation develops completely, bubbles begin to propagate towards the trailing edge of the blade pressure surface. When the cavitation develops completely, the cavitation occupies almost the entire blade surface. At this time, the cavitation will block the flow passage and destroy the continuity of liquid flow in the impeller, resulting in the decrease of pump efficiency and head. In summary, according to the classification of cavitation types in the pump, it can be seen that the main type of blade cavitation shown in the figure is airfoil cavitation.

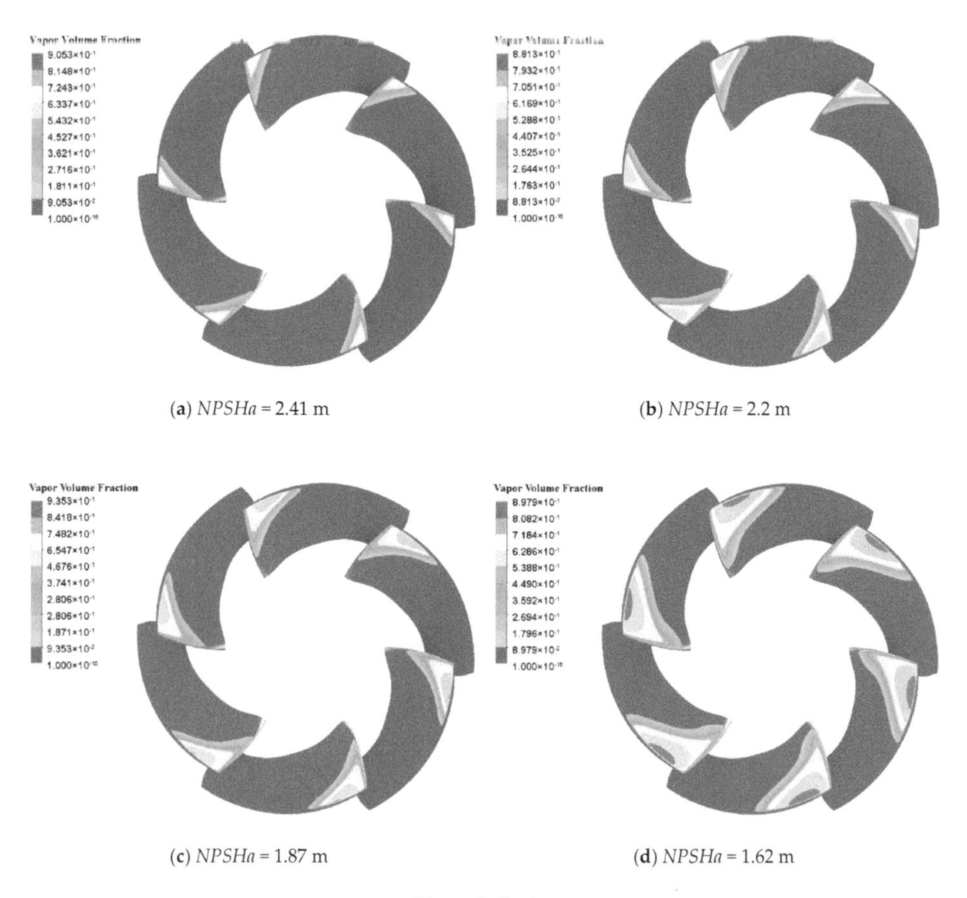

(**a**) *NPSHa* = 2.41 m (**b**) *NPSHa* = 2.2 m

(**c**) *NPSHa* = 1.87 m (**d**) *NPSHa* = 1.62 m

Figure 9. *Cont.*

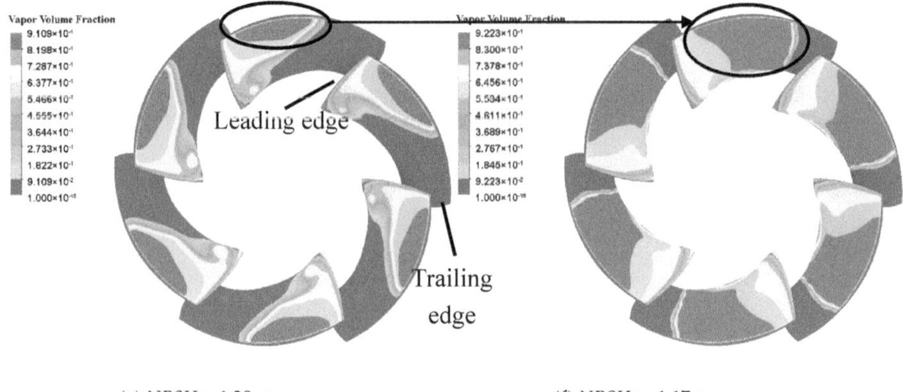

(**e**) *NPSHa* =1.38 m (**f**) *NPSHa* = 1.17 m

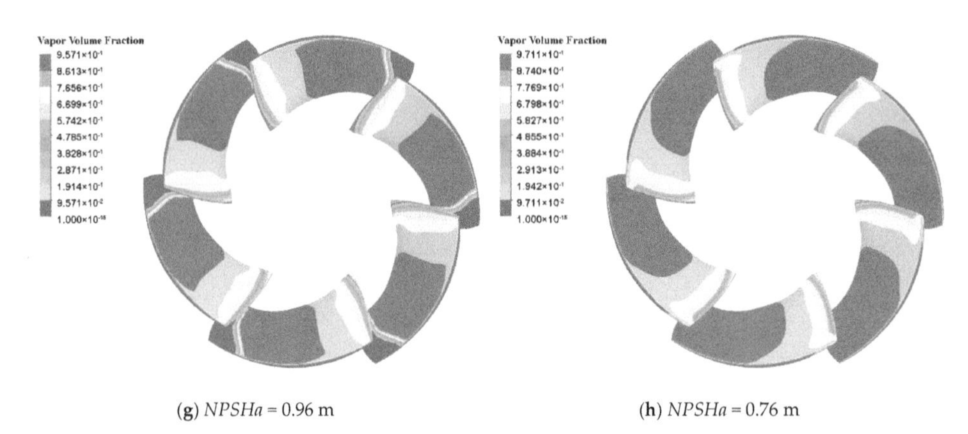

(**g**) *NPSHa* = 0.96 m (**h**) *NPSHa* = 0.76 m

Figure 9. Vapor volume fraction distribution on suction surface of blade.

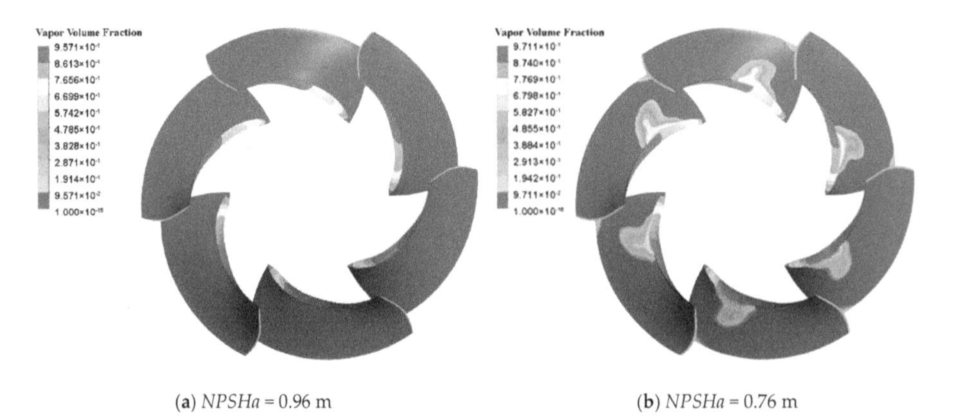

(**a**) *NPSHa* = 0.96 m (**b**) *NPSHa* = 0.76 m

Figure 10. Vapor volume fraction distribution on pressure surface of blade.

Figure 11 presents the static pressure distribution along the chord on the suction and pressure surfaces of the blades under three spans at different *NPSHa* values. As can be seen from the figure, the static pressure on the hub side is lower than that on the rim side, and the pressure distribution at the span of 0.9 times shows that the rim produces a lower pressure zone lower than the vaporization pressure (Pv = 3574 Pa) earlier than the hub. As the value of *NPSHa* decreases, the continuous development of cavitation causes the velocity and pressure distribution in the impeller passage to change, resulting in a decrease in the pressure of the blade working face and an increase in the area of the low pressure region.

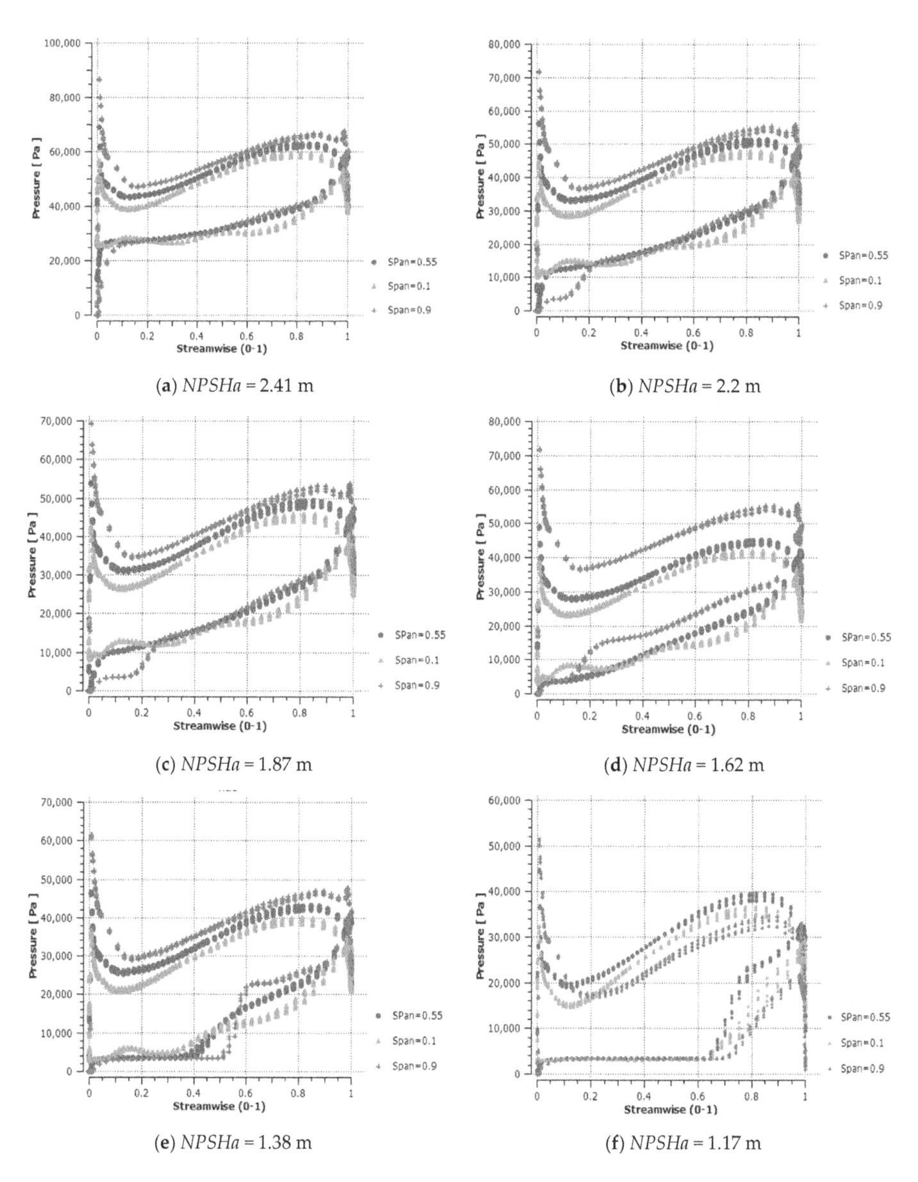

Figure 11. *Cont.*

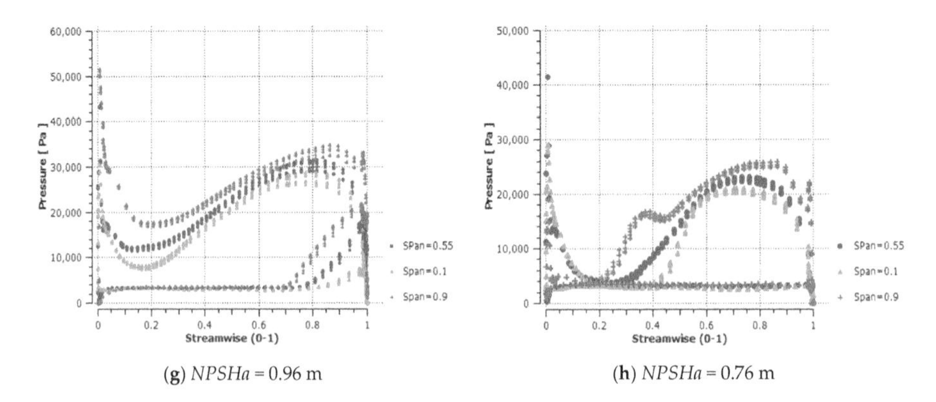

(**g**) *NPSHa* = 0.96 m (**h**) *NPSHa* = 0.76 m

Figure 11. Static pressure distribution at different spans on blade surface.

Figure 12 shows the pressure variation from hub to rim with different *NPSHa* values. The calculation results show that the pressure close to the hub edge is less than the rim edge. The pressure at the same point of the blade surface increases as the value of *NPSHa* increases.

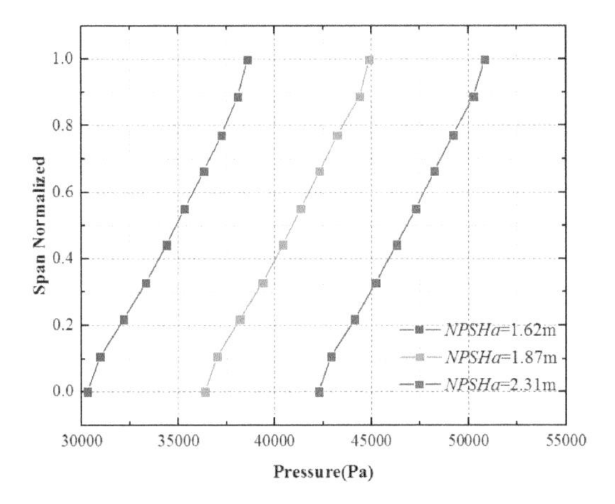

Figure 12. Pressure variation from hub to rim with different *NPSHa* values.

As shown in Figure 13, the figure presents the distribution of the vapor volume fraction in the impeller passage (facing the incoming flow direction). The figure shows that the vapor in the impeller flow passage first appears at the rim. Moreover, the vapor mainly occurs on the leading edge of the blade, and the vapor gradually extends toward the trailing edge as the value of *NPSHa* decreases. As shown in Figure 13c–e, the vapor at the hub first appears near the trailing edge of the airfoil root, and its gas content is significantly higher than the rim. When the value of *NPSHa* is 0.76m, the maximum vapor volume fraction of the rim is about 51.69%, and the maximum value of the gas content of the hub is 97.11%. The reason may be due to the large distortion of the blade root airfoil and the flow separation of the water flow, resulting in a local low pressure region.

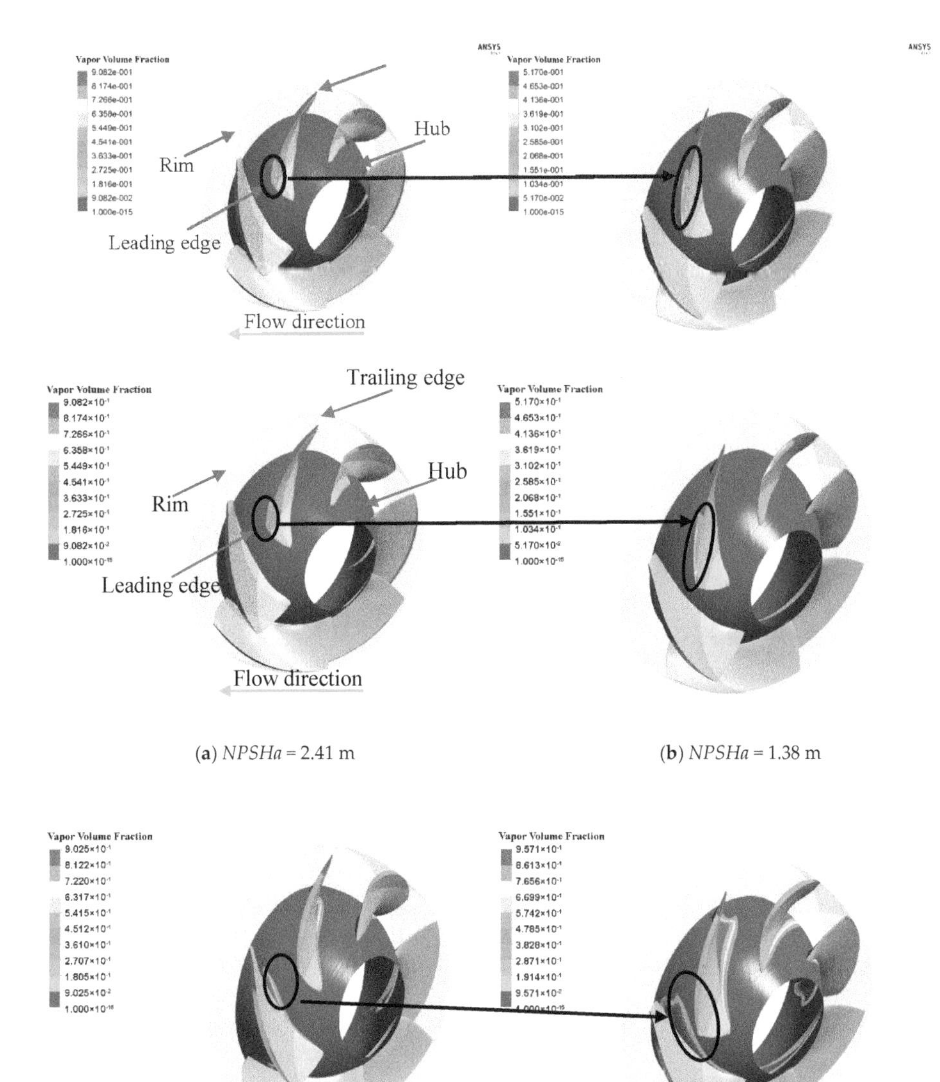

(**a**) *NPSHa* = 2.41 m (**b**) *NPSHa* = 1.38 m

(**c**) *NPSHa* = 1.21 m (**d**) *NPSHa* = 0.96 m

Figure 13. *Cont.*

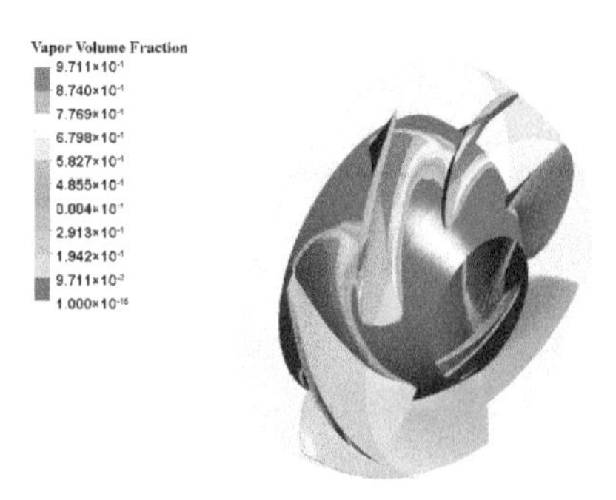

(**e**) *NPSHa* = 0.76 m

Figure 13. Vapor volume distribution of impeller rim and hub.

The pump efficiency η, shaft power N, and head H were used to define the hydraulic characteristics of the waterjet propulsion system. The calculation formula is as follows:

$$\eta = \frac{30\rho gQH}{\pi nM} \times 100\% \tag{11}$$

where T is the torque of blades, N·m, n is the rotating speed of the impeller, r/min, Q is the flow rate, m^3/s, and N is the shaft power, kW.

The cavitation performance curve of the pump under the designed flow rate is shown in Figure 14. As shown in Figure 10, the main type of blade cavitation is airfoil cavitation. When the *NPSHa* value decreases from 2.41 m to 1.47 m, cavitation can be found on the blade surface of the impeller. However, compared with Figure 14, it is found that the head and efficiency curves do not decrease sharply but increase slightly in this process, which is caused by the complexity and instability of cavitation flow. When the pump operates between the initial cavitation condition and the critical cavitation condition, the lift coefficient of the supercavitation will increase slightly as *NPSHa* gradually decreases toward the critical value, and the development of blade cavitation will lead to a certain degree of increase of the pump head before the breakdown cavitation condition. However, the increase of lift and the development of cavitation will also cause cavitation oscillation and damage the flow passage parts of the pump. Hence, pumps are generally not allowed to operate between these two conditions. As shown in Figure 14, the *NPSHa* corresponding to the critical cavitation point K is 1.29 m, and the head decreases by 3.28%. In fact, when the value of *NPSHa* is 1.27 m, it can be seen from the vapor fraction distribution of Figure 9 that cavitation has developed to a certain extent. When the value of *NPSHa* is less than 1.29 m, the head and efficiency of the pump drops sharply as the total pressure at the inlet decreases further. When the *NPSHa* value is reduced from 1.29 m to 1.21 m, the pump head is reduced by 23.23%. When the *NPSHa* value is less than 1.29 m, with the further decrease of *NPSHa* value, the cavitation rapidly covers the suction surface of the blade and gradually extends to the pressure surface, thus blocking the impeller passage and making the pump unable to work normally.

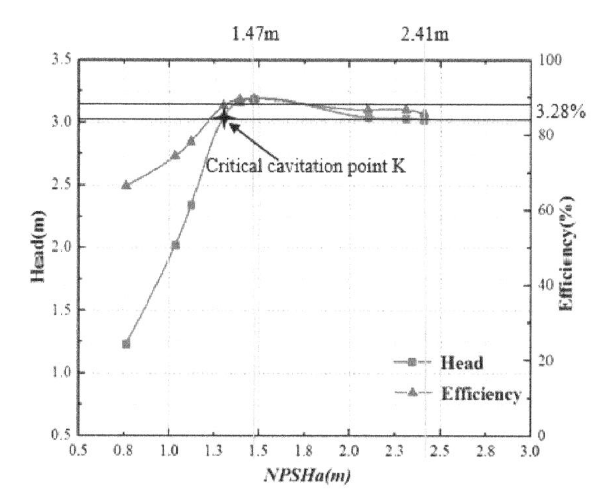

Figure 14. Cavitation characteristic curve of pump under design flow condition.

4.2. Cavitation Characteristics of Impeller in Waterjet Propulsion Pump System

In order to understand the cavitation characteristics of the water jet propulsion pump system more comprehensively, the waterjet propulsion pump equipped with inlet passage was calculated and analyzed. Figure 15 is the comparison between the cavitation characteristic curves of waterjet propulsion system and that of the waterjet propulsion pump at the designed flow rate. The figure shows that the head and efficiency of the pump have decreased after assembling the inlet passage. The flow rate of the pump has not changed under the same inlet condition, so the *NPSHr* of the pump remains unchanged. After assembling the inlet passage, the hydraulic loss and installation height increase, which reduces the *NPSHa* of the pump system and makes the cavitation performance of the pump worse.

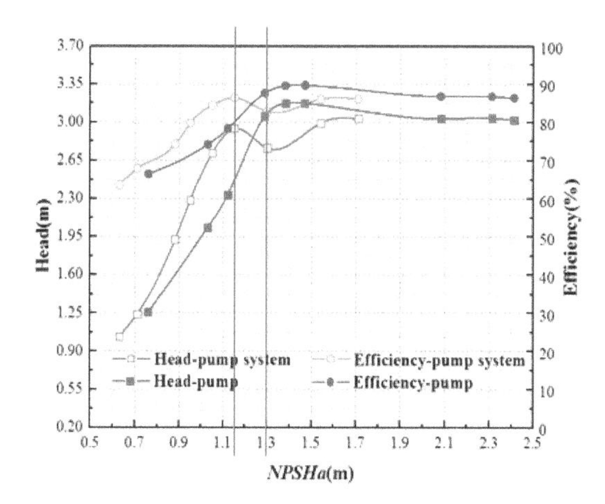

Figure 15. Cavitation characteristic curve of the waterjet propulsion system.

Figure 16 shows the distribution of the vapor volume fraction on the suction surface of the blade in the waterjet propulsion pumping system. Compared with Figure 10, the cavitation form in the impeller is still airfoil cavitation after assembling the inlet passage. When the inlet passage was assembled, the distribution of cavitation region on the blade surface becomes uneven due to the influence of the flow pattern of the inlet passage. As shown in Figure 16, under the same *NPSHa* conditions the area of cavitation region on different blade surfaces is different, and the degree of cavitation development is also different. Comparing Figure 16a,b, it can be seen that when the *NPSHa* value is 1.71 m, the cavitation of the No. 3 blade develops the fastest, and the cavitation was first discovered at the rim of the No. 3 blade near the leading edge. This phenomenon is mainly due to the impeller rotation and the uneven distribution of the flow pattern at the outlet of the inlet passage. With the decrease of *NPSHa* value, cavitation began to appear in the rim position near the leading edge of the blade, except for blade No. 3. It can be seen that the cavitation characteristics of impeller blades are changed after assembling the inlet passage due to the influence of the passage. With the decrease of *NPSHa*, the cavitation on the suction surface of the blades gradually extends from the leading edge to the trailing edge. When the *NPSHa* value decreases from 2.03 m to 1.30 m, it can be seen from Figure 16 that cavitation gradually occurs in the impeller blade during this process. When the *NPSHa* value is 1.3 m, 50% of the suction surface of the No. 2 and No. 3 blades are covered by vapors. At the same time, it can be seen from Figure 15 that the head and efficiency of the pump system are declining in this process. However, the head and efficiency of the pump device in Figure 13 have a certain upward trend when the *NPSHa* value decreases from 1.30 m to 1.15 m. Compared with Figure 16d,e, it can be seen that the distribution of vapors on suction surface of the No. 2, No. 3, and No. 4 blades becomes uneven. As can be seen from Figure 16h, when the cavitation is fully developed, the regions with large vapor volume fraction are mainly concentrated in the middle and trailing edges of the blades. Therefore, cavitation erosion is more serious in the middle and trailing edges of suction surfaces of blades.

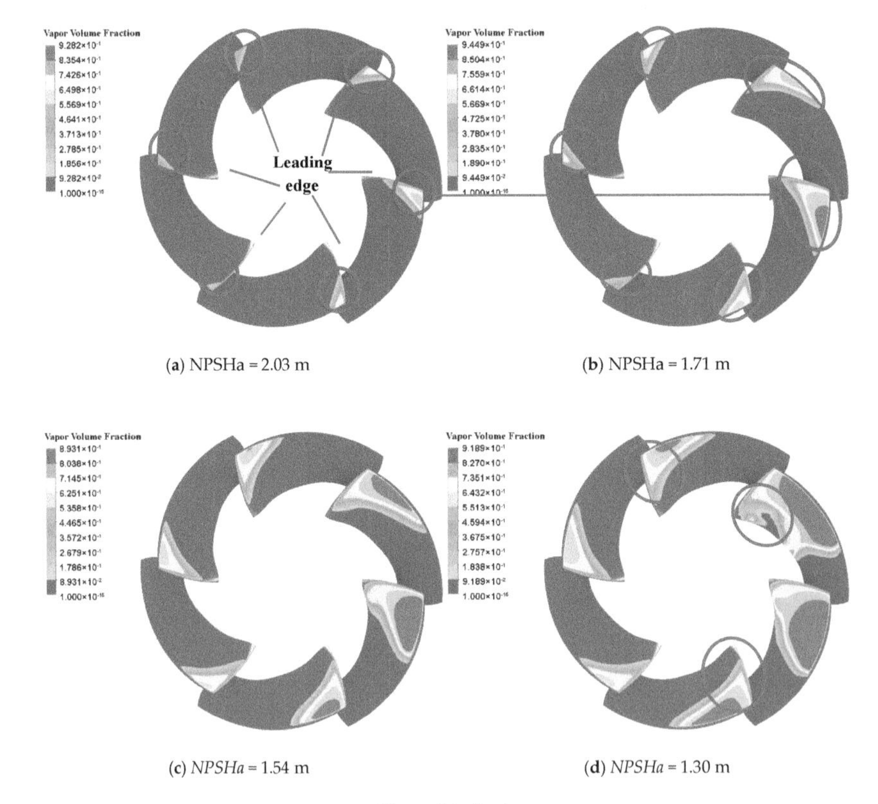

(**a**) NPSHa = 2.03 m

(**b**) NPSHa = 1.71 m

(**c**) *NPSHa* = 1.54 m

(**d**) *NPSHa* = 1.30 m

Figure 16. *Cont.*

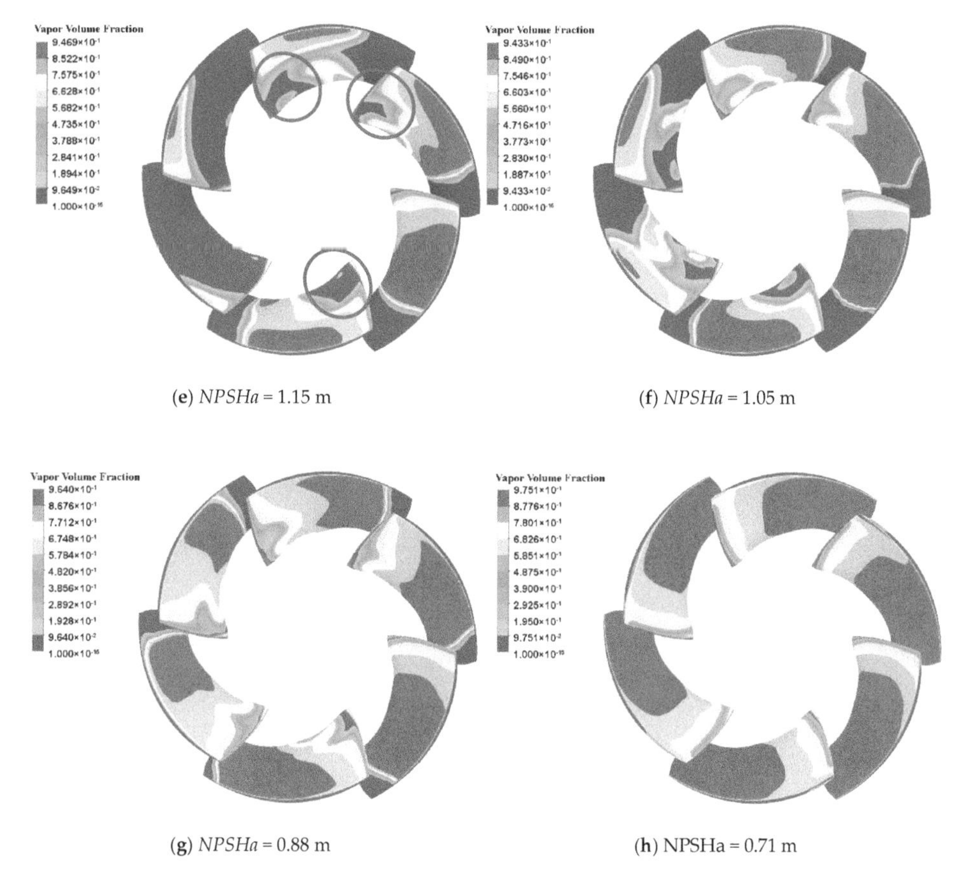

(e) *NPSHa* = 1.15 m (f) *NPSHa* = 1.05 m

(g) *NPSHa* = 0.88 m (h) NPSHa = 0.71 m

Figure 16. Distribution of vapor volume fraction on suction surface of blade.

The uneven distribution of cavitation region in impeller is related to the flow pattern at the inlet of impeller. The cavitation development in impeller is affected by the change of inlet structure after assembling the inlet passage. Figure 17 shows the velocity distribution contour of the passage outlet with different *NPSHa* values. It can be seen form the figure that the flow velocity distribution at the outlet of the flow passage is not uniform. The flow velocity around the wall of the flow passage is high, while that near the drive shaft is low. According to the law of conservation of energy, the flow velocity around the wall of the passage is high, so the corresponding low pressure region will first occur at the rim of the blade, which results in the initial position of cavitation near the leading edge of the blade. The red color region in Figure 17a represents the high-speed region, so the region where the blade under high-speed impact first produces cavitation in Figure 16d.

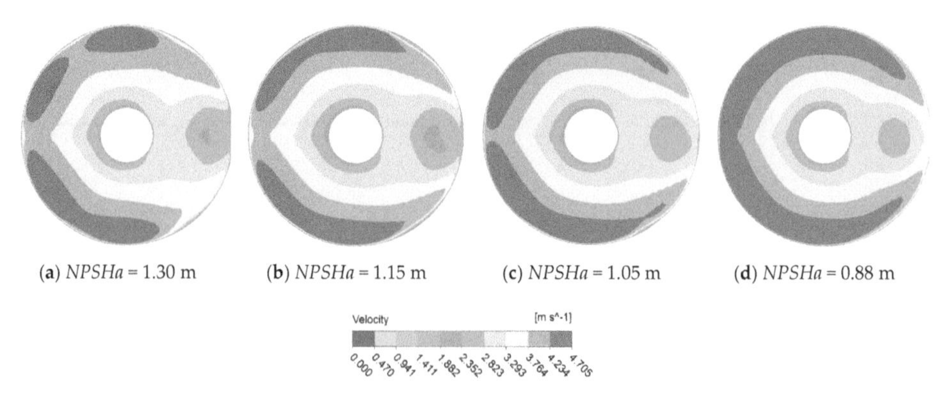

Figure 17. Velocity distribution contour of the passage outlet with different *NPSHa* values.

The inlet passage has a great influence on the flow pattern of impeller inlet, mainly because the inlet shape of the passage is irregular, especially the lip. The lip curvature changes greatly, so the flow pattern at the flow passage inlet is uneven, and the lip is the most prone part of cavitation in the passage. Moreover, there is a diffusion section between the impeller inlet and the outlet of the runner, and the diffusion angle affects the influent flow pattern. At the same time, the curved section in front of the impeller inlet causes a change in the flow rate, and the curved section is also a potential region where cavitation occurs. It can be seen from Figures 18 and 19 that critical cavitation occurs at the lip location. As *NPSHa* decreases, the cavitation region of the curved connecting section on the lower side of the drive shaft gradually increases, and the cavitation region begins to appear on the upper and lower sides of the drive shaft.

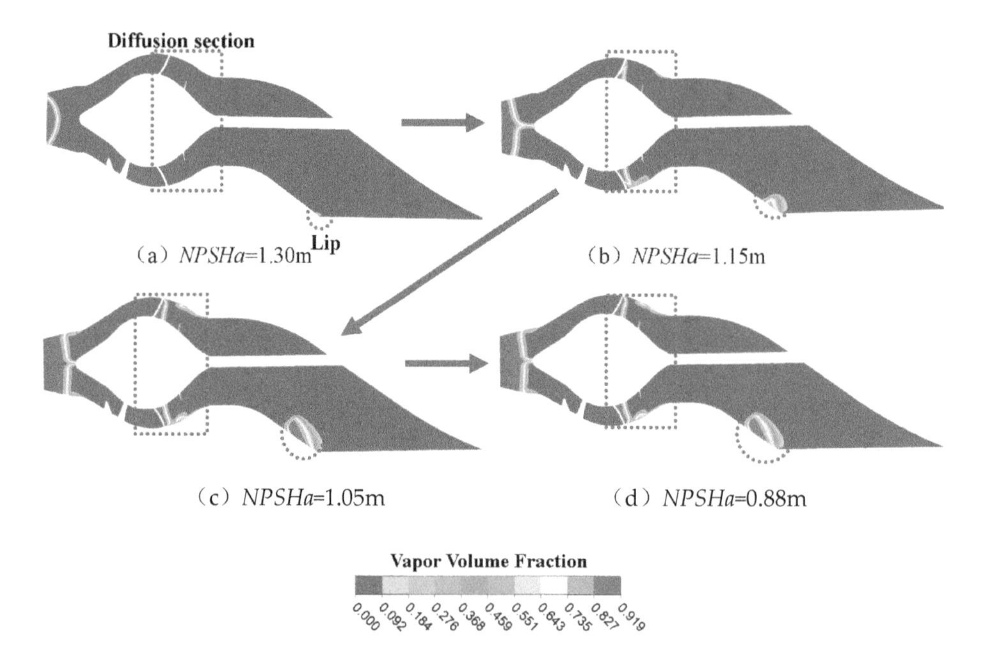

Figure 18. Vapor volume distribution in the inlet passage with different *NPSHa* values.

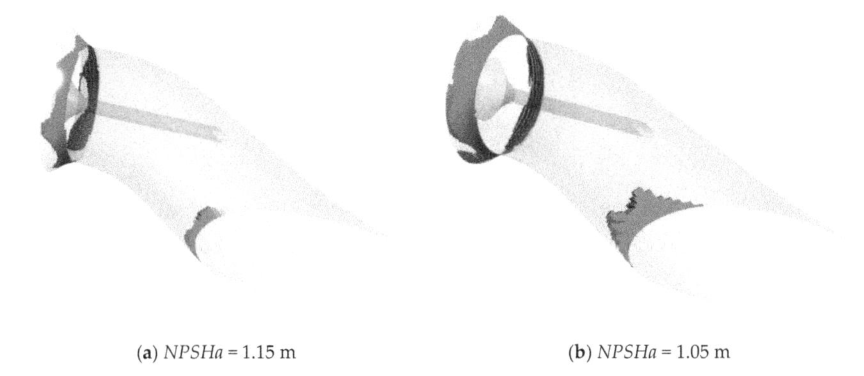

(a) *NPSHa* = 1.15 m (b) *NPSHa* = 1.05 m

Figure 19. Cavitation volume distribution in the inlet passage with different *NPSHa* values.

5. Conclusions

1. Under the designed flow rate condition, vapors in the process of cavitation initiation first accumulate on the leading edge of the blade's suction surface, which is close to the impeller rim. With the decrease of inlet total pressure, the cavitation region extended towards the trailing edge and the vapor fraction volume become gradually larger. When the value of *NPSHa* is equal to 1.38 m, vapors begin to appear in the middle of the blade surface, and the area of cavitation region accounts for about half of the surface area of the blade suction surface. When the value of *NPSHa* is equal to 0.76 m, the cavitation is fully developed. The suction surface and pressure surface of the blade are covered by vapors and the impeller passage is blocked, which makes the pump unable to work properly.
2. After assembling the inlet passage, the flow pattern at the impeller inlet is affected by the passage, resulting in the uneven distribution of the cavitation region of the impeller. This is more disadvantageous to the cavitation performance of the waterjet propulsion pump. The potential danger regions of cavitation are the lip of inlet passage and the upper and lower connecting curved section of the inlet passage. The occurrence time of cavitation in impeller is prior to that in the inlet passage.

Author Contributions: Data curation, L.C.; Formal analysis, W.J. and L.C.; Methodology, W.J.; Writing – original draft, W.J.; Writing – review & editing, L.C., C.W. and J.X.

Funding: This research was funded by the National Natural Science Foundation of China (Grant No.51779214 and No.51609105), the Peak Six Talents Plan in the Jiangsu Province (Grant No.2015-JXQC-007), the Jiangsu Province 333 High-Level Talents Training Project (BRA2018382). And the APC a project funded by the Priority Academic Program Development of Jiangsu Higher Education Institutions (PAPD).

Conflicts of Interest: The authors declare no conflict of interest.

References

1. Bulten, N.W.H. Numerical Analysis of a Waterjet Propulsion System. Ph.D. Thesis, Library Eindhoven University of Technology, Eindhoven, The Netherlands, 2006.
2. Park, W.G.; Jang, J.H.; Chun, H.H.; Kim, M.C. Numerical flow and performance analysis of waterjet propulsion system. *Ocean. Eng.* **2004**, *32*, 1740–1761. [CrossRef]
3. Xia, C.; Cheng, L.; Luo, C.; Jiao, W.; Zhang, D. Hydraulic Characteristics and Measurement of Rotating Stall Suppression in a Waterjet Propulsion System. *Trans. FAMENA* **2018**, *4*, 85–100. [CrossRef]
4. Laberteaux, K.; Ceccio, S.; Mastrocola, V.; Lowrance, J. High speed digital imaging of cavitating vortices. *Exp. Fluids* **1998**, *24*, 489–498. [CrossRef]
5. Cheng, H.; Long, X.; Ji, B.; Zhu, Y.; Zhou, J. Numerical investigation of unsteady cavitating turbulent flows around twisted hydrofoil from the Lagrangian viewpoint. *J. Hydrodyn.* **2016**, *28*, 709–712. [CrossRef]

6. Stephen, C.; Kumaraswamy, S. Experimental Determination of Cavitation Characteristics of Low Specific Speed Pump using Noise and Vibration. *J. Inst. Eng. (India)* **2019**, *100*, 64–74. [CrossRef]
7. Wang, C.; He, X.; Shi, W.; Wang, X.; Wang, X.; Qiu, N. Numerical study on pressure fluctuation of a multistage centrifugal pump based on whole flow field. *AIP Adv.* **2019**, *9*, 035118. [CrossRef]
8. Zhang, J.; Xia, S.; Ye, S.; Xu, B.; Song, W.; Zhu, S.; Tang, H.; Xiang, J. Experimental investigation on the noise reduction of an axial piston pump using free-layer damping material treatment. *Appl. Acoust.* **2018**, *139*, 1–7. [CrossRef]
9. Ye, S.; Zhang, J.; Xu, B.; Zhu, S.; Xiang, J.; Tang, H. Theoretical investigation of the contributions of the excitation forces to the vibration of an axial piston pump. *Mech. Syst. Signal Proc.* **2019**, *129*, 201–217. [CrossRef]
10. Zhu, Y.; Qian, P.; Tang, S.; Jiang, W.; Li, W.; Zhao, J. Amplitude-frequency characteristics analysis for vertical vibration of hydraulic AGC system under nonlinear action. *AIP Adv.* **2019**, *9*, 035019. [CrossRef]
11. Long, X.; Zhang, J.; Wang, Q.; Xiao, L.; Xu, M.; Qiao, L.; Ji, B. Experimental investigation on the performance of jet pump cavitation reactor at different area ratios. *Exp. Therm. Fluid Sci.* **2016**, *78*, 309–321. [CrossRef]
12. Tan, L.; Zhu, B.; Cao, S.; Wang, Y.; Wang, B. Influence of Prewhirl Regulation by Inlet Guide Vanes on Cavitation Performance of a Centrifugal Pump. *Energies* **2014**, *7*, 1050–1065. [CrossRef]
13. Christopher, E. *Brennen. Cavitation and Bubble Dynamics*; Oxford University Press: Oxford, UK, 1995.
14. He, X.; Jiao, W.; Wang, C.; Cao, W. Influence of surface roughness on the pump performance based on Computational Fluid Dynamics. *IEEE Access* **2019**, *7*, 105331–105341. [CrossRef]
15. Jayaprakash, A.; Kapahi, A.; Choi, J.; Chahine, G. Modelling of material pitting from cavitation bubble collapse. *J. Fluid Mech.* **2014**, *755*, 142–175.
16. Wang, C.; Chen, X.; Qiu, N.; Zhu, Y.; Shi, W. Numerical and experimental study on the pressure fluctuation, vibration, and noise of multistage pump with radial diffuser. *J. Braz. Soc. Mech. Sci. Eng.* **2018**, *40*, 481. [CrossRef]
17. Zhang, D.; Shi, L.; Shi, W.; Zhao, R.; Wang, H.; van Esch, B.P.M. Numerical analysis of unsteady tip leakage vortex cavitation cloud and unstable suction-side-perpendicular cavitating vortices in an axial flow pump. *Int. J. Multiph. Flow* **2015**, *77*, 244–259. [CrossRef]
18. Monte Verde, W.; Biazussi, J.; Sassim, N.; Bannwart, A. Experimental study of gas-liquid two-phase flow patterns within centrifugal pumps impellers. *Exp. Therm. Fluid Sci.* **2017**, *85*, 37–51. [CrossRef]
19. Ji, B.; Luo, X.; Peng, X.; Wu, Y. Three-dimensional large eddy simulation and vorticity analysis of unsteady cavitating flow around a twisted hydrofoil. *J. Hydrodyn.* **2013**, *25*, 510–519. [CrossRef]
20. Wang, C.; Hu, B.; Zhu, Y.; Wang, X.; Luo, C.; Cheng, L. Numerical study on the gas-water two-phase flow in the self-priming process of self-priming centrifugal pump. *Processes* **2019**, *7*, 330. [CrossRef]
21. Li, C.; Qi, W. Rotating stall region of water-jet pump. *Trans. FAMENA* **2014**, *38*, 31–40.
22. Duplaa, S.; Coutierdelgosha, O.; Dazin, A.; Roussette, O.; Bois, G.; Caignaert, G. Experimental Study of a Cavitating Centrifugal Pump During Fast Startups. *J. Fluids Eng.* **2010**, *132*, 365–368. [CrossRef]
23. Ji, B.; Long, Y.; Long, X.; Qian, Z.D.; Zhou, J. Large eddy simulation of turbulent attached cavitating flow with special emphasis on large scale structures of the hydrofoil wake and turbulence-cavitation interactions. *J. Hydrodyn. Ser. B* **2017**, *29*, 27–39. [CrossRef]
24. Rayleigh, L., VIII. On the pressure developed in a liquid during the collapse of a spherical cavity. *Philos. Mag.* **1917**, *34*, 94–98. [CrossRef]
25. Bal, S.; Kinnas, S.; Lee, H. Numerical Analysis of 2-D and 3-D Cavitating Hydrofoils Under a Free Surface. *J. Ship Res.* **2001**, *45*, 34–49.
26. Brennen, C.; Acosta, A. Fluid-induced Rotordynamic Forces and Instabilities. *Struct. Control Health Monit.* **2005**, *13*, 10–26. [CrossRef]
27. Singhal, A.; Athavale, M.M.; Li, H.; Jiang, Y. Mathematical basis and validation of the full cavitation model. *J. Fluids Eng.* **2002**, *124*, 617–624. [CrossRef]
28. Schmidt Steffen, J.; Sezal Ismail, H.; Schnerr Günter, H. Compressible simulation of high-speed hydrodynamics with phase change. In Proceedings of the European Conference on Computational Fluid Dynamics, Egmond aan Zee, The Netherlands, 5–8 September 2006.
29. Gopalan, S.; Katz, J. Flow structure and modeling issues in the closure region of attached cavitation. *Phys. Fluids* **2000**, *12*, 895–911. [CrossRef]

30. Escaler, X.; Egusquiza, E.; Farhat, M.; Avellan, F.; Coussirat, M. Detection of cavitation in hydraulic turbines. *Mech. Syst. Signal Proc.* **2006**, *20*, 983–1007. [CrossRef]
31. Johnson, M.; Moore, J. The development of wake flow in a centrifugal impeller. *J. Eng. Gas Turbines Power* **1980**, *102*, 382–389. [CrossRef]
32. Wang, C.; He, X.; Zhang, D.; Hu, B.; Shi, W. Numerical and experimental study of the self-priming process of a multistage self-priming centrifugal pump. *Int. J. Energy Res.* **2019**, 1–19. [CrossRef]
33. Zhu, Y.; Tang, S.; Quan, L.; Jiang, W.; Zhou, L. Extraction method for signal effective component based on extreme-point symmetric mode decomposition and Kullback-Leibler divergence. *J. Braz. Soc. Mech. Sci. Eng.* **2019**, *41*, 100. [CrossRef]
34. Qian, J.; Chen, M.; Liu, X.; Jin, Z. A numerical investigation of the flow of nanofluids through a micro Tesla valve. *J. Zhejiang Univ. Sci. A* **2019**, *20*, 50–60. [CrossRef]
35. Zhou, L.; Wang, Z. Numerical simulation of cavitation around a hydrofoil and evaluation of a RNG k-ε model. *J. Fluids Eng.* **2008**, *130*, 1–7. [CrossRef]
36. Ji, B.; Luo, X.; Arndt, R.E.; Wu, Y. Numerical simulation of three dimensional cavitation shedding dynamics with special emphasis on cavitation–vortex interaction. *Ocean Eng.* **2014**, *87*, 64–77. [CrossRef]
37. Qian, J.; Gao, Z.; Liu, B.; Jin, Z. Parametric study on fluid dynamics of pilot-control angle globe valve. *J. Fluids Eng.* **2018**, *140*, 111103. [CrossRef]
38. Wang, C.; Shi, W.; Wang, X.; Jiang, X.; Yang, Y.; Li, W.; Zhou, L. Optimal design of multistage centrifugal pump based on the combined energy loss model and computational fluid dynamics. *Appl. Energy* **2017**, *187*, 10–26. [CrossRef]
39. Zwart, P.; Gerber, A.; Belamri, T. A two-phase model for predicting cavitation dynamics. In Proceedings of the ICMF 2004 International Conference on Multiphase Flow, Yokohama, Japan, 30 May–4 June 2004; pp. 1–11.
40. Kunz, R.; Boger, D.; Stinebring, D.; Chyczewski, T.; Lindau, J.; Gibeling, H. A preconditioned Navier-Stokes method for two-phase flows with application to cavitation prediction. *Comput. Fluids* **2000**, *29*, 849–875. [CrossRef]
41. Schnerr, G.H.; Sauer, J. Physical and numerical modeling of unsteady cavitation dynamics. In Proceedings of the ICMF 2001 International Conference on Multiphase Flow, New Orleans, LA, USA, 27 May–1 June 2001; pp. 1–8.
42. Ji, B.; Luo, X.; Wu, Y.; Peng, X.; Xu, H. Partially-Averaged Navier-Stokes method with modified k–ε model for cavitating flow around a marine propeller in a non-uniform wake. *Int. J. Heat Mass Transf.* **2012**, *55*, 6582–6588. [CrossRef]
43. Li, X.; Yuan, S.; Pan, Z.; Yuan, J.; Fu, Y. Numerical simulation of leading edge cavitation within the whole flow passage of a centrifugal pump. *Sci. China Technol. Sci.* **2013**, *56*, 2156–2162. [CrossRef]
44. Ji, B.; Luo, X.; Wu, Y.; Peng, X.; Duan, Y. Numerical analysis of unsteady cavitating turbulent flow and shedding horse-shoe vortex structure around a twisted hydrofoil. *Int. J. Multiph. Flow* **2013**, *51*, 33–43. [CrossRef]
45. Shi, S.; Wang, G.; Chen, G.; Zhang, M. Experimental investigation of the thermal effect on the unsteady cavitating flow structure. *J. Ship Mech.* **2013**, *17*, 327–335.

Article

Numerical Analysis of the Diaphragm Valve Throttling Characteristics

Yingnan Liu [1], Liang Lu [1,*] and Kangwu Zhu [2,3]

[1] School of Mechanical Engineering, Tongji University, Shanghai 200092, China; 1451722@tongji.edu.cn
[2] Shanghai Institute of Aerospace Control Technology, Shanghai 201109, China; zjuzkw@zju.edu.cn
[3] Shanghai Servo System Engineering Technology Research Center, Shanghai 201109, China
* Correspondence: luliang829@tongji.edu.cn

Received: 4 September 2019; Accepted: 19 September 2019; Published: 28 September 2019

Abstract: The throttling characteristics of the diaphragm valve are numerically studied in this paper. Firstly, the diaphragm deformation performance is analyzed by a finite element method, while the upper boundary morphology of the internal flow field under different valve openings was obtained. Then the two-dimensional simulation of the weir diaphragm valve flow field is carried out in order to explore the optimal design of flow path profile. The study shows that the throttling characteristics can be improved by flatting the ridge side wall, widening the top of the ridge and gently flatting the internal protruding of the flow path. In addition, using the local grid encryption techniques based on velocity gradient adaptive and y^+ adaptive can improve the accuracy of simulation results. Finally, a cavitation two-phase flow simulation is carried out. The results show that cavitation may occur below 50% opening of diaphragm valve in ultra-pure water system, which becomes more intense with the increase of inlet pressure and even leading to flow saturation on the micro-orifice state.

Keywords: diaphragm valve; throttling characteristics; profile design optimize; computational fluid dynamics

---

1. Introduction

The diaphragm valves are extensively used in various industrial applications, including breweries, pharmaceutical chemical, dairy, food, petroleum and mining etc. [1]. This is due to several unique advantages: the ability to handle solids, the low resistance coefficient, no dead areas where particles may collect, easy maintenance, and automation [1–3]. Traditionally, the diaphragm valves are commonly used as shut-off valves for non-Newtonian fluids with variable viscosities [4], such as suspensions, mixed solutions, and the like, because diaphragm valves in such applications do not need and are not easy to achieve accurate flow control. With the rapid development of information technology, diaphragm valves have been widely used in the precise control of Newtonian fluids with downsizing technology trend [5,6]. In order to obtain higher integration and smaller feature sizes of the chip, immersion lithography is the cutting-edge technology and key step in the semiconductor industry [7,8]. The fluid control requirements for ultra-pure water treatment and liquid injection systems are becoming more stringent, which mainly reflects the water purity of ultra-pure water and the stability of immersion flow field [9,10]. Diaphragm valves are the key components of ultra-pure water systems due to their excellent cleanliness and low resistance loss, ensuring a stable flow balance of the infusion flow and the immersion flow field [11].

The diaphragm valve structure and diaphragm material affect the throttling characteristics and sealing performance [12]. From the structural point of view, the most widely used diaphragm valves are straight-through type and weir type [1,13]. Although the flowing resistance in straight-through type valve body is lower and the flow performance is better than that in weir type valve body, the diaphragm travels longer in straight-through type, which means that under the condition of frequent opening and closing control, the life of the straight-through type valve body diaphragm will be much lower than that of the weir type valve body diaphragm. In addition, during the closing process in weir type valve body, the diaphragm forms a vertical coincident seal with the sealing table without any mechanical friction, thereby meeting the sanitary level cleaning requirements. Based on the special advantages of the weir type, the diaphragm valves used in the semiconductor industry are mostly weir type valve bodies. From the material point of view, the properties of rubber diaphragm such as high temperature resistance and oil resistance in the field of chemical industry are not the key points to be considered in ultra-pure water system. The surface quality, tensile strength, fatigue strength and shape retention of rubber are more suitable for evaluating diaphragm valve diaphragm in ultra-pure water system. Currently NBR, HNBR, ACM, FKM, etc. are the most widely used rubber material [14]. Nickel or silicon or bimorph materials are also used as diaphragm materials [15,16]. In addition, The diaphragm made of fluoroplastics (PTFE as an example) or rubber and fluoroplastics composite materials has better elasticity and strength, and its surface is smoother [17].

The throttling characteristics and flow control performance of the diaphragm valve are expressed in the form of valve flow coefficient C_v (or K_v) value, local resistance coefficient and valve flow characteristic curve [18,19]. The valve flow coefficient K_v value is the unit of measurement in metric units (the valve flow coefficient in English units is called the C_v value) and is defined as: under the standard pressure drop of 1 bar, the volume flow rate of water (the fluid severity at 1 N/m^3) through the valve per hour, as expressed in Equation (1).

$$K_v = Q\sqrt{\frac{\rho_0}{\Delta p}} \tag{1}$$

where, K_v—valve flow coefficient; Q—volume flow rate (m^3/h); Δp—pressure drop (1 bar); ρ_0—density of water (kg/m^3).

The resistance coefficient is the index to measure the pressure loss of the valve. It is a dimensionless coefficient representing the pressure loss of the valve. It depends on the type, diameter, structure and cavity shape of the valve, and is calculated by the experimental data of the valve. According to the Bernoulli equation, the formula for calculating the drag coefficient can be deduced as Equation (2) [20],

$$\zeta = \frac{2\Delta p}{\rho v^2} \tag{2}$$

where, ρ—density of medium (kg/m^3); v—average velocity of inlet (outlet) (m/s); Δp—pressure drop (Pa).

The flow characteristic curve of the valve shows the flow through the valve at different openings in the process of opening and closing [21]. It is an index to describe the flow control performance of the valve. The higher the linearity of the curve, the larger the linearity range, the better the flow control performance of the valve. The curve was obtained experimentally.

At present, there are many experimental studies and valve body structure improvement studies on diaphragm valves, mainly focusing on sealing performance and operability. There are also studies on throttling characteristics of diaphragm valves and simulation studies on flow field [6,22,23] and pressure characteristics [24,25], but not in the ultra-pure water system of the semiconductor industry. Various studies based on other types of valves have confirmed that the change of flow path shape and the use of pilot valves can improve flow control characteristics, reduce vibration and cavitation [26–29]. The weir type diaphragm valve has become the mainstream for the sake of better cleanliness, longer diaphragm life, and better flow control characteristics. However, due to its large flow resistance and lack of design criteria for internal flow channel surface structure, it is difficult to obtain optimization methods to further improve valve control performance. In addition, the outlet of diaphragm valves in ultra-pure water treatment system is generally atmospheric back pressure environment, but the pressure in the pipeline of ultra-pure water treatment system is generally about 5 times atmospheric pressure, so it is easy to produce cavitation and jet under the condition of diaphragm valve micro-orifice, leading to flow saturation and further increase pressure loss.

In order to solve the above problems, based on the structure of the weir type diaphragm valve body, the numerical simulation analysis of two-dimensional flow field is carried out. By comparing the flow coefficient, resistance coefficient and flow characteristic curve, the flow control performance of valves with different profiles is evaluated, and the structural optimization scheme is proposed and verified. Using the local grid encryption techniques based on velocity gradient adaptive and y^+ adaptive the accuracy of simulation results are improved [30]. The flow field of diaphragm valve was simulated and analyzed, and the cavitation phenomenon of small valve opening of diaphragm valve was explored under the condition of ultra-pure water system.

2. Finite Element Analysis of Diaphragm Deformation

The flow path surface structure of weir type diaphragm valve is determined not only by the shape of the bottom ridge and the tube wall itself, but also by the deformation of the diaphragm bottom surface and the angle of the installation joint at different openings. A number of shape elements determine the upper and lower wall boundary of the two-dimensional flow field of weir type diaphragm valve.

However, the rubber diaphragm is hyper elastic and irregular in shape. It is difficult to calculate the deformation by theoretical calculation [31,32]. At present, there are few experimental or simulation results for the diaphragm deformation when the diaphragm valve works. Therefore, in order to obtain the inner surface shape of the diaphragm at different openings and determine the changing two-dimensional flow field boundary, firstly, finite element method analysis on the diaphragm of diaphragm valve is carried out.

2.1. Structure and Grid

According to the design method of diaphragm size of diaphragm valve, based on the nominal size DN20 series of diaphragm valve, the shape and size of diaphragm are preliminarily determined (Figure 1). The effects of the diaphragm thickness (B_0 = 4 mm, B_0 = 5 mm, B_0 = 6 mm) and the stem loading surface diameter on the diaphragm deformation were investigated.

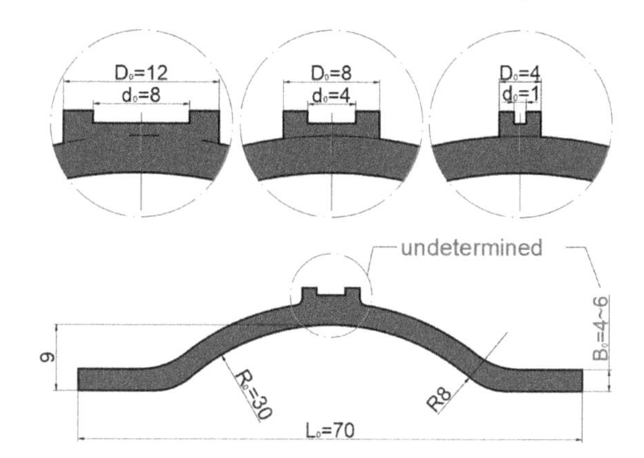

Figure 1. Shape and size of the diaphragm with three kinds of the stem loading surface parameter.

The three-dimensional model of rubber diaphragm is imported into ANSYS static structure and meshed into model interface. Based on the mesh automatic meshing basic set value, the size function is selected adapted, and the relevance center selects course, and the mesh size is 1 mm. The hyper elastic material was selected as the model material. The neoprene rubber was extracted from ANSYS engineering data, and the density (ρ_1 = 1.25 g/cm^3) and tensile strength (σ > 10 MPa) of the material were manually added. Figure 2a shows the model mesh with a diaphragm thickness of B_0 = 6 mm.

The results show that the maximum skewness is 0.81, 0.81 and 0.82 under the three thicknesses, and the average is less than 0.5. The grid quality is good.

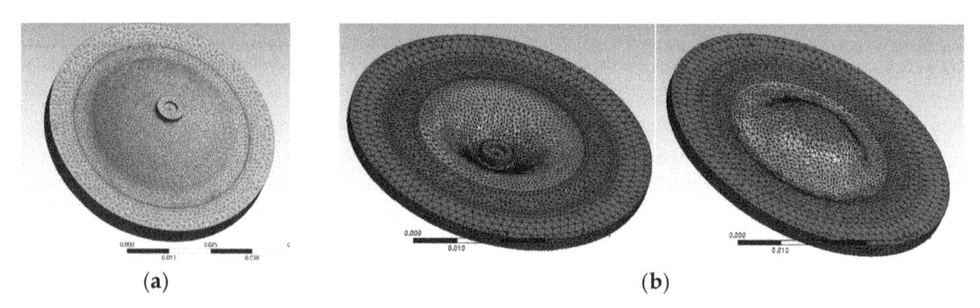

Figure 2. Numerical model of the diaphragm (**a**) the model mesh with a diaphragm thickness of B_0 = 6 mm and (**b**) the total deformation of the diaphragm with a thickness of B_0 = 4 mm.

Load setting (Table A1): Select large deformation in analysis setting, apply fixed load (flange of diaphragm upper and lower surface, simulate assembly fixing), standard gravity, uniform pressure (arc under diaphragm, simulate medium back pressure), fixed displacement (stem action plane, direction downward, x_0 = 15 mm, solve maximum deformation), and add no friction load (loading the side wall, limiting the deformation of the cylindrical surface). The resolver outputs the total displacement map, the longitudinal section displacement map and the longitudinal section stress map. Figure 2b shows the total deformation of the diaphragm with a thickness of B_0 = 4 mm.

2.2. Analysis and Conclusion

2.2.1. The Influence of Diaphragm Thickness

Contrastive displacement diagram (Figure 3) shows that with the decrease of diaphragm thickness, the maximum strain region increases gradually, the strain distribution is more uniform and the strain transmission is better. That is, the displacement input of valve stem can be well maintained to the surface of diaphragm, so that the diaphragm can produce a certain amount of deformation with the input requirements, and the runner can produce the corresponding opening. From this point of view, the thinner diaphragm is more conducive to the diaphragm valve flow accurate control. In addition, the thinner the diaphragm, the smoother the deformation at the bottom, the approximate planes at the maximum displacement, the larger the contact area with the top of the ridge, and the better the sealing performance.

However, this does not mean that the thinner the diaphragm, the better. Contrast stress diagram (Figure 3) shows that the thinner the diaphragm valve is, the greater the maximum stress is, and the greater the maximum stress area is, especially the upper and lower surface of the diaphragm. The thicker the diaphragm, the lower the stress is in most areas. The maximum stress is only concentrated in the sealing contact area of the diaphragm bottom surface, and sticks to the ridge top. After tightening, part of the stress is offset, which means that the possibility of diaphragm damage is reduced and the life of the diaphragm is longer.

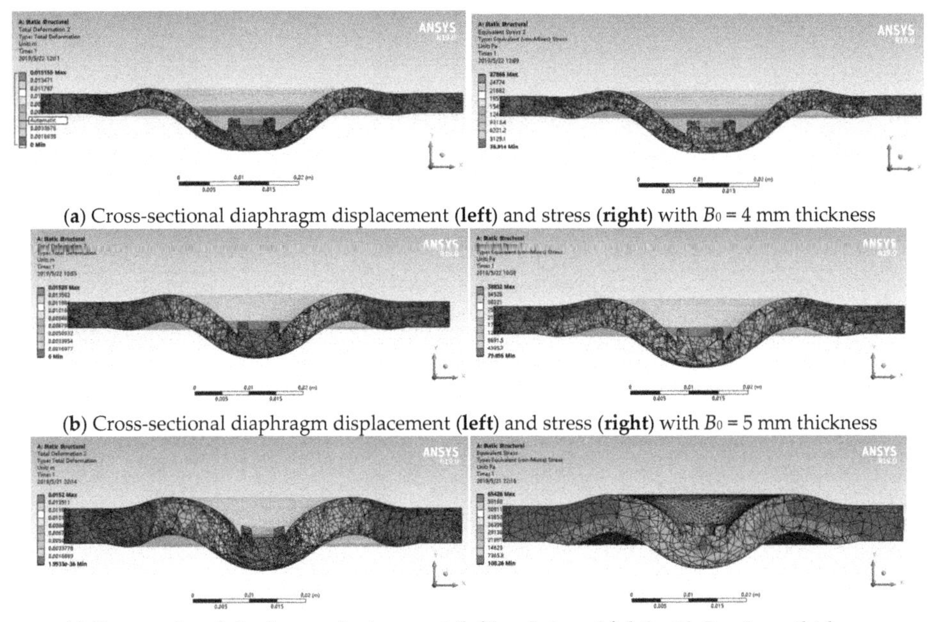

(**a**) Cross-sectional diaphragm displacement (**left**) and stress (**right**) with B_0 = 4 mm thickness

(**b**) Cross-sectional diaphragm displacement (**left**) and stress (**right**) with B_0 = 5 mm thickness

(**c**) Cross-sectional diaphragm displacement (**left**) and stress (**right**) with B_0 = 6 mm thickness

Figure 3. Displacement Diagram and Stress Diagram of Diaphragms with Different Thicknesses.

2.2.2. Effect of Stem Loading Area

By analyzing the deformation of diaphragm (Figure 4) under different loading areas, it can be seen that when the valve stem is loaded in a large area, there is almost no stress at the center of the diaphragm and the maximum displacement is not reached. On the contrary, the maximum displacement occurs directly below the edge of the convex platform. When the maximum displacement of the valve stem is input, the center of the diaphragm bottom surface is depressed. It shows that the displacement input cannot be transmitted well. Moreover, due to the concave surface under the diaphragm, the diameters above the ridge have undergone two shrinkage and expansion processes. The continuous change of the diameters will also lead to increased pressure loss and increase the resistance coefficient of the valve. In addition, this kind of depression will also lead to incomplete sealing and easy to leak.

When the diaphragm is loaded in a moderate area, the central area of the lower surface is close to the plane, the boundary of the runner is more uniform, and the sealing performance is better. The diaphragm deflection cannot meet the working requirement when the minimum area load or approximately concentrated load occurs. Even some element grids have been seriously damaged during the simulation iteration process, which leads to the termination of the simulation, indicating that the loading area of the valve stem should not be too small.

By exploring the deformation of diaphragm at different opening, using FEM module in ANSYS WORKBENCH, the geometric model of the deformed diaphragm is obtained, and the shape and size of the diaphragm under different opening are obtained. The flow field boundary at different opening can be obtained by matching the shape of the runner.

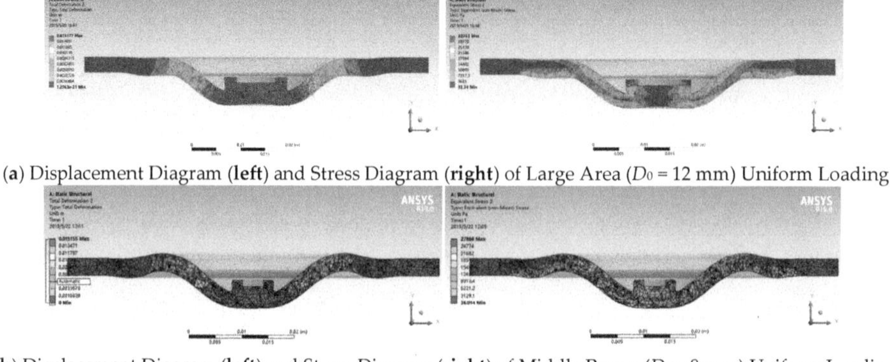

(**a**) Displacement Diagram (**left**) and Stress Diagram (**right**) of Large Area (D_0 = 12 mm) Uniform Loading

(**b**) Displacement Diagram (**left**) and Stress Diagram (**right**) of Middle Range (D_0 = 8 mm) Uniform Loading

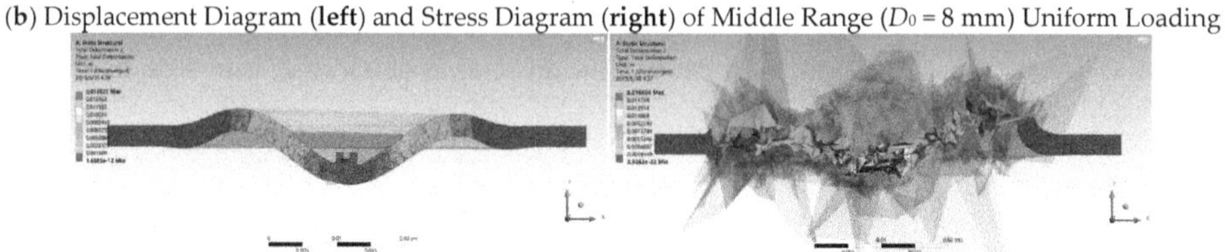

(**c**) Deformation deflection (**left**) or even damage (**right**) of diaphragm under small area (D_0 = 4 mm) loading (When the grid size is not so small (e.g., 2 mm), the state of the diaphragm is shown in the left figure. When the grid size is set 1 mm or smaller, the calculation reports error telling which grid is seriously damaged and stops iteration. The right figure shows the state when the iteration stops, indicating that the diaphragm has been destroyed. The two figures show the two probable situations in reality.)

Figure 4. Diaphragm Deformation under Different Loading Area.

3. CFD Flow Structure and Grid Independence Analysis

3.1. Two-Dimensional Flow Field Model of Weir Diaphragm Valve

The diaphragm of Weir-Type diaphragm valve is deformed and displaced by the stem loading, and the diameter of flow passage is compressed to achieve throttling effect until it is fitted with the sealing table at the top of the ridge to realize the complete closure of the valve (Figure 5).

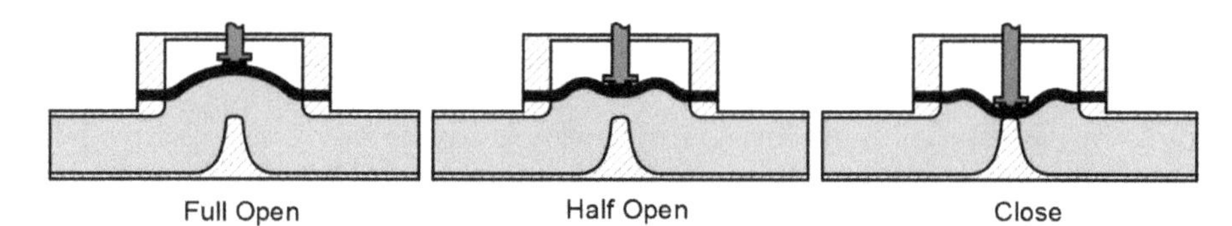

Figure 5. Working Principle of Weir Diaphragm Valve.

In this paper, the LiquidlensTMLLG4 ultra-pure water treatment system of ENTEGRIS Company is taken as the reference of working condition of diaphragm valve. The relevant size parameters of DS12-2M-12F-3 manual diaphragm valve are taken as the reference for modeling. The flow field structure size can be obtained by simplification (Figure 6). The shape of the ridge has a major influence on the flow state and loss of the medium in the weir diaphragm valve. Five different sizes of the ridge structures (Figure 6) are designed to explore their effects on the flow characteristics of the diaphragm valve.

Determine the relevant flow parameters according to the reference operating conditions (Table 1).

The diaphragm valve is used at the inlet and outlet of the system, so the back pressure at the outlet of the simulated diaphragm valve is generally set to p_{out} = 400 kPa. When the nominal diameter of diaphragm valve is $DN = 20$ mm and the system flow rate Q is set to 8 L/min, the average velocity of flow can be calculated from Equation (3),

$$v = \frac{Q}{A} = \frac{4Q}{\pi(DN)^2} \tag{3}$$

The average velocity of flow can be calculated to be about $v = 0.42$ m/s. According to the table above, the flow rate varies in the vicinity of 8 L/min. Therefore, a set of different flow velocities is set up in simulation to explore the characteristics of diaphragm valves under different Reynolds numbers.

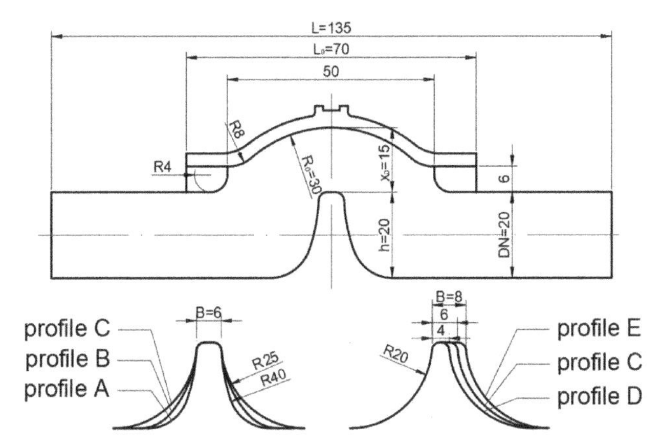

Figure 6. Flow Path Structure Dimensions and Five Ridge Profiles.

Table 1. Liquidlens^{TM}LLG4 System Flow Parameter Index.

	Parameters	Units	Index
System inlet index	Flow rate range	L/min	>8
	Pressure range	kPa	350~550
	Temperature rage	°C	17~26
System outlet index	Flow rate range	L/min	1~8
	Pressure range	kPa	100~500
	Temperature rage	°C	17~24

In addition, according to the deformation results of diaphragm at different openings, the two-dimensional flow field model of valve at different openings can be obtained (profile E as an example in Figure 7).

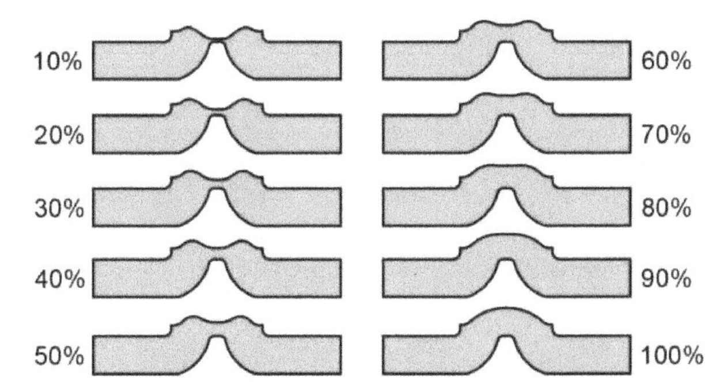

Figure 7. The two-dimensional flow field model of valve at different openings (profile E).

3.2. Boundary Conditions and Simulation Settings

After the flow field model is established, the mesh module is used to mesh the flow field and discrete flow fields. Select the adaptation in the size function, select the course in the relevant center, and set the grid size to 1 mm, 0.5 mm, and 2 mm respectively to generate the mesh. The meshing results are as shown in Figure 8 for the profile E. The simulation parameters in detail are shown in Table A1.

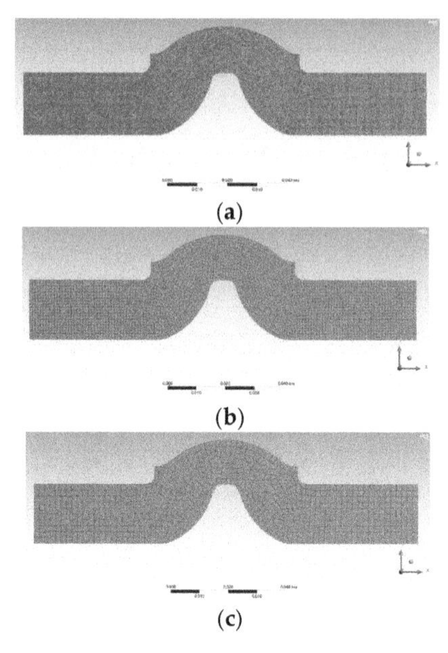

Figure 8. Mesh of different sizes under profile E. (**a**) for 0.5 mm; (**b**) for 1 mm; (**c**) for 2 mm.

The skewness is checked in the grid quality column, and the maximum skewness of the three meshes is 0.593 (1 mm), 0.539 (0.5 mm) and 0.504 (2 mm), respectively. All of these can meet the requirements of this simulation. In order to reduce the computational complexity of the simulation iteration and ensure a certain degree of accuracy, this simulation is other. The grid size of each basin shape is set to 1 mm.

The grid file is imported into FLUENT software, and the environment is two-dimensional flow field simulation. The acceleration of gravity is added in general. Model models select standard k-ε model, and model parameters follow the default values of the system. Water-liquid is added to the liquid medium material, and the default parameters are obtained. The wall material is made of steel as the prototype. Some parameters are modified, such as the absolute roughness value is changed to $\Delta = 0.02$. The other parameters can be maintained by default because heat exchange is not involved in this time.

Boundary conditions: The inlet is the velocity inlet. When the relationship between Reynolds number and valve flow coefficient is simulated, the velocity input range is from v = 0.3 m/s to v = 0.8 m/s, and the interval is 0.05 m/s. When investigating the influence of ridge shape, the input is v = 0.5 m/s; the outlet is a pressure outlet, and the back pressure is p_{out} = 400 kPa; the characteristic parameters such as turbulent intensity, hydraulic diameter (DN = 0.02 m), turbulent energy k, turbulent dissipation rate ε, turbulent viscosity ratio in the setting of inlet and outlet can be calculated according to the following formulas.

$$i = 0.16\mathrm{Re}^{(-1/8)} \tag{4}$$

$$k = 1.5(vi)^2 \tag{5}$$

$$\varepsilon = 0.75 C_\mu k^{1.5}/l = 0.75 C_\mu k^{1.5}/(0.07d) \tag{6}$$

$$\mu_t = \rho_0 C_\mu \frac{k^2}{\varepsilon} \tag{7}$$

Among them, $C_\mu = 0.09$, v is the velocity of flow, ρ_0 is the density of water, μ_t/μ is the turbulent viscosity ratio, and μ is the dynamic viscosity of water.

Solver method sets pressure as standard, calculates second-order upwind, and sets reference cross-section at the center of the valve orifice, so as to output characteristic parameters at the valve orifice in the calculation results. The simulation step size is set to 5000 steps, and other parameters such as relaxation factor and convergence residual range are all defaulted by the system.

The surface integrals are selected from the simulation results to output the results. By calculating the total pressure at the inlet, outlet and the middle section of the valve, the pressure loss between the outlet and the inlet can be obtained, and the flow coefficient C_v (K_v) and local resistance coefficient ζ can be calculated according to Equations (1) and (2).

3.3. Mesh Refinement and Grid Independence Verification Based on Velocity Gradient and y^+ Adaptive

When simulating the flow of medium fluid in the valve at different opening, the velocity field cloud atlas shows that the velocity gradient at the valve opening is very large when the valve is half-open. The smaller the opening, the more concentrated the maximum velocity at the valve opening center, the more obvious the increase of the velocity gradient (Figure 9). In addition, when the opening is less than 40%, the height of the valve mouth is less than 6 mm, which can only accommodate 0 to 6 grids. It is necessary to take into account the insufficient development of turbulence in the boundary layer of the river basin. Especially when the opening is 1.5 mm at 10%, the scale of the turbulent viscous bottom layer is generally less than 1 mm. At this time, the influence of the flow boundary layer on the flow state cannot be ignored. However, it is not unnecessary to increase the overall grid density. Therefore, the velocity gradient adaptive method and y^+ adaptive method are adopted to solve the local mesh refinement based on the initial operation results (Figure 10).

It can be seen that with the alternating adaptive process, the mesh is continuously optimized, and the simulation results converge gradually. Therefore, the hybrid adaptive method of velocity gradient and y^+ adaptive has good applicability for reducing the error of simulation results of small opening state of valve body, and has guiding significance for grid optimization.

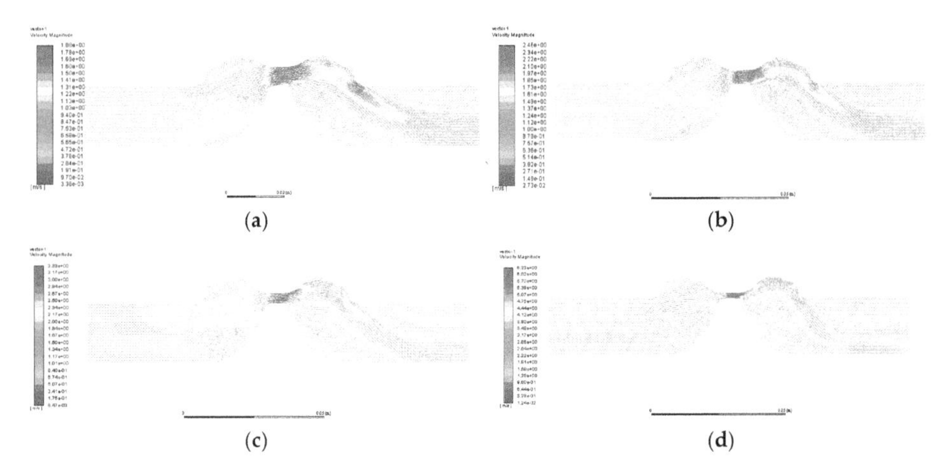

Figure 9. Cloud Map Display of Velocity Field at Different Openings shows that as the valve opening decreases, the velocity gradient increases. (**a**) for 40% opening; (**b**) for 30% opening; (**c**) for 20% opening; (**d**) for 10% opening.

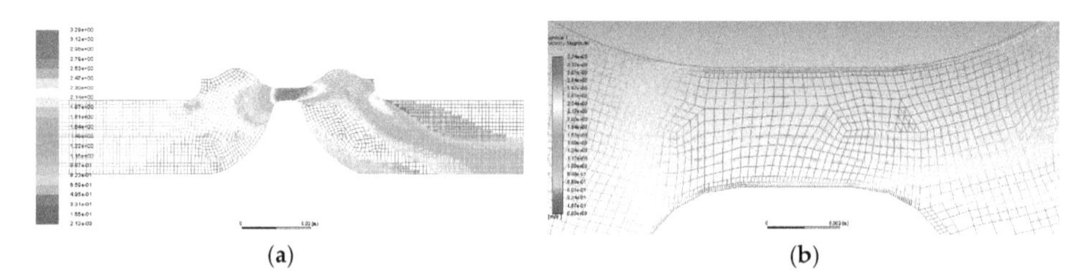

Figure 10. The meshes of three-time velocity adaptation (**a**) velocity gradient and y$^+$ hybrid adaptation; (**b**) and their results of calculation (Tables 2 and 3), respectively.

Table 2. Result of Velocity gradient adaptive calculation.

	Initial Result	One-Time Adaption	Two-Time Adaption	Three-Time Adaption	Four-Time Adaption
Δp/Pa	3560.4	3274.2	3268.5	3226.8	3243.4
C_v	3.46473	3.61230	3.61613	3.63941	3.63009
K_v	94.685	98.737	98.823	99.459	99.215

Table 3. y$^+$ adaptive results at different mesh densities.

Mesh Size	Initial Grid Number	Index	Initial Result	One-Time Adaption	Two-Time Adaption	Three-Time Adaption
1 mm	2778	C_v	3.4647	3.6774	3.7093	3.7072
		K_v	94.685	100.497	101.370	101.313
0.5 mm	10850	C_v	3.5182	3.6819	3.6748	3.6733
		K_v	96.146	100.620	100.426	100.386
0.25 mm	42664	C_v	3.5393	3.6769	3.6624	3.6708
		K_v	96.722	100.483	100.088	100.346
0.1 mm	266735	C_v	3.5560	3.6679	3.6680	3.6701
		K_v	97.179	100.239	100.240	100.305

3.4. Cavitation Model

The geometric model is based on watershed model with different valve opening under profile E. The two-phase flow material is set as water-liquid and water-vapor in the material library of the system (The simulation parameters are listed in Table A1). For the space discretization of the governing equations, simple algorithm is used to couple the velocity and pressure. The pressure term discretization format is PRISTO! In other terms, the second-order upwind model is used to simulate the river basin grid model with different openings in fluent. In LiquidlensTMLLG4 ultra-pure water treatment system, the inlet pressure is more than 500 kPa, and the outlet pressure ranges from p_{out} = 100 kPa to p_{out} = 500 kPa. That is to say, the maximum pressure difference between the two ends of the diaphragm valve in the system is about 400 kPa. Therefore, the simulated inlet is set as p_{in} = 500 kPa pressure inlet, and the outlet is set as p_{out} = 100 kPa pressure outlet.

4. Results and Discussion

4.1. Throttling Characteristics

In general, the flow coefficients of profile B, C, and E decrease with the increase of flow velocity, which indicates that the greater the flow velocity, the greater the turbulent fluctuation, the more pressure loss is formed, which is in accordance with the general law (Figure 11). The flow coefficients of profile A increase obviously with the increase of Reynolds number, because the ridge of profile A is too steep and the flow velocity is lower. As a result of the blocking effect, it is not easy for the fluid to flow through the ridge. On the contrary, when the flow velocity is high, the fluid momentum is large and easy to pass, so the pressure drop decreases and the flow coefficient increases. The flow coefficient of profile d does not change obviously with the increase of Reynolds number. The possible reason is that the side of the ridge of profile d is gentle and easy for the fluid to pass through, but the width of the ridge top is wide. The narrow passage leads to an approximate sharp angle, which leads to an excessive bending angle of the flow passage, approaching 180 degrees. The fluid is not easy to turn, resulting in pressure loss. The two factors restrict each other, resulting in the flow coefficient almost unchanged.

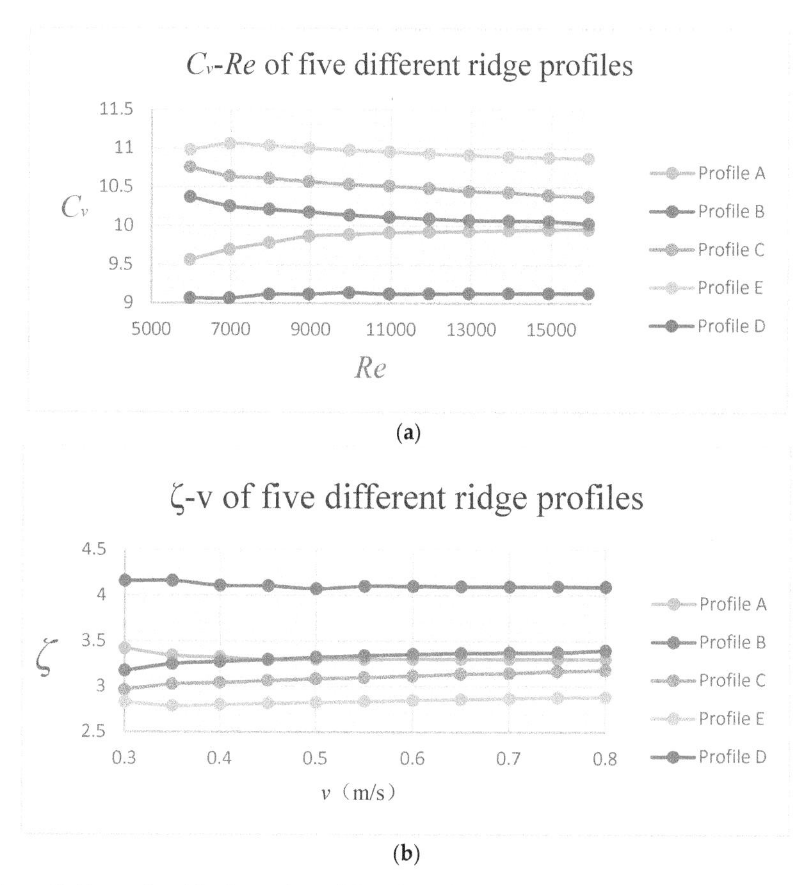

Figure 11. The results of the Comparison of Throttling Characteristics of Five Profiles. (**a**) C_v-Re of five different ridge profiles; (**b**) ζ-v of five different ridge profiles.

By comparing the curves of profile A, B and C, it can be known that with the gradual increase of the flatness of the side wall of the ridge, the flow coefficient becomes larger and the flow performance is better at the same flow velocity; by comparing the curves of profile C, D and E, we can know that the appropriate increase of the width of the top of the ridge is beneficial to the fluid steering, flow passage and flow coefficient. The circulation performance is better. Therefore, the top of the ridge should be too narrow, and the side walls of the ridge should be too steep.

Comparing the streamline diagrams of four types of surfaces (Figure 12), it can be seen that there are large vortices near the right side of the ridge and the upper wall of the exit, which is the direct cause of pressure loss. The former is determined by the shape of the ridge, and the latter is caused by the protrusion of the pipe wall on the right side of the upper wall hindering the flow. In addition, there are small eddies on the edge of the diaphragm and valve body installation. In addition, although there are two large eddy zones, the shape and size of the eddy zones are also different. When the ridge is too steep or the ridge top is too narrow, leading to too large bending, the scale of the right-side eddy on the ridge is very large, and the shape of the eddy is nearly circular, resulting in a great pressure loss. In contrast, the right-side eddy zones of the ridge of B and E are relatively flat. It can be inferred that increasing the flatness of the side of the ridge can reduce the vortices to a certain extent; for the vortices near the upper wall at the exit, the roundness of the protrusion of the inner wall of the upper wall can be appropriately increased or a more suitable shape can be designed to weaken the vortices.

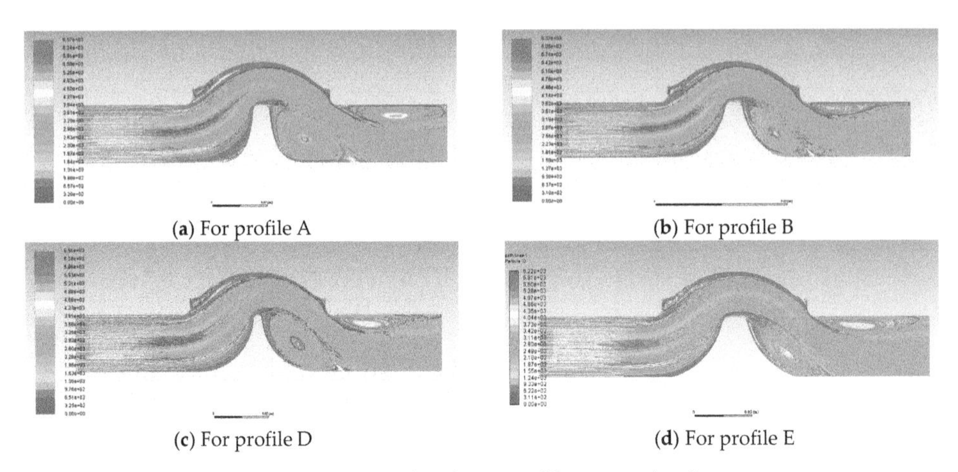

(**a**) For profile A (**b**) For profile B

(**c**) For profile D (**d**) For profile E

Figure 12. The streamline diagrams of four types of surfaces.

4.2. *Opening and Closing Characteristics*

According to the flow coefficient diagram (Figure 13a), with the increase of opening, the flow performance increases, pressure loss decreases, and flow coefficient increases. However, in the range of 80–100% opening, the flow coefficient has experienced a process of first rising and then falling, and the flow performance is the best around 90% opening (Figure 13b). By comparing the inner surface shapes of 80%, 90% and 100% open diaphragms, it can be seen that the central area of the inner surface of 90% open diaphragms is close to the plane, while the central area is concave when the valve orifice is fully open. It can be inferred that the flow performance is better when the central area of the inner surface of diaphragms is horizontal than that when the valve orifice is concave. It can be guessed that the central area will be approximately planar when the diaphragm is manufactured, or a 10% pre-clearance will be given when the diaphragm is installed to ensure the best flow performance when the diaphragm is fully open.

By analyzing the local resistance coefficient diagram (Figure 13c), we can know that the resistance effect of the valve is obvious when the valve is small. The relationship between outlet flow rate and valve opening shows that the linearity of the middle section of flow control is better but the linearity area is not large. In the two opening areas of 0–20% and 80–100%, the flow characteristics of the valve are not good. The reason is that the small opening can easily cause jet flow or even cavitation, and blocking flow is easy to occur, which results in a small flow variation range and a large flow rate. When opening, the throttling effect caused by the change of valve opening is not obvious and the flow control characteristics are not good because the diameter of valve opening is similar to the diameter of pipeline at the entrance of river basin. In the middle opening range, we can know that the flow characteristics are approximately linear, and the flow control performance is good.

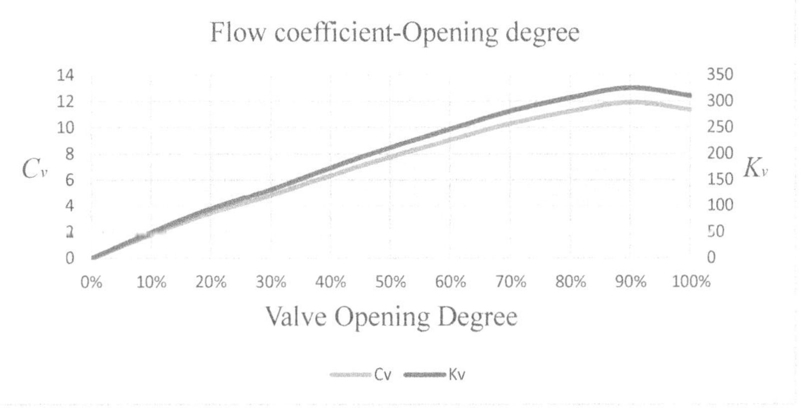

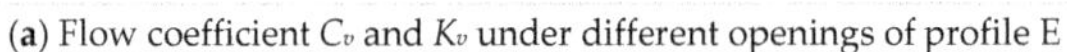

(**a**) Flow coefficient C_v and K_v under different openings of profile E

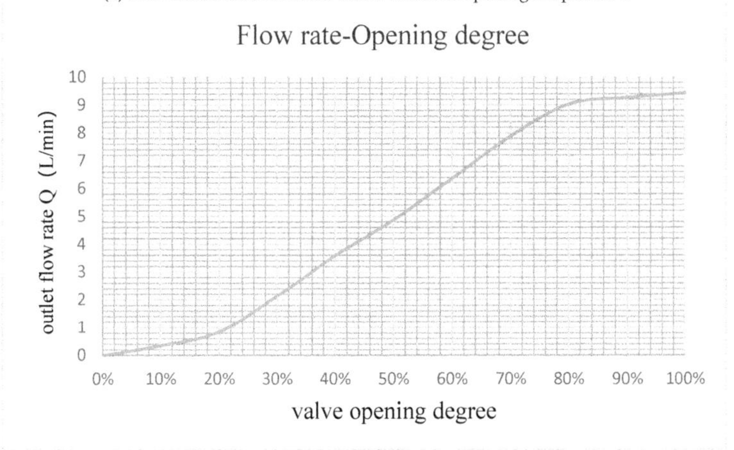

(**b**) Flow rate at different openings of profile E

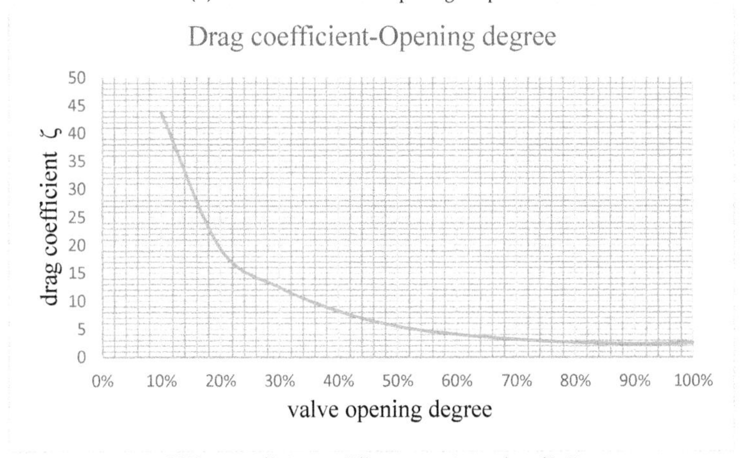

(**c**) Drag coefficient at different openings of profile E

Figure 13. The results of the flow characteristics at different openings of profile E.

4.3. Cavitation Characteristics

4.3.1. The Influence of Opening Degree

Under the given boundary conditions, cavitation begins to appear near the lower boundary of the valve orifice and the corner of the outlet bending when the opening of the diaphragm valve drops to 50%. With the opening decreasing, the cavitation area gradually expands and develops to the upper boundary of the valve orifice (Figure 14). When the opening is below 20%, the cavitation area shrinks to the valve orifice and cavitates near the corner of the outlet bending. When the opening is less than 10%, the cavitation area continues to shrink near the center of the valve opening.

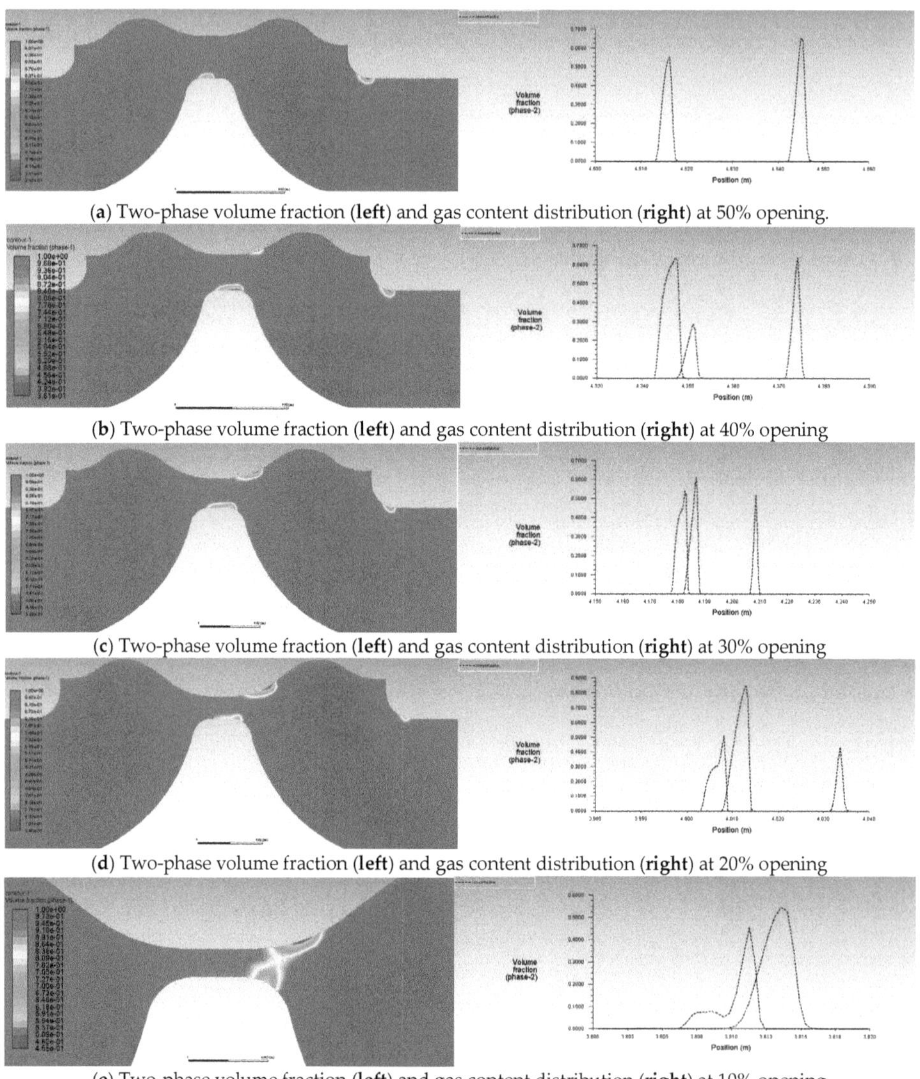

(**a**) Two-phase volume fraction (**left**) and gas content distribution (**right**) at 50% opening.

(**b**) Two-phase volume fraction (**left**) and gas content distribution (**right**) at 40% opening

(**c**) Two-phase volume fraction (**left**) and gas content distribution (**right**) at 30% opening

(**d**) Two-phase volume fraction (**left**) and gas content distribution (**right**) at 20% opening

(**e**) Two-phase volume fraction (**left**) and gas content distribution (**right**) at 10% opening

Figure 14. *Cont.*

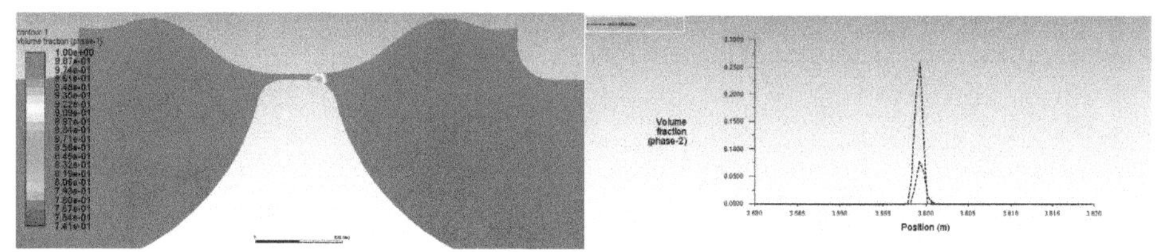

(**f**) Two-phase volume fraction (**left**) and gas content distribution (**right**) at 5% opening

Figure 14. Volume fraction α of two phases and void fraction distribution of cavitation phase in valve flow field.

4.3.2. Effect of Entrance Boundary Conditions (at 10% Opening)

The inlet is still set as pressure inlet, the setting value increases from p_{in} = 500 kPa to p_{in} = 100 kPa intervals, and the outlet is set as pressure outlet with p_{out} = 100 kPa back pressure. Two-phase volume fraction nephogram of observed time-averaged cavitation results can be obtained by processing and calculating the data of outlet flow rate, pressure difference at both ends and inlet flow rate, and the change of mass flow rate at both ends of inlet and outlet can be obtained with the increase of pressure boundary conditions (Figure 15).

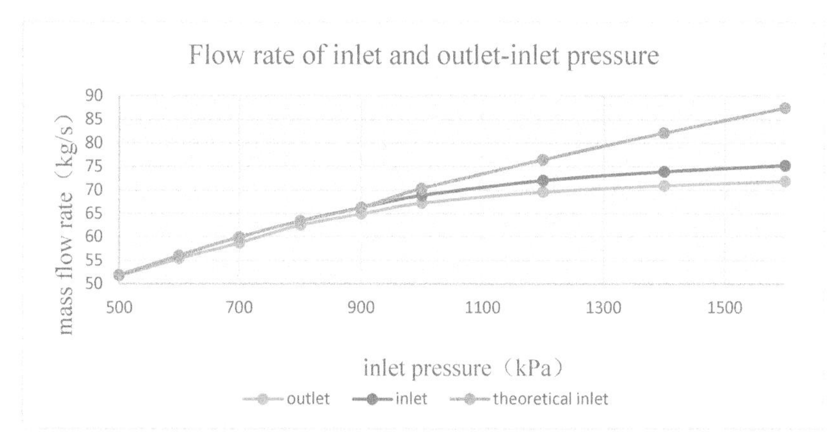

Figure 15. With the increase of inlet pressure, the inlet flow rate and outlet flow rate increase almost linearly, but when the inlet pressure p_{in} increases to more than 900 kPa, the increase of outlet flow rate slows down, resulting in a larger flow loss.

The reason is that when the pressure difference between the two ends Δp is greater than 800 kPa, the cavitation phenomenon becomes serious and more cavitation bubbles are blocked at the valve mouth, which leads to flow saturation and deviates from the theoretical flow input. And Figure 16 shows a volume fraction nephogram at different inlet pressures.

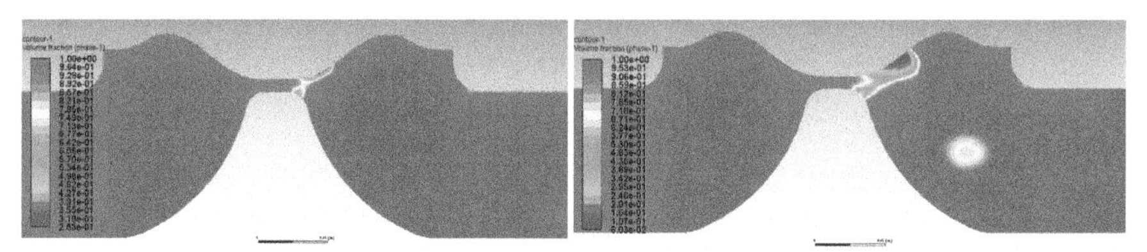

(**a**) Aqueous phase volume fraction at inlet pressure of p_{in} = 600 kPa (**left**) and p_{in} = 900 kPa (**right**)

Figure 16. *Cont.*

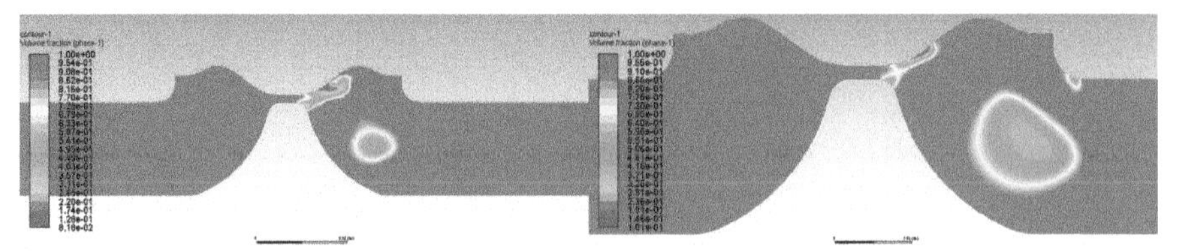

(**b**) Aqueous phase volume fraction at inlet pressure of p_{in} = 1200 kPa (**left**) and p_{in} = 1500 kPa (**right**)

Figure 16. Volume fraction nephogram at different inlet pressures.

4.4. Shape Optimization and Verification

According to the conclusion analysis of the influence of the structure of the runner profile on the throttling characteristics and the flow control characteristics, the following optimization ideas can be obtained: Increasing the lateral flatness of the ridge, appropriately increasing the width of the ridge top and reducing the protrusion of the inner wall of the valve at the diaphragm installation can increase the flow coefficient and reduce the local resistance coefficient of the valve, and appropriately reduce the diaphragm stroke, the central area of the inner surface of the diaphragm approximates the plane at the maximum opening, and give the diaphragm a pre-stroke during installation. The flow control characteristics of the diaphragm valve can be improved appropriately and the linear range of the flow control zone can be enlarged. The improvement of various surface structures should be carried out on the premise of ensuring the overall size ratio of diaphragm valves.

The optimized prototype is based on the well-behaved surface e, and further flattens the ridge ridge, reduces the obstacles at the protrusions, and modifies the shape of the inner surface of the diaphragm so that the central area of the inner surface of the diaphragm is planar in the fully open state, thus reducing the diaphragm travel. The comparison between the optimized profile F and profile E is shown in Figure 17.

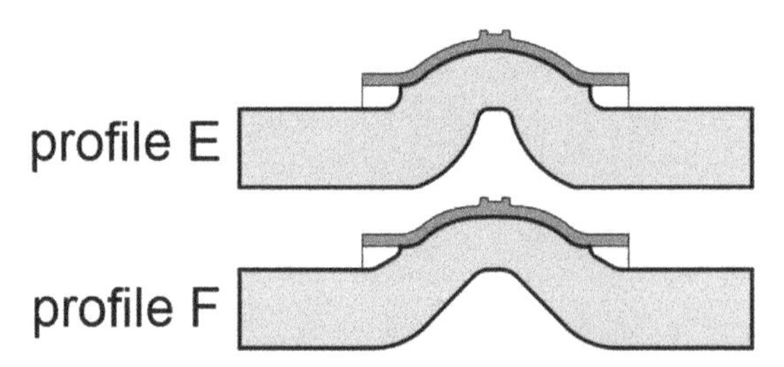

Figure 17. The optimized surface F and the optimized surface E.

The flow coefficient and local resistance coefficient (Figure 18) of the optimized profile are improved to a certain extent compared with the prototype profile E before optimization; with the increase of the opening of the profile F valve, the flow coefficient shows an increasing trend, and no longer decreases with the increase of the opening when approaching full opening (Figure 19a). In addition, some improvements have been made in the flow control characteristics of the valve under profile F. In addition to the good linearity, the linearity range has also been enlarged at large valve ports (Figure 19b).

As can be seen from the streamline diagram (Figure 20), the vortices on the right side of the ridge are compressed to a great extent, which hardly hinders the main flow. The vortices near the upper wall of the valve outlet are almost non-existent, indicating that the reduction of the flow path hindrance has a great effect on the reduction of the vortices.

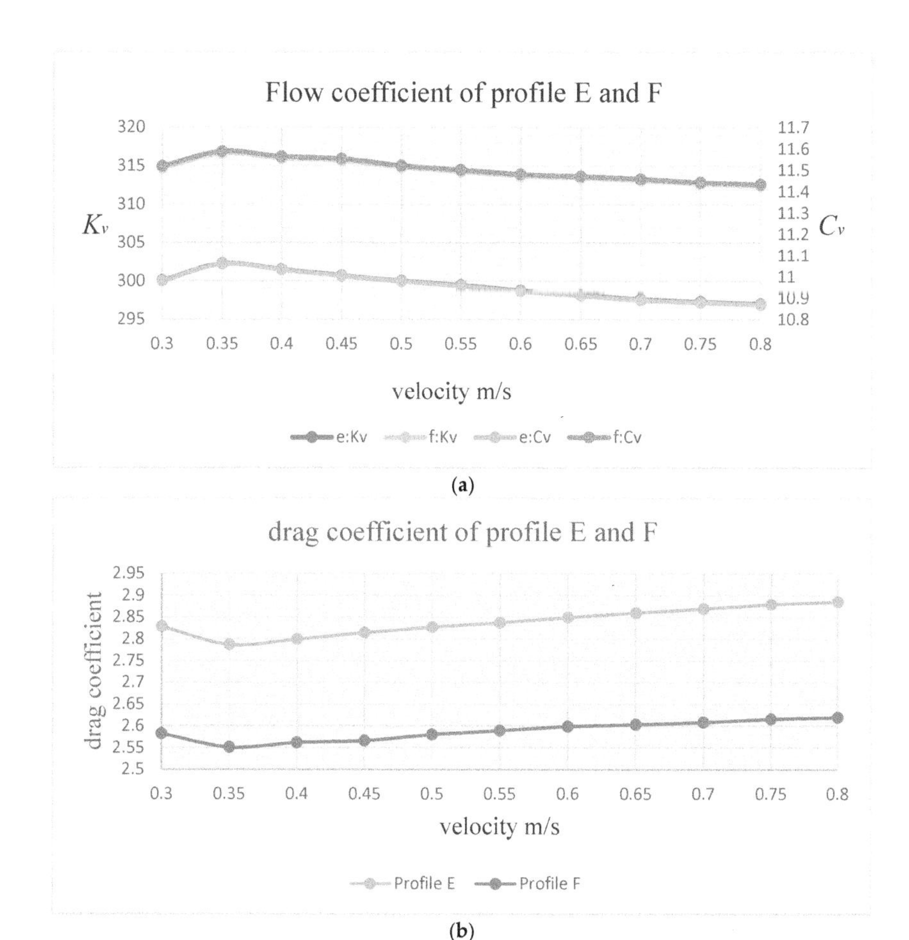

(b)

Figure 18. Comparison on flow characteristics (**a**) and drag characteristics (**b**) between profile E and the optimized profile F.

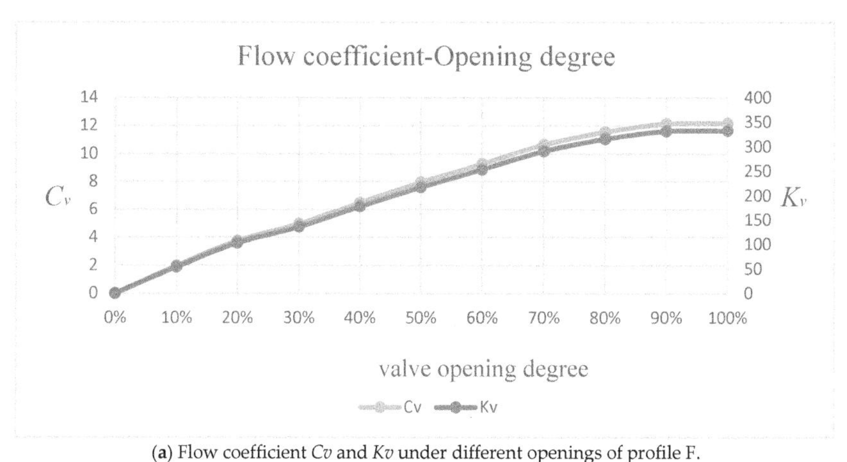

(**a**) Flow coefficient *Cv* and *Kv* under different openings of profile F.

Figure 19. *Cont.*

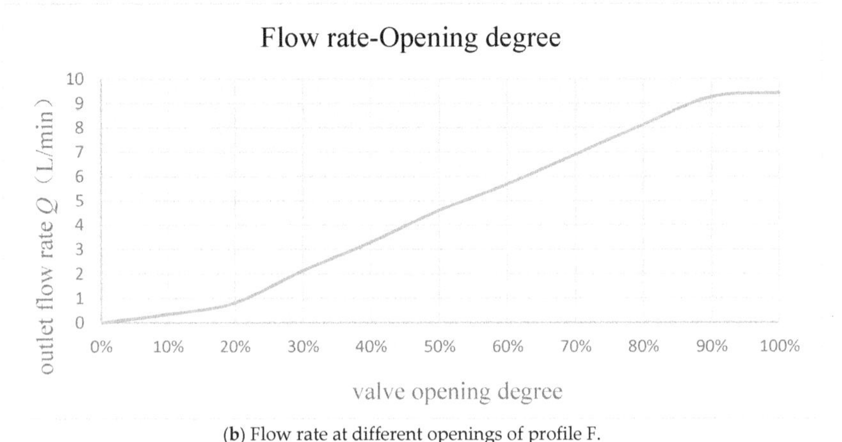

(**b**) Flow rate at different openings of profile F.

Figure 19. The results of the flow characteristics at different openings of profile F.

Figure 20. Streamline diagram of profile F at full opening.

5. Conclusions

The deformation mechanism of diaphragm valve diaphragm is analyzed and simulated by the finite element method. The influence of diaphragm thickness and the loading mode of stem loading on diaphragm deformation are explored. The conclusion is drawn that the diaphragm with thin and moderate loading area had better sealing and control characteristics. The effect of velocity gradient adaptive and y^+ adaptive methods on mesh optimization in some regions and the effect of refined mesh on simulation accuracy are verified by means of mesh adaptive technology.

In the way of design comparison, the influence of the sill-shaped diaphragm structure on the flow coefficient and the drag coefficient of the valve is discussed. Combined with the simulation of the cloud map and the simulation of the flow field under different opening degrees of the valve, the optimized diaphragm flow is obtained. The theoretical idea of the pavement surface structure: smoothing the side wall of the ridge, increasing the width of the ridge, reducing the obstruction of the pipeline and smoothing the lower surface of the diaphragm. Optimize the design in this direction and further simulate to analyze the correctness of the verification optimization.

Through the analysis of gas-liquid two-phase flow nephogram, it is concluded that with the valve closing cavitation gradually occurs and expands, and with the increase of inlet boundary conditions, the small valve opening cavitation becomes more and more serious.

In this paper, the factors influencing the flow characteristics of diaphragm valves and the method of optimizing the flow profile are systematically studied. The main technical indexes of diaphragm valves, such as controllability, flow performance, diaphragm life, cleanliness, and sealing, are comprehensively studied. The improvement of Throttling Characteristics and control characteristics is beneficial to ultra-pure water system and submerged photolithography. The improvement of technology is of great benefit and constructive and forward-looking to the development of semiconductor industry.

Author Contributions: Investigation Y.L. and L.L; Simulation and Analysis Y.L.; Methodology L.L.; Software Y.L.; Writing and Editing Y.L. and L.L.; Validation L.L. and K.W.

Funding: The authors are grateful to the National Natural Science Foundation of China (no. 51605333 and no. 51805317) for financial support.

Conflicts of Interest: The authors declare no conflict of interest.

Nomenclature

C_v	valve flow coefficient in British units
ζ	resistance coefficient/drag coefficient
Δp	pressure drop between valve inlet and outlet
ρ	density of medium
D_0	outer diameter of diaphragm convex
L_0	diaphragm diameter
ρ_1	density of diaphragm
σ	tensile strength
DN	nominal diameter/hydraulic diameter
h	roof ridge height
p_{out}	pressure of outlet
ε	turbulent dissipation rate
μ_t/μ	turbulent viscosity ratio
μ	dynamic viscosity of water
i	turbulence intensity
K_v	valve flow coefficient in metric units
Q	valve volume flow rate
G	density ratio of medium to water at 15.6 °C
v	average velocity of flow
d_0	diameter of diaphragm convex
B_0	diaphragm thickness
R_0	spherical diameter of sub-diaphragm surface
x_0	deformation travel of diaphragm
L	flow field length
B	ridge top width/loading table width
p_{in}	pressure of inlet
k	turbulent energy
ρ_0	density of water
Δ	absolute roughness of pipe wall
α	vapor volume fraction number

Appendix A

Table A1. Simulation settings and parameters.

The FEM Simulation of the Diaphragm			
Grid size	1 mm	Engineering data sources	Neoprene rubber
Density	1.25 g/cm^3	Tensile ultimate strength	100 MPa
Design modeller	Add extrude	Mesh	Size function:adaptive
Relative center	Coarse	Static structure	Structural
Analysis settings	Program controlled	Large deflection	On
Standard earth gravity	All bodies	Fixed support	Upper and lower ring extruded surfaces
Fixed displacement	1.5–15 mm	Frictionless support	Loading stage inner wall
Solution	Total deformation	Solution	Equivalent stress
The Two-Dimension Simulation of the Weir-Type Diaphragm Valve Flow Filed			
Basic grid size	1 mm	General: type	Pressure-based
General: time	Steady	Velocity formulation	Absolute
Viscous model	Standard k-epsilon	Near-wall treatment	Standard wall functions
Model constants	Default	Materials	Fluid:water-liquid
Velocity inlet	0.5–0.8 m/s	Initial gauge pressure	400 kPa
Turbulence parameters	Based on calculation	Pressure outlet	Total pressure
Wall roughness constant	0.02	Solution methods	Simple
Pressure	Standard	Momentum	Second order upwind
Turbulent kinetic energy	Second order upwind	Solution controls	Default
Iterations	50,000	Monitor check absolute criteria	0.001
The cavitation simulation of small valve orifice state			
Grid size	1 mm	General: type	Pressure-based
General: time	Steady	Velocity formulation	Absolute
Multiphase model	Mixture	Phase interaction	Mass: mechanism
Cavitation model	Zwart-Gerber-Belamri	Cavitation properties	Constant 3540
Model constants	Default	Phase 1	Water-liquid
Phase 2	Water-vapor	Viscous model	Realizable k-epsilon
Near wall treatment	Standard wall functions	Pressure inlet	500–1600 kPa
Turbulent intensity	5%	Hydraulic diameter	0.02
Pressure outlet	100 kPa	Outlet backflow pressure	Total pressure
Inlet volume fraction	Phase 2: 0 (constant)	Solution methods	Scheme: simple
Spatial discretization	Pressure: PRESTO!	Other settings	Second order upwind
Solution controls	All (0.1–0.2)	Number of interactions	50,000

References

1. Mbiya, B.M.; Fester, V.G.; Slatter, P.T. Evaluating resistance coefficients of straight through diaphragm control valves. *Can. J. Chem. Eng.* **2009**, *10*, 704–714. [CrossRef]
2. Cheng, Y.; Liu, M. Selection of common valves in chemical engineering design. *Guangdong Chem. Ind.* **2009**, *8*, 244–245.
3. Wu, H. Requirements for selection and installation of diaphragm valve in clean fluid conveying system. *Chem. Pharm. Eng.* **2016**, *37*, 50–54.
4. Nguyen, T.; van der Meer, D.; van den Berg, A.; Eijkel, J.C.T. Investigation of the effects of time periodic pressure and potential gradients on viscoelastic fluid flow in circular narrow confinements. *Microfluid Nanofluid* **2017**, *21*, 37. [CrossRef]
5. Jerman, J.H. *Electrically-Activated, Micromachined Diaphragm Valves. Micro System Technologies 90*; Springer: Berlin/Heidelberg, Germany, 1990; pp. 806–811.
6. Yang, F.; Imin, K. Analysis of fluid flow and deflection for pressure-balanced MEMS diaphragm valves. *Sens. Actuators A Phys.* **2000**, *79*, 13–21. [CrossRef]
7. Lin, B.J. Immersion lithography and its impact on semiconductor manufacturing. *J. Micro/Nanolithogr. MEMS MOEMS* **2004**, *3*, 377–501. [CrossRef]
8. Owa, S.; Nagasaka, H. Immersion lithography: its potential performance and issues. In *Optical Microlithography XVI*; International Society for Optics and Photonics: Bellingham, WA, USA, 2003; pp. 724–733.
9. Cao, F. Design and Implementation of the State Control of Immersion System of Immersion Lithography Machine based on DSL. Master's Thesis, Zhejiang University, Hangzhou, Zhejiang, China, 15 January 2017.
10. Chen, W. Study on Liquid Supply and Sealing of Immersion Control Unit in Immersion Lithography. Master's Thesis, Zhejiang University, Hangzhou, Zhejiang, China, 30 June 2010.
11. Kim, T.H.; Sunkara, V.; Park, J.; Kim, C.J.; Woo, H.K.; Cho, Y.K. A lab-on-a-disc with reversible and thermally stable diaphragm valves. *Lab Chip* **2016**, *16*, 3741–3749. [CrossRef]
12. Ohlsson, P.A. Diaphragm valve development–challenging traditional thinking. *Pharm. Eng.* **2013**, *33*, 1–4.
13. Tanikawa, T.; Yamaji, M.; Yakushijin, T.; Fukuchi, O. Direct Touch Type Metal Diaphragm Valve. U.S. Patent 8,256,744, 4 September 2012.
14. Lv, B. *Practical Rubber Handbook*; Chemical Industry Press: Beijing, China, 2009.
15. Jerman, H. Electrically-activated, normally-closed diaphragm valves. In Proceedings of the International Conference on Solid-State Sensors and Actuators, San Francisco, CA, USA, 24–27 June 1991; pp. 1045–1048.
16. Sun, X.; Gu, X.; Carr, W.N. Lateral in-plane displacement microactuators with combined thermal and electrostatic drive. In Proceedings of the 8th International Conference on Solid-State Sensors and Actuators and Eurosensors IX, Stockholm, Sweden, 25–29 June 1995.
17. Li, Y.; Sun, L. Manufacture and Application of Rubber Compound—Fluoroplastics Composite Diaphragm for Pumps and Valves. *Spec. Purp. Rubber Prod.* **2010**, *31*, 35–38.
18. Lu, P. *Practical Valve Design Manual*; Machinery Industry Press: Beijing, China, 2012.
19. Cui, B.; Ma, G.; Wang, H.; Lin, Z. Influence of valve core structure on flow resistance characteristics and internal flow field of throttling stop valve. *J. Mech. Eng.* **2005**, *51*, 178–184. [CrossRef]
20. Zhang, Y. *Hydromechanics*; Higher Education Press: Beijing, China, 2016.
21. Cai, S. Diaphragm flow regulating valve. *China Pet. Mach.* **2006**, *2*, 31–32.
22. Fester, V.G.; Kazadi, D.M.; Mbiya, B.M.; Slatter, P.T. Loss coefficients for flow of Newtonian and non-Newtonian fluids through diaphragm valves. *Chem. Eng. Res. Des.* **2007**, *85*, 1314–1324. [CrossRef]
23. Wang, Z. Research on Flow Field Simulation and Optimum Design Method for Process Valve. Master's Thesis, University of Electronic Science and Technology of China, Chengdu, Sichuan, China, 19 June 2017.
24. Wu, D.; Li, S.; Wu, P. CFD simulation of flow-pressure characteristics of a pressure control valve for automotive fuel supply system. *Energy Convers. Manag.* **2015**, *101*, 658–665. [CrossRef]
25. Liu, X.; Liu, Y.; Guo, X.; Zhang, K. Research on the simulation of inflow control valve based on FLUENT. *Mech. Eng.* **2018**, *10*, 3–6.
26. Zhang, J.; Wang, D.; Xu, B.; Gan, M.Y.; Pan, M.; Yang, H.Y. Experimental and numerical investigation of flow forces in a seat valve using a damping sleeve with orifices. *J. Zhejiang Univ. Sci. A* **2018**, *19*, 417–430. [CrossRef]
27. Qian, J.; Gao, Z.; Liu, B.; Jin, Z.J. Parametric study on fluid dynamics of pilot-control angle globe valve. *J. Fluids Eng.* **2018**, *140*, 111103. [CrossRef]

28. Ye, S.; Zhang, J.; Xu, B.; Zhu, S.; Xiang, J.; Tang, H. Theoretical investigation of the contributions of the excitation forces to the vibration of an axial piston pump. *Mech. Syst. Signal Process.* **2019**, *129*, 201–217. [CrossRef]
29. Qian, J.; Chen, M.; Liu, X.; Jin, Z.J. A numerical investigation of the flow of nanofluids through a micro Tesla valve. *J. Zhejiang Univ. Sci. A* **2019**, *20*, 50–60. [CrossRef]
30. Zhou, X.; Wang, Z.; Zhang, Y. Flow Coefficient calculation method based on CFD and mesh adaptive method. *J. Univ. Electron. Sci. Technol.* **2017**, *2*, 475–480.
31. Javorik, J.; Stanek, M. The numerical simulation of the rubber diaphragm behavior. *Situations* **2011**, *3*, 4.
32. He-xiang, W.; Yong-ming, W.; Jia-zhen, P. Optimizing the configuration of the diaphragm valve film by ANSYS. *Mech. Res. Appl.* **2009**, *3*, 63–64.

Article

Analysis of Air–Oil Flow and Heat Transfer inside a Grooved Rotating-Disk System

Chunming Li [1], Wei Wu [2,*], Yin Liu [2], Chenhui Hu [2] and Junjie Zhou [2]

[1] China North Vehicle Research Institute, Beijing 100072, China; lichunming201@163.com
[2] National Key Laboratory of Vehicular Transmission, Beijing Institute of Technology, Beijing 100081, China; 3220180250@bit.edu.cn (Y.L.); huchenhui_bit@126.com (C.H.); zhoujunjie@bit.edu.cn (J.Z.)
* Correspondence: wuweijing@bit.edu.cn

Received: 18 August 2019; Accepted: 11 September 2019; Published: 18 September 2019

Abstract: An investigation on the two-phase flow field inside a grooved rotating-disk system is presented by experiment and computational fluid dynamics numerical simulation. The grooved rotating-disk system consists of one stationary flat disk and one rotating grooved disk. A three-dimensional computational fluid dynamics model considering two-phase flow and heat transfer was utilized to simulate phase distributions and heat dissipation capability. Visualization tests were conducted to validate the flow patterns and the parametric effects on the flow field. The results indicate that the flow field of the grooved rotating-disk system was identified to be an air–oil flow. A stable interface between the continuous oil phase and the two-phase area could be formed and observed. The parametric analysis demonstrated that the inter moved outwards in the radial direction, and the average oil volume fraction over the whole flow field increased with smaller angular speed, more inlet mass flow of oil, or decreasing disk spacing. The local Nusselt number was remarkably affected by the oil volume fraction and the fluid flow speed distributions in this two-phase flow at different radial positions. Lastly, due to the change of phase volume fraction and fluid flow speed, the variation of the average Nusselt number over the whole flow field could be divided into three stages.

Keywords: CFD; two-phase flow; volume of fluid; Nusselt number; grooved disk; flow pattern

---

1. Introduction

The viscous flow between rotating disks has received special attention for its wide applications, including wet clutches in automatic transmission [1], in hydro-viscous drive [2], and in turbomachinery [3]. The theoretical research of the two-disk flow was first proposed by Von Karman in 1921 [4]. Later, the fundamental flow behaviors of the counter-rotating and the rotor-stator disks were investigated theoretically by Batchelor [5] and Stewartson [6]. Most of the literature on the two-disk flow dealt with the shrouded disk systems, forming the theoretical foundation of the turbomachinery [7]. However, studies on an open disk system are also important, since the theory is the foundation of many applications, such as wet clutches. In the open disk configuration, the flow field becomes a Stewartson-type flow [8]. It has been found that annular or spiral rolls and an Ekman layer exist on a rotor disk [9]. Moreover, traditional research of a flat disk flow field is also focused on the laminar–turbulent transition [10].

In applications of the open disk system, such as mechanical seals and thrust bearings, the disks are featured with complicated surface texture. The load-carrying capacity and the cooling effect are improved by various groove structures on the surface [11–15]. The radial single-phase flow between a grooved rotating disk and a flat stationary disk was investigated analytically and experimentally [16]. Additionally, the effect of grooves on lubricant flow and thermal characteristics was examined for engagement of wet clutches [17,18]. The cooling capacity was notably affected by the single or the two phase flow patterns of the flow field with grooved plates [19,20]. The influence of flow dynamic behavior inside the grooved rotating-disk system on the disk temperature distribution was also demonstrated [21]. A rotating disk model, which is the topological geometry of bearing rotating parts, was established, and the volume of fluid (VOF) method was used to instantaneously track the interface between the oil film and the air on the disk surface [22]. For open wet clutches, the flow changed from single-phase to multi-phase flow in the gap at high speed. The commonly used simplified analytical model was constituted by the multi-phase flow theory of open grooved two-disk systems [23,24]. The flow field of the oil film based on the VOF model was simulated numerically by fluid dynamics simulation software to study the two-phase flow of a hydro-viscous drive [25]. The flow behavior was found to have an enormous effect on mass and heat transfer in the multiphase flows [26,27]. The effects of radial grooves and waffle-shape grooves on disk thermal and torque responses were studied [28]. Several other models were formulated to determine the flow impact on the heat exchange and the temperature distribution in the grooved two-disk system [29,30], including a computational fluid dynamics (CFD) model [31]. The drag torque and the flow pattern in the grooved rotating-disk system also received special focus [1,32–34]. It was found that the efficiency of the grooved rotating-disk system varied with flow field parameters [35–37]. From the above review, it could be concluded that the flow field would be dramatically affected by the disk configuration and the operation status in the grooved rotating-disk system. The variation of the flow field has a tremendous effect on the operation performance of the grooved rotating-disk system. However, the details of the flow pattern variation under different parametric conditions are still not clear. The heat dissipation characteristics of the flow field in a two-disk system remains to be further studied. Thus, a deeper insight into the interaction between the flow field characteristics and the grooved two-disk system enables potential improvements in practical applications, such as wet clutches.

In this paper, the flow pattern and the heat dissipation of the flow field inside the grooved rotating-disk system is studied. Different from the single phase studies before, the air–oil two-phase flow is investigated. The air–oil phase distribution in the flow field is presented in detail and compared with visualization experimental results. The effects of angular speed of grooved disk, oil flow rate at inlet, and disk spacing on the flow pattern and the oil phase distribution are analyzed. The local and the average Nusselt number was utilized to investigate the heat dissipation capacity of the two-disk system. This research aims to propose a quantitative method for the advanced precision cooling mechanism design of the clutch disk.

2. Grooved Rotating-Disk System Configuration

The schematic of a simplified open grooved two-disk system model is presented in Figure 1. The flow field in the gap is described by a cylindrical coordinate system (r, θ and z), indicating the radial, the azimuthal, and the axial coordinates, respectively. The disk with the radial grooves rotates axially with an angular velocity ω, and the stationary disk is ungrooved. There is a gap H between two disks. The groove area is defined by the groove number Ng and the circumferential angle of each single groove. The groove depth of the textured disk is h, and $r1$ and $r2$ represent the inner and the outer radii of the disks, respectively. A dimensionless radial location $r/r2$ is defined to analyze the parameter effects.

An oil volumetric flow rate Q, which is assumed to be a constant temperature (320 K), is prescribed at the inner radius surface as the inlet boundary condition, while an ambient pressure is prescribed at the outer radius surface as the outlet boundary condition. For the heat transfer process, the thermal effect of the system is coupled in the model by considering the heat conduction of disks and the heat convection between fluid and disks. In spite of a rather small gap and the viscous shear of the fluid, the viscous heating effect of oil is omitted, since the oil is supplied continuously at the inner radius, and the heat of viscous friction is taken away. The stationary ungrooved disk is assumed to be a heat source with a constant heat flux uniformly distributed on the surface. A stationary wall boundary condition is specified on the surface of the stationary disk. The surface of the rotating grooved disk is specified as an adiabatic wall boundary condition with a constant angular speed.

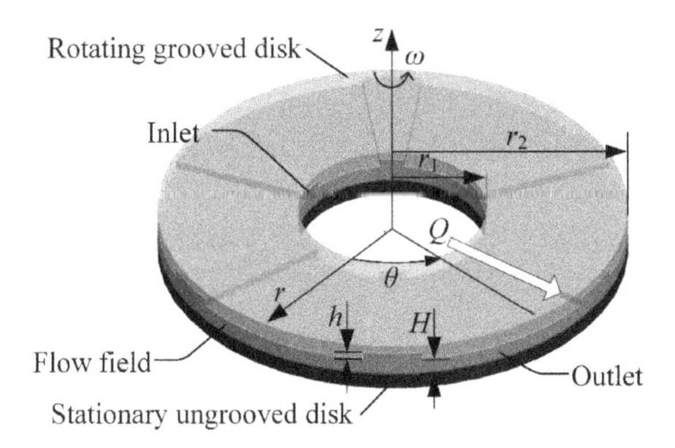

Figure 1. Schematic of the grooved rotating-disk system.

3. Grooved Rotating-Disk System Modeling

3.1. Governing Equations

In the present study, the flow field transits from single-phase flow to air–oil two-phase flow with the increasing speed of the grooved disk. Thus, a three-dimensional CFD approach dealing with a multiphase flow problem presented for the open grooved two-disk system [31] was utilized in this flow field simulation. In the visualization experiment, it was found that the air–oil interface between the inner and the outer radii of the grooved disk was rather stable after the rotating speed of the grooved disk was reached and kept constant. Thus, a steady-state model was established. The model was developed using the commercial software ANSYS FLUENT 15.0. In this model, the VOF approach developed by Hirt and Nichols [38] was adopted to track the two-phase interface of the air and the oil. A volume fraction φ_{oil} was used to mark the volume fraction of the oil phase. In the model, $\varphi_{\text{oil}} = 1$ represents the pure oil phase in a cell, while $\varphi_{\text{oil}} = 0$ represents the cell full of air phase. If $0 < \varphi_{\text{oil}} < 1$, it describes the air–oil two-phase interface. The volume fractions of two phases satisfy the following constraint:

$$\varphi_{air} + \varphi_{oil} = 1 \tag{1}$$

In the VOF method, the flow properties are averaged by phase volume fraction [39]. The density, the dynamic viscosity, and the effective thermal conductivity coefficients of the air–oil two-phase flow are given by:

$$\rho = \varphi_{oil}\rho_{oil} + (1 - \varphi_{oil})\rho_{air} \tag{2}$$

$$\mu = \varphi_{oil}\mu_{oil} + (1 - \varphi_{oil})\mu_{air} \tag{3}$$

$$k_{eff} = \varphi_{oil}k_{oil} + (1 - \varphi_{oil})k_{air} \tag{4}$$

The governing equations of the CFD method are the momentum equation, the continuity equation, and the energy equation. These equations are solved in the VOF multiphase model. The VOF model solves one single momentum equation, and the velocity field results are shared among both phases and result in a direct velocity coupling of the phases at the interface [34]. This computational model selects steady reference frame. The simulation is steady-state, and the transient terms in the governing equations are omitted.

The continuity equation is:

$$\nabla \cdot \left(\rho \vec{v}\right) = 0 \tag{5}$$

The momentum equation is:

$$\nabla \cdot \left(\rho \vec{v} \vec{v}\right) = \nabla \cdot \left(\mu(\nabla \vec{v} + (\nabla \vec{v})^{T})\right) - \nabla \cdot p + \rho \vec{g} + \vec{F} \tag{6}$$

Equations (2) and (3) were implemented as the averaged dynamic viscosity and the density of the fluid in the above equations. $\vec{v}$ denotes the velocity vector of fluid, p is the pressure of fluid, $\vec{g}$ is the gravity acceleration vector, and $\vec{F}$ denotes the surface tension source term. The surface tension force is given as:

$$\vec{F_i} = \sigma \frac{\rho \kappa_i \nabla \varphi_i}{\left(\rho_i + \rho_j\right)/2} \tag{7}$$

where i and j denote the air and the oil phase, respectively. σ is the surface tension coefficient assumed to be constant 0.03 N/m in this simulation. κ_i is the surface curvature at the interface defined as the divergence of the unit normal. For two phases presented in a cell, $\kappa_i = -\kappa_j$.

$$\kappa_i = \nabla \cdot \hat{n}_i \tag{8}$$

where:

$$\hat{n}_i = \hat{n}_w \cos \theta_w + \hat{t}_w \sin \theta_w \tag{9}$$

θ_w is contact angle at the wall. θ_w = 9.3° and 10.1° on the surfaces of the stationary disk made of quartz glass and the rotating aluminum disk, respectively. The data were obtained from the contact angle test results described in reference [40]. $\hat{n}_w$ and $\hat{t}_w$ are the unit vectors normal and tangential to the interface, respectively. Since the surface tension coefficient was assumed to be constant, the Marangoni effect was neglected due to zero surface tension gradient.

In the simulation, the air phase was set as the primary phase. The continuity equation of the air phase was solved first, given as:

$$\nabla \cdot \left(\alpha_{air} \rho_{air} \vec{v}_{air}\right) = 0 \tag{10}$$

After obtaining the volume fraction of the air phase in a cell, the volume fraction of the oil was determined from the constraint (1) as:

$$\varphi_{oil} = 1 - \varphi_{air} \tag{11}$$

The energy equation omitting radiation as a source term for the fluid domain was deduced as [41]:

$$\nabla \cdot (\vec{v} \rho E) = \nabla \cdot (k_{eff} \nabla T) \tag{12}$$

where:

$$E = \frac{\varphi_{oil} \rho_{oil} E_{oil} + (1 - \varphi_{oil}) \rho_{air} E_{air}}{\varphi_{oil} \rho_{oil} + (1 - \varphi_{oil}) \rho_{air}} \tag{13}$$

The energy equation for the solid domain is:

$$\nabla \cdot (k_s \nabla T) = 0 \tag{14}$$

where T denotes the temperature shared between the air and the oil phases, k_{eff} represents the effective thermal conductivity of the two-phase flow given by Equation (4), φ is the fluid volume fraction, E is the specific sensible enthalpy, and k_s denotes the thermal conductivity of the solid structure. The oil and air subscripts identify the oil and the air parameters, respectively.

3.2. Computational Model

Since the flow was assumed to be axisymmetric, one single-groove section (a 36° section) of disk was selected, and a periodic boundary condition was adopted in the simulation. In terms of boundary conditions, the moving disk was the rotating wall surface, and the static disk was the stationary wall surface. The standard wall function method was adopted to deal with the flow near the wall surface. The model inner diameter was the inlet of the flow field and was set as the mass-flow-inlet condition. The outer diameter was the outlet of the flow field. Since the outlet of the flow field was connected with the atmosphere, it was set as the pressure-outlet, and the pressure was the standard atmospheric pressure. In order to better simulate the flow details of the disk clearance flow field, the mesh needed to be encrypted near the disk wall and the radial slot wall. The flow field was discretized by a structured mesh scheme. The Green Gauss node-based method was used for the evaluation of gradients. The momentum and the energy equations were solved by the QUICK scheme [42]. The PRESTO! (Pressure staggering option) scheme was employed for the discretization of the pressure. A continuum surface force model proposed by Brackbill et al. [43] was applied in this calculation. Interface reconstruction was performed using the explicit piecewise linear interface construction (PLIC) scheme [44]. To handle the two-phase flow condition, it was assumed that the inlet was filled with oil, thus the oil volume fraction was one at the inlet. In this paper, the convergence of the calculation was determined by monitoring the residual error and the flow parameters. The numerical results were obtained when the drag torque, the volume average oil phase, and the area average wall surface heat transfer coefficient were stable. To ensure the accuracy and the validity of numerical results, a careful check for the grid independence of the numerical solutions was conducted. Three sets of mesh resolutions were adopted for the flow field with 307,644 (Mesh 1), 365,796 (Mesh 2), and 439,432 (Mesh 3). In Table 1, the calculated results of the average oil volume fraction with different angular velocities are given. The differences between the mesh density of 307,644 elements and the other two were smaller than 5%, indicating that a higher resolution of mesh than Mesh 1 limited improvement on the calculation result. Mesh 1 basically satisfied the requirement of accuracy. Considering the reduction of computational time consumption, Mesh 1 was selected in the simulation model.

Table 1. Oil volume fraction results under different mesh resolutions and speeds.

Speed	15.7 rad/s	18.8 rad/s	21.9 rad/s
Mesh 1	0.4841	0.3608	0.3020
Mesh 2	0.4914	0.3663	0.3062
Mesh 3	0.4935	0.3721	0.3091

4. Experimental Apparatus

A flow field test platform for the grooved rotating-disk system was built based on the Bruker's Universal Mechanical Tester, as shown in Figure 2. The platform consisted of a rotating grooved disk, a stationary ungrooved disk, a peristaltic oil pump, a data acquisition system, and a control system. The rotating disk was driven by an electric motor, and the precision of the controlled disk speed was 1.0 r/min. The gap between the stationary disk and the rotating disk could be adjusted continuously with the precision of 10 μm. In the test, the gap was set to be zero at first. The measured force in the axial direction was used to detect the zero-gap position. Then, the gap was increased gradually from zero position to a prescribed position. The pump could supply a maximum volumetric flow rate of 1000 mL/min with the accuracy of 1.0 mL/min. When the flow field became uniform and stable, the visualized image of the flow field was captured. The specifications of the grooved rotating-disk system in the test are shown in Table 2.

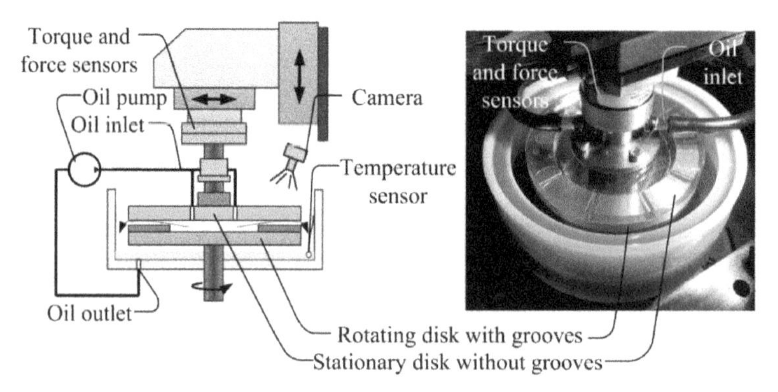

Figure 2. Test apparatus of the grooved rotating-disk system.

Table 2. Specifications of the grooved rotating-disk system model.

Parameter	Value
r_1	42 mm
r_2	60 mm
h	0.4 mm
H	1.6–2.0 mm
N_g	10

5. Results and Discussion

5.1. Flow Field Visualization

The flow field simulation and the visualization test were conducted under the condition of the rotating speed of the grooved disk ω = 18.8 rad/s, the disk spacing H = 1.8 mm, and the inlet volumetric flow rate of oil Q = 40 mL/min. The comparison of oil volume fraction distribution between the simulation result and the visualization test of the flow field is shown in Figure 3. The 3D contour of the oil phase distribution is presented in Figure 4. In general, a two phase flow was exhibited in the flow field in the simulation. The simulation demonstrated that the oil film was continuous and complete near the inlet of the flow field. The air entrainment and the decrease of oil volume fraction took place radially between the inlet and the outlet, presenting an air–oil two-phase flow. There was a clear interface formed between the continuous oil phase and the two-phase area. Near the outlet surface, the oil volume fraction became close to zero, indicating an air phase flow. Previous studies also showed that a sufficient flow rate could ensure that the oil phase filled the entire flow field at a given disk speed and gap [24]. However, in practice, the flow rate was not enough when the rotating speed increased continuously. Then, the air entered the gap from the periphery of the disk gap to ensure the continuity of the flow, which could be observed from the streamline simulation of air phase in Figure 5. In the visualization test result, the inner and the outer radii of the grooved disk were specified as the inlet and the outlet, as marked in Figure 3. A relatively stable boundary between the continuous oil phase and the aerated part could also be clearly observed. The boundary was considered to be the air–oil interface, as shown in Figure 3. Obviously, the visualization test also showed the same flow. In conclusion, the simulated oil volume fraction distribution correlated well with the visualization test result. The CFD model was therefore validated through this comparison.

Figure 5 illustrates the contour surface of φ_{oil} = 0.75 and the streamline of fluid flow. From the contour surface of φ_{oil} = 0.75 and the 3D contour of oil phase distribution in Figure 4, the oil phase occupied more area on the stationary disk surface than the rotating disk surface. The flow of oil gathered around the grooved zone. From the flow streamline in Figure 5, it can be noted that the oil flowed circumferentially along the ungrooved surface within the continuous film. The oil phase flowed into the radial groove and then came out, retaining circumferential flow status. Outside the interface of the continuous oil film, the oil phase was aerated and became a two-phase flow. The two-phase flow mainly discharged through the radial groove. Near the outer edge of disks, the air phase entered through the radial groove and flowed circumferentially in the ungrooved area. From the above observations, it can be deduced that a better cooling effect around the groove could be achieved due to a stronger mixing and heat exchange of oil and air around the grooved area. In practice, the oil flow was mainly used to provide sufficient cooling capability, especially in the high-speed operation. Thus, a sufficient assessment of the flow inside the groove was helpful for the heat dissipation.

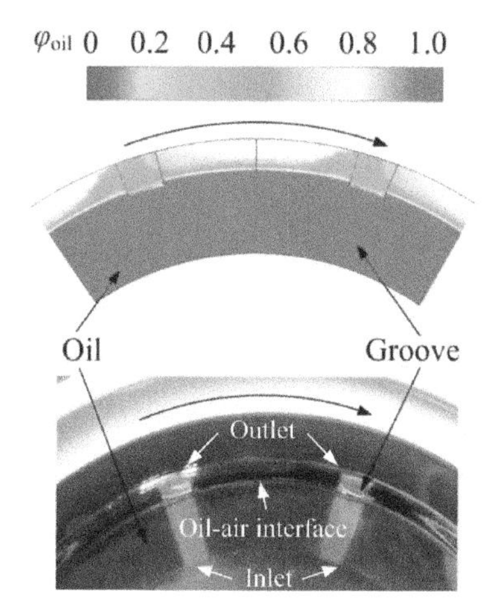

Figure 3. Comparison of the oil volume distribution between simulation and visualization test.

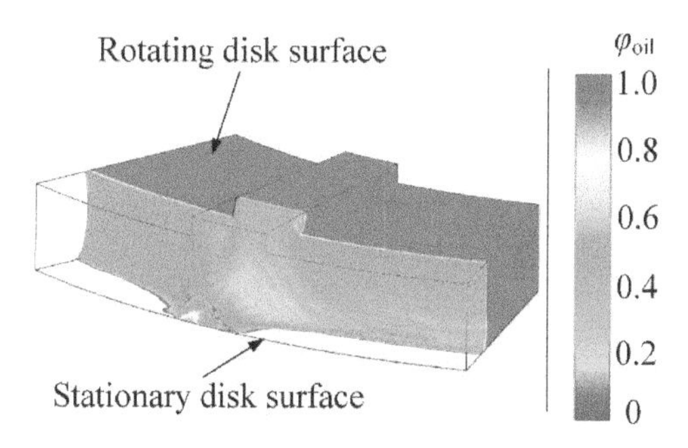

Figure 4. 3D contour of oil phase distribution for a 36° section of flow field.

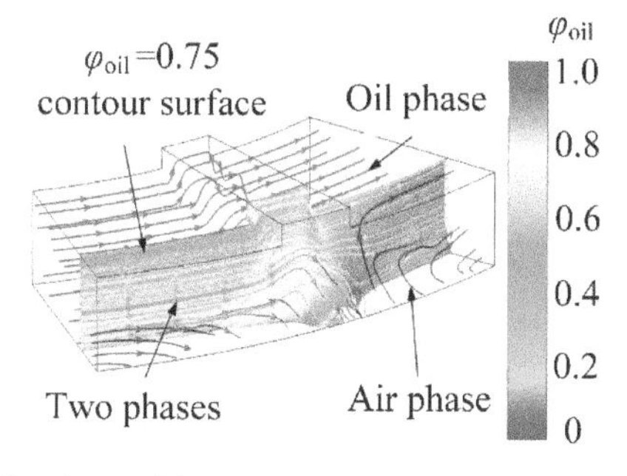

Figure 5. Streamline figure of fluid flow with the $\varphi_{oil} = 0.75$ contour surface shown.

5.2. Parametric Analysis

The angular speed, the flow rate, and the disk spacing are sensitive system parameters. In the visualization test results, the radial position of the continuous oil film boundary (which indicates the air–oil interface) could be utilized to demonstrate the effect of each parameter on the phase distribution in the flow field. Figure 6 presents the effects of different angular velocities on the oil volume fraction along the radial direction. The horizontal value of the coordinate was a dimensionless radial location r/r_2. It is noteworthy that the oil volume fraction at one radial position was the averaged value of all the cells at this same radius. The following figures with different parameters were obtained in the same approach. For a certain angular velocity, the oil volume fraction tended to be 1.0 near the inlet area, indicating a pure oil flow. Around the middle radius region, the oil volume fraction decreased rapidly, indicating that the oil phase was aerated and a two-phase flow existed. As the radial position approached the outlet, the decline slope of the oil volume fraction reduced, and an air phase flow played a dominate role near the outer edge.

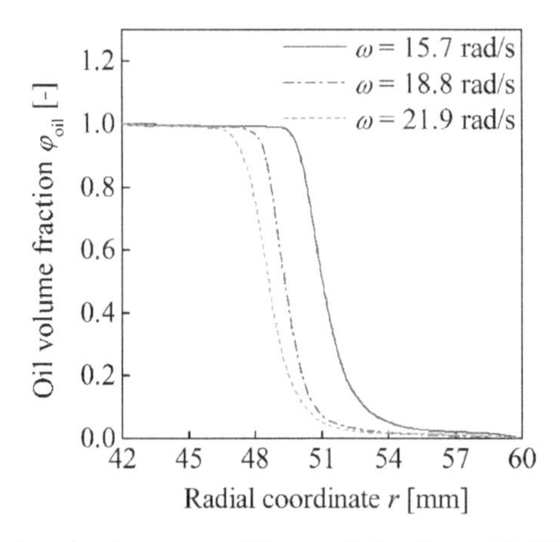

Figure 6. Average oil volume fraction curves at different radial positions with different angular velocities.

For different angular velocities, it could be seen that the transition location of the full oil phase to two-phase flow in the radial coordinate became closer to the inner radius with the angular speed increasing. In the two-phase flow zone, the oil volume fraction of the flow field became lower with the increasing angular speed at the same radial position. Figure 7 shows the effect of angular velocity on the flow field in the visualization test. It was observed that the interface between the continuous oil phase and the aerated part moved radially inwards with the increasing angular velocity. The continuous oil film shrunk at higher angular speeds. Thus, the simulation results agreed well with those of the visualization test.

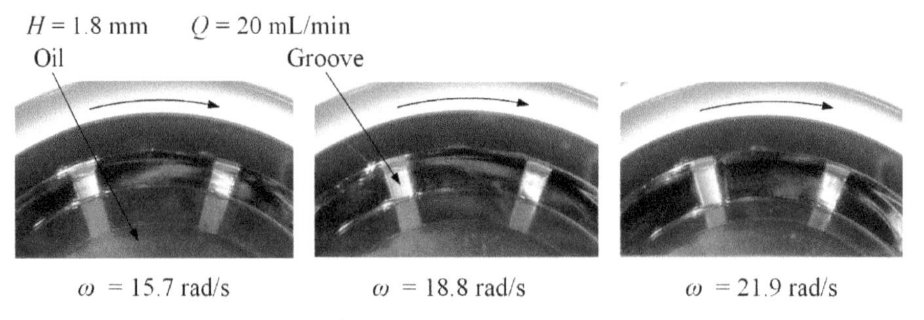

Figure 7. Comparison of the air–oil interface radial position at different angular velocities.

Figure 8 presents the variations of the oil volume fraction along the radial direction at different inlet volumetric flow rates. With the increase of the flow rate, the transition location of the full oil phase to two-phase flow in the radial coordinate moved gradually outwards to the outer radius. At the same radial location in the two-phase flow zone, the oil volume fraction became larger with higher flow rate. Furthermore, the decline slope of the average oil volume fraction was reduced with larger flow rate. As a validation, the visualization test results of various flow rates are shown in Figure 9. The boundary of the continuous oil phase area expanded with the increasing flow rate at the same speed. Since a larger flow rate is beneficial for heat dissipation in practical application, the flow rate is an important operating parameter. However, a larger flow rate requires a greater oil pump, leading to greater power consumption. Besides, it also results in higher drag torque at the same speed. Thus, a suitable flow rate needs to be confirmed according to the variation of the flow pattern.

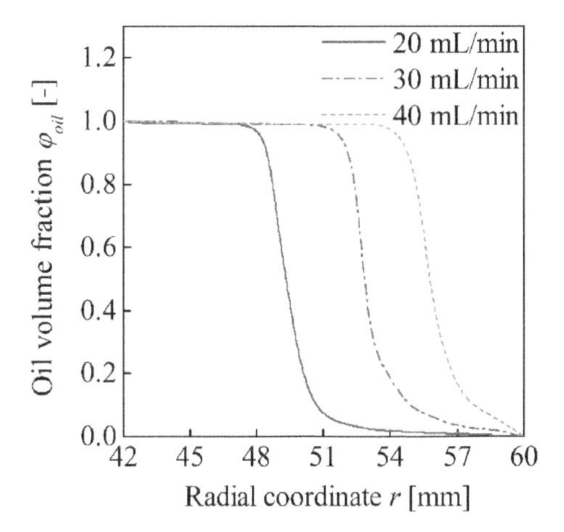

Figure 8. Average oil volume fraction curves at different radial positions with different oil flow rates.

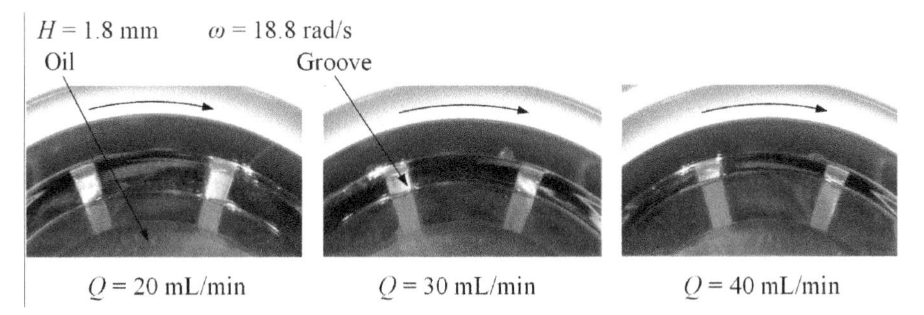

Figure 9. Comparison of the air–oil interface radial position with different inlet oil flow rates.

The effects of the disk spacing on the flow field are shown in Figure 10. The boundary of the continuous oil phase remarkably moved inwards with higher disk spacing. At the same radial position in the two-phase zone, the oil volume fraction became much smaller with a higher disk spacing. The visualization test results of different disk spacing are shown in Figure 11. The continuous oil phase area shrunk with larger disk spacing. The numerical simulation results agreed well with the experimental results. The disk spacing is also an important system geometric parameter, since it significantly influences the viscous drag. Thus, the disk spacing needs to be optimized according to the practical restriction.

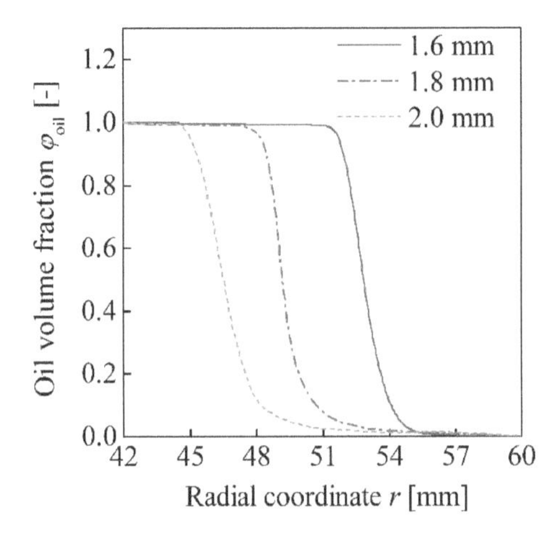

Figure 10. Average oil volume fraction curves at different radial positions with different disk spacing.

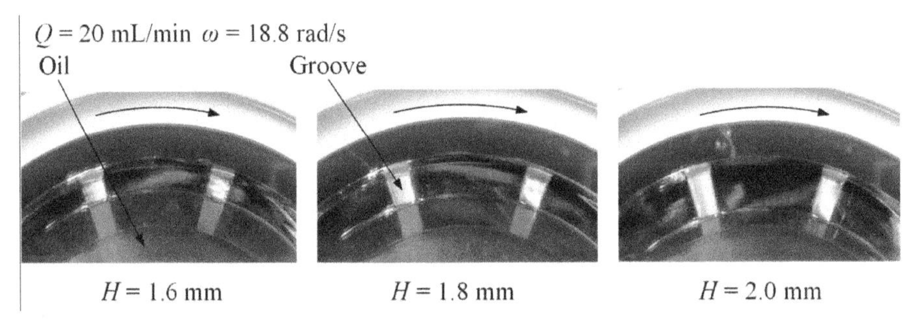

Figure 11. Comparison of the air–oil interface radial position with different disk spacing.

For the flow field inside the grooved two-disk system, both the flow rate and the disk spacing had direct effects on the radial flow velocity. The variation of the flow rate or the disk spacing changed the radial flow velocity. A more detailed investigation on the radial flow velocity is necessary. Furthermore, the flow rate and the disk spacing mentioned above are usually applied as the control parameters in engineering.

The average oil volume fractions of the flow field within the whole calculation domain under different angular velocities, oil flow rates, and disk spacing are shown in Figure 12. The simulated and the measured values of the average oil volume fraction were compared. It seems that the average oil volume fraction became lower with the increase of the angular speed, as shown in Figure 12a. When the angular speed was relatively low, the average oil volume fraction decreased nearly linearly. At relatively high speed, the slope of the average oil volume fraction curve got smaller and tended to be a constant value. In Figure 12b, both the simulated and the measured average oil volume fractions increased linearly with the flow rate before the flow rate reached 50 mL/min, which filled the flow field between the two disks. Lastly, in Figure 12c, the average oil volume fraction decreased with the increase of the disk spacing in a linear way. In the application of the disengaged wet clutch, the oil phase distribution is of crucial importance, since it affects the heat dissipation and the drag loss significantly. Thus, the two-phase flow behavior should be considered carefully in the clutch disk design.

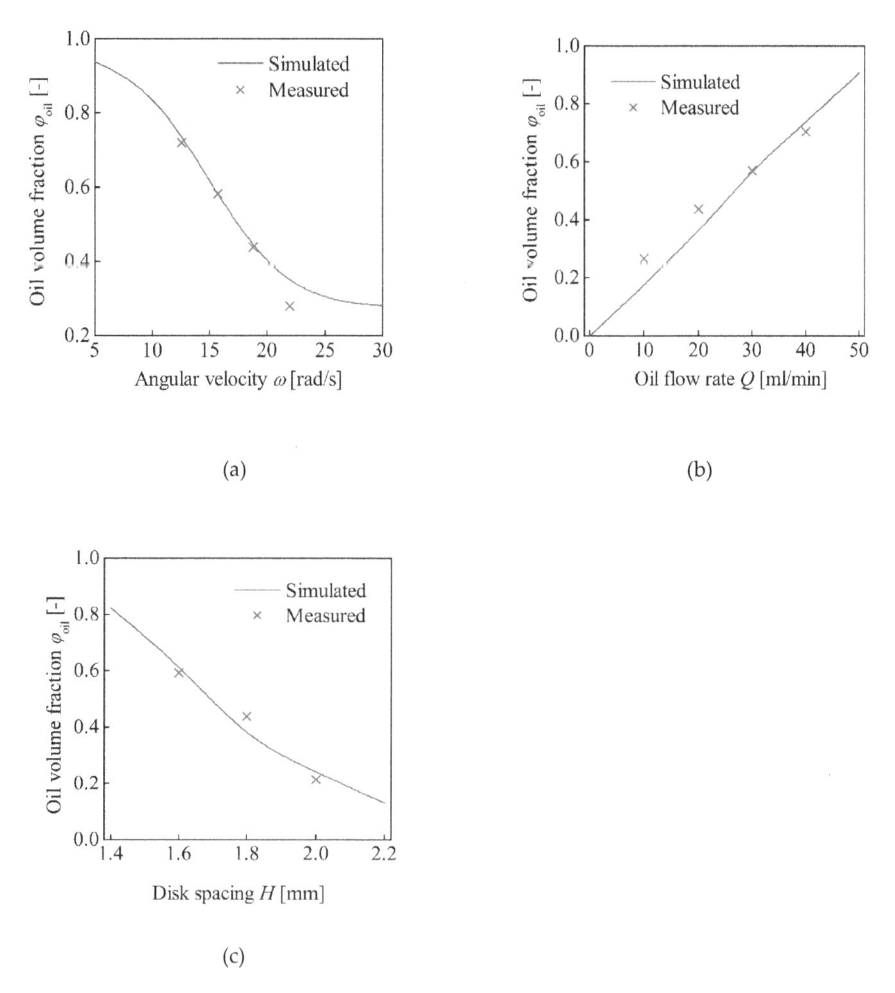

Figure 12. Parametric effects on the average oil volume fraction. (**a**) effect of the angular velocity; (**b**) effect of the inlet oil flow rate; (**c**) effect of the disk spacing.

5.3. Heat Dissipation Capability

The heat dissipation capability was investigated using the method presented in reference [45]. The Nusselt number is an important non-dimensional quantity used in heat transfer, which relates the convective heat transfer to the conductive heat transfer by a fluid across a surface. In this study, local and average Nusselt numbers were utilized to describe the heat dissipation capability under different conditions. The local and the average Nusselt numbers on the surface of the stationary flat disk are respectively defined as:

$$Nu_l = \frac{qr}{(T_W - T)k} \tag{15}$$

$$Nu_{av} = \frac{qr_2}{(T_{Wav} - T)k} \tag{16}$$

where q denotes the heat flux, T_W denotes the local temperature of the stationary disk surface, T is the temperature of the two-phase flow, k is the effective thermal conductivity of the fluid, and T_{Wav} indicates the average temperature of the stationary disk surface.

Figure 13 shows the radial distribution of the local Nusselt number along the center line of the stationary disk. At a certain angular velocity, it is noted from the figure that the local Nusselt number was high close to the inner radius position of the disk. It dropped rapidly and fluctuated remarkably outward along the radial direction. Near the outer diameter position of the disk, the Nusselt number increased rapidly to a great value. The reasons for the variation of the Nusselt number along the radial direction were as follows. The inner radius region was filled with the continuous oil phase. The oil temperature was low, and the heat convection of the oil was much stronger than the air, leading to a high local Nusselt number near the inlet. In the middle radius region, the oil volume fraction decreased quickly around the interface of oil film and two-phase flow. The heat convection of the two-phase flow was much weaker than the pure oil phase. The heat dissipation was weakened, and the fluid temperature rose. Furthermore, compared with the flow within the full oil phase zone, the flow of the air–oil two-phase flow was much more unstable around the interface of the oil film and the two-phase flow, which led to drastic fluctuations of the Nusselt number. Near the outer diameter position of the disk, though the oil volume fraction was very small, the velocity of the air flow was relatively much higher than the air–oil flow. Besides, the temperature of the air phase was lower, thus the cooling intensity of the disk surface was higher. This made the local Nusselt number increase near the outlet boundary.

With the angular speed of the disk increasing, it was concluded from Figure 13 that the fluctuation of the local Nusselt number moved towards the inner diameter, and its range notably increased. This was because the oil volume fraction decreased at higher speeds, and the area of the air phase expanded towards the inner diameter, which resulted in an increase in the mixing area of the two-phase flow. The higher the speed of the disk was, the more unstable the two-phase flow became, and the higher the flow velocity was. Thus, the fluctuation was aggravated. In the regions near the inner and the outer diameter, a higher local Nusselt number was attributable to a greater shear rate of the oil phase and the air phase, respectively, which led to stronger heat dissipation.

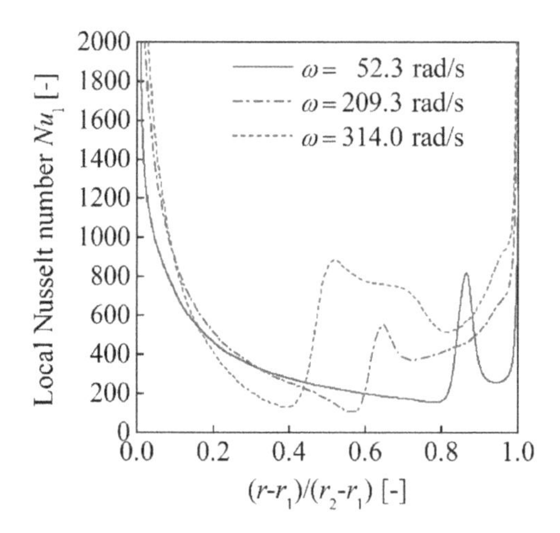

Figure 13. Results of the local Nusselt number on the stationary disk surface along the radial direction.

Lastly, the variation of the average Nusselt number on the stationary disk surface with increasing angular velocity is shown in Figure 14. The change of the average Nusselt number with angular velocity could be divided into three stages. For stage I, the angular velocity of the rotating disk was small, and the gap was filled with the oil phase. As the disk speed increased, the shear rate of the oil phase in the gap increased, resulting in stronger heat dissipation. Therefore, the average Nusselt number rose remarkably in stage II. In stage II, the oil volume fraction decreased rapidly due to the aeration effect. The heat dissipation capability of the mixture was weaker than that of the full oil phase, which led to a rapid decrease of the average Nusselt number. In stage III, the oil volume fraction in the flow field decreased much more slowly. However, the speed of the air–oil two-phase flow and the pure air phase inside the flow field increased with a higher disk speed. The mixture flow around the inner diameter region and the pure air phase flow around the outer diameter region played major roles in the heat convection. Thus, the average Nusselt number increased gradually with the angular velocity of the disk.

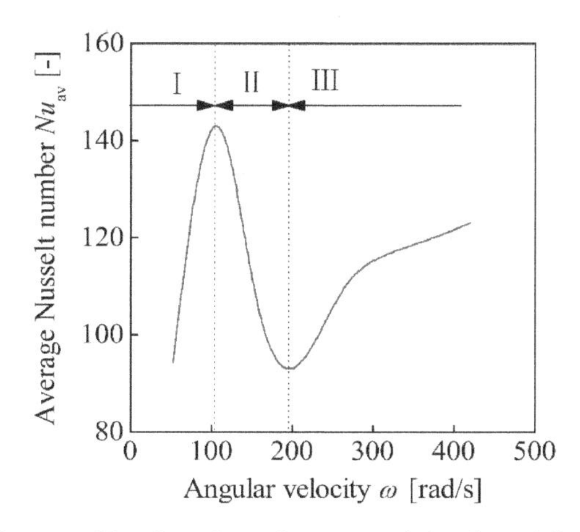

Figure 14. Curve of the average Nusselt number on the stationary disk surface at different angular velocities.

6. Conclusions and Future Work

In this paper, the viscous flow field inside a grooved rotating-disk system was studied quantitatively. A 3D CFD model considering two-phase flow and heat transfer was utilized to simulate phase distributions and heat dissipation. Visualization tests were conducted to validate the flow pattern and the parametric effect on the flow field. Several conclusions were drawn as follows.

(1) The viscous flow field of the grooved rotating-disk system was identified to be an air–oil two-phase flow. A stable interface between the continuous oil phase and the two-phase area could be formed and observed in the flow field. The proposed 3D CFD model with VOF model and heat transfer consideration was validated by the comparison of simulation with the visualization test result of the flow field.

(2) The effects of angular velocity, inlet volumetric flow rate of oil, and disk spacing on the oil volume fraction along the radial direction were analyzed and compared with the visualization test results. In general, the interface between the continuous oil phase and the two-phase flow moved outwards with smaller angular speed, more inlet flow of oil, or decreasing disk spacing. Besides, for the whole flow field, the average oil volume fraction decreased with increasing angular speed, less inlet flow rate of oil, or increasing disk spacing.

(3) The local Nusselt number was remarkably affected by the air–oil volume fraction and the flow velocity distributions in this two-phase flow. The local Nusselt number was much higher near the disk edge—both the inner and the outer ones. Fluctuations of the local Nusselt number appeared near the two-phase interface due to the flow speed changes. At higher angular velocity, the area of fluctuation expanded. The variation of average Nusselt number over the whole flow field could be divided into three stages resulting from the variation of fluid volume fraction and fluid speed.

The results can be used to optimize the design of the grooved rotating-disk system disk fluid machine. The energy transfer characteristics of the air–oil two-phase flow need more study. Work on these topics is currently underway in the National Key Laboratory of Vehicular Transmission at the Beijing Institute of Technology.

Author Contributions: Conceptualization, C.L. and W.W.; Methodology, W.W.; Validation, C.L., W.W. and Y.L.; Formal Analysis, J.Z.; Investigation, C.H.; Writing—Original Draft Preparation, W.W. and Y.L.; Writing—Review & Editing, C.L.; Project Administration, C.L.; Funding Acquisition, C.L. and W.W.

Funding: This work is supported by the National Natural Science Foundation of China (Grant No. U1864210 and 51975045) and National Key R&D Program of China (Grant No. 2018YFB2001300).

Acknowledgments: The authors thank referees for improving the quality of the paper. If readers are interested in this paper, please contact the authors. It's our pleasure to provide readers with relevant model, grid and setup files.

Conflicts of Interest: The authors declare no conflict of interest.

Nomenclature

E	Specific sensible enthalpy, J/kg
F	External force, N
g	Gravity acceleration, m/s^2
h	Depth of radial grooves, m
H	Disk spacing, m
k	Effective thermal conductivity coefficient of fluid, W/(m·K)
k_s	Thermal conductivity coefficient of solid structure, W/(m·K)
N_g	Groove number
Nu_{av}	Average Nusselt number
Nu_l	Local Nusselt number
p	Pressure, Pa
Q	Oil flow rate, L/min
q	Heat flux, W/m^2
r	Radial coordinate
r_1	Inner radius, m
r_2	Outer radius, m
T	Temperature of two-phase flow, K
T_W	Local temperature of stationary disk surface, K
T_{Wav}	Average temperature of stationary disk surface, K
t	Time, s
z	Axial coordinate
Greek symbols	
θ	Azimuthal coordinate
μ	Dynamic viscosity, Pa·s
ρ	Density, kg/m^3
v	Velocity, m/s
φ	Volume fraction
ω	Angular velocity, rad/s

Subscripts

air	Subscript for the air phase parameterparameter
oil	Subscript for the oil phase parameter

References

1. Liu, F.; Wu, W.; Hu, J.; Yuan, S. Design of multi-range hydro-mechanical transmission using modular method. *Mech. Syst. Signal Process.* **2019**, *126*, 1–20. [CrossRef]
2. Cui, J.; Wang, C.; Xie, F.; Xuan, R.; Shen, G. Numerical investigation on transient thermal behavior of multidisk friction pairs in hydro-viscous drive. *Appl. Therm. Eng.* **2014**, *67*, 409–422. [CrossRef]
3. Guha, A.; Sengupta, S. The fluid dynamics of the rotating flow in a Tesla disc turbine. *Eur. J. Mech. B/Fluids* **2013**, *37*, 112–123. [CrossRef]

4. Von Kármán, T. Über laminare und turbulente reibung. *J. Appl. Math. Mech./Z. Für Angew. Math. Und Mech.* **1921**, *1*, 233–252.
5. Batchelor, G.K. Note on a class of solutions of the Navier-Stokes equations representing steady rotationally-symmetric flow. *Q. J. Mech. Appl. Math.* **1951**, *4*, 29–41. [CrossRef]
6. Stewartson, K. On the flow between two rotating coaxial disks. *Math. Proc. Camb. Philos. Soc.* **1953**, *49*, 333–341. [CrossRef]
7. Srinivasacharya, D.; Kaladhar, K. Analytical solution for Hall and Ion-slip effects on mixed convection flow of couple stress fluid between parallel disks. *Math. Comput. Model.* **2013**, *57*, 2494–2509.
8. Brady, J.F.; Durlofsky, L. On rotating disk flow. *J. Fluid Mech.* **1987**, *175*, 363–391. [CrossRef]
9. Soong, C.Y.; Wu, C.C.; Liu, T.P.; Liu, T.P. Flow structure between two co-axial disks rotating independently. *Exp. Therm. Fluid Sci.* **2003**, *27*, 295–311. [CrossRef]
10. Schouveiler, L.; Le Gal, P.; Chauve, M.P. Instabilities of the flow between a rotating and a stationary disk. *J. Fluid Mech.* **2001**, *443*, 329–350. [CrossRef]
11. Bai, S.; Peng, X.; Li, Y.; Sheng, S. A hydrodynamic laser surface-textured gas mechanical face seal. *Tribol. Lett.* **2010**, *38*, 187–194. [CrossRef]
12. Liu, J.; Shao, Y. An improved analytical model for a lubricated roller bearing including a localized defect with different edge shapes. *J. Vib. Control* **2018**, *24*, 3894–3907. [CrossRef]
13. Fesanghary, M.; Khonsari, M.M. On the modeling and shape optimization of hydrodynamic flexible-pad thrust bearings. *Proc. Inst. Mech. Eng. Part J J. Eng. Tribol.* **2013**, *227*, 548–558. [CrossRef]
14. Chao, Q.; Zhang, J.; Xu, B.; Wang, Q. Discussion on the Reynolds equation for the slipper bearing modeling in axial piston pumps. *Tribol. Int.* **2018**, *118*, 140–147. [CrossRef]
15. Huang, J.H.; Fan, Y.R.; Qiu, M.X.; Fang, W.M. Effects of groove on behavior of flow between hydro-viscous drive plates. *J. Cent. South Univ.* **2012**, *19*, 347–356. [CrossRef]
16. Missimer, J.R.; Johnson, W.S. Flow between a smooth stationary disk and grooved rotating disk. *J. Tribol.* **1982**, *104*, 248–254. [CrossRef]
17. Li, M.; Khonsari, M.M.; McCarthy, D.M.C.; Lundin, J. Parametric analysis for a paper-based wet clutch with groove consideration. *Tribol. Int.* **2014**, *80*, 222–233. [CrossRef]
18. Razzzaque, M.M.; Kato, T. Effects of a groove on the behavior of a squeeze film between a grooved and a plain rotating annular disk. *J. Tribol.* **1999**, *121*, 808–815. [CrossRef]
19. Xie, F.; Wu, D.; Tong, Y.; Zhang, B.; Zhu, J. Effects of structural parameters of oil groove on transmission characteristics of hydro-viscous clutch based on viscosity-temperature property of oil film. *Ind. Lubr. Tribol.* **2017**, *69*, 690–700. [CrossRef]
20. Said, I.A.; Abdel-Aziz, M.H.; El-Taweel, Y.A.; Sedahmed, G.H. Mass and heat transfer behavior of a rotating disc with parallel rectangular grooves. *Chem. Eng. Process. Process Intensif.* **2016**, *105*, 110–116. [CrossRef]
21. Beretta, G.P.; Malfa, E. Flow and heat transfer in cavities between rotor and stator disks. *Int. J. Heat Mass Transf.* **2003**, *46*, 2715–2726. [CrossRef]
22. Wang, D.M.; Chen, B.; Gu, Z.T.; Zhou, C. Flow characteristics of oil film on surface of a rotating disk. *J. Propuls. Technol.* **2019**, *40*, 53–60.
23. Aphale, C.R.; Schultz, W.W.; Ceccio, S.L. The influence of grooves on the fully wetted and aerated flow between open clutch plates. *J. Tribol.* **2010**, *132*, 011104. [CrossRef]
24. Iqbal, S.; Al-Bender, F.; Pluymers, B.; Desmet, W. Model for predicting drag torque in open multi-disks wet clutches. *J. Fluids Eng.* **2014**, *136*, 021103. [CrossRef]
25. Xie, F.; Zheng, X.; Sheng, G.; Sun, Q.; Agarwal, R.K. Numerical investigation of cavitation effect on two-phase oil film flow between a friction pair in hydro-viscous drive. *P. I. Mech. Eng. C.-J. Mec.* **2018**, *232*, 4626–4636. [CrossRef]
26. Hu, J.; Wu, W.; Wu, M.; Yuan, S. Numerical investigation of the air–oil two-phase flow inside an oil-jet lubricated ball bearing. *Int. J. Heat Mass Transf.* **2014**, *68*, 85–93. [CrossRef]
27. Payvar, P. Laminar heat transfer in the oil groove of a wet clutch. *Int. J. Heat Mass Transf.* **1991**, *34*, 1791–1798. [CrossRef]
28. Jang, J.Y.; Khonsari, M.M.; Maki, R. Three-dimensional thermohydrodynamic analysis of a wet clutch with consideration of grooved friction surfaces. *J. Tribol.* **2011**, *133*, 011703. [CrossRef]
29. Jen, T.C.; Nemecek, D.J. Thermal analysis of a wet-disk clutch subjected to a constant energy engagement. *Int. J. Heat Mass Transf.* **2008**, *51*, 1757–1769. [CrossRef]

30. Marklund, P.; Mäki, R.; Larsson, R.; Höglund, E.; Khonsari, M.M.; Jang, J. Thermal influence on torque transfer of wet clutches in limited slip differential applications. *Tribol. Int.* **2007**, *40*, 876–884. [CrossRef]
31. Wu, W.; Xiong, Z.; Hu, J.B.; Yuan, S.H. Application of CFD to model oil–air flow in a grooved two-disc system. *Int. J. Heat Mass Transf.* **2015**, *91*, 293–301. [CrossRef]
32. Zhang, B.; Wang, Z.; Wang, T. Drag Torque Model with Decoupling of Temperature and Radius. *J. Comput. Theor. Nanosci.* **2016**, *13*, 1337–1342. [CrossRef]
33. Wu, W.; Xiao, B.; Hu, J.; Yuan, S.; Hu, C. Experimental investigation on the air-liquid two-phase flow inside a grooved rotating-disk system: Flow pattern maps. *Appl. Therm. Eng.* **2018**, *133*, 33–38. [CrossRef]
34. Neupert, T.; Bartel, D. High-resolution 3D CFD multiphase simulation of the flow and the drag torque of wet clutch discs considering free surfaces. *Tribol. Int.* **2019**, *129*, 283–296. [CrossRef]
35. Pahlovy, S.A.; Mahmud, S.F.; Kubota, M.; Ogawa, M.; Takakura, N. Prediction of Drag Torque in a Disengaged Wet Clutch of Automatic Transmission by Analytical Modeling. *Tribol. Online* **2016**, *11*, 121–129. [CrossRef]
36. Xie, H.; Gong, H.; Hu, L.; Yang, H. Coriolis effects on torque transmission of hydro-viscous film in parallel disks with imposed throughflow. *Tribol. Int.* **2017**, *115*, 100–107. [CrossRef]
37. Gong, H.; Xie, H.; Hu, L.; Yang, H. Combined effects of Coriolis force and temperature-viscosity dependency on hydro-viscous transmission of rotating parallel disks. *Tribol. Int.* **2018**, *117*, 168–173. [CrossRef]
38. Hirt, C.W.; Nichols, B.D. Volume of fluid (VOF) method for the dynamics of free boundaries. *J. Comput. Phys.* **1981**, *39*, 201–225. [CrossRef]
39. Ganapathy, H.; Shooshtari, A.; Choo, K.; Dessiatoun, S.; Alshehhi, M.; Ohadi, M. Volume of fluid-based numerical modeling of condensation heat transfer and fluid flow characteristics in microchannels. *Int. J. Heat Mass Transf.* **2013**, *65*, 62–72. [CrossRef]
40. Hu, J.; Peng, Z.; Wei, C. Experimental Research on Drag Torque for Single-Plate Wet Clutch. *J. Tribol. Trans. ASME* **2012**, *134*, 014502.
41. Da Riva, E.; Del Col, D. Numerical simulation of laminar liquid film condensation in a horizontal circular minichannel. *J. Heat Transf.* **2012**, *134*, 051019. [CrossRef]
42. Wang, F.J. *Principle and Application of CFD Software*; Tsinghua University Press: Beijing, China, 2015; pp. 40–44.
43. Brackbill, J.U.; Kothe, D.B.; Zemach, C. A continuum method for modeling surface tension. *J. Comput. Phys.* **1992**, *100*, 335–354. [CrossRef]
44. Benson, D.J. Volume of fluid interface reconstruction methods for multi-material problems. *Appl. Mech. Rev.* **2002**, *55*, 151–165. [CrossRef]
45. Yuan, Z.X.; Saniei, N.; Yan, X.T. Turbulent heat transfer on the stationary disk in a rotor–stator system. *Int. J. Heat Mass Transf.* **2003**, *46*, 2207–2218. [CrossRef]

Article

Fault Diagnosis Method for Hydraulic Directional Valves Integrating PCA and XGBoost

Yafei Lei [1], Wanlu Jiang [1,*], Anqi Jiang [2], Yong Zhu [3,*], Hongjie Niu [4] and Sheng Zhang [1,*]

1 College of Mechanical Engineering, Yanshan University, Qinhuangdao 066004, China
2 College of Electrical Engineering, Yanshan University, Qinhuangdao 066004, China
3 Research Center of Fluid Machinery Engineering and Technology, Jiangsu University, Zhenjiang 212013, China
4 Qinhuangdao Shouqin Metal Materials Co., Ltd., Qinhuangdao 066009, China
* Correspondence: wljiang@ysu.edu.cn (W.J.); zhuyong@ujs.edu.cn (Y.Z.); zsqhd@ysu.edu.cn (S.Z.); Tel.: +86-0335-806-1729 (W.J.); +86-0511-8879-9918 (Y.Z.); +86-0335-805-7073 (S.Z.)

Received: 18 August 2019; Accepted: 29 August 2019; Published: 3 September 2019

Abstract: A novel fault diagnosis method is proposed, depending on a cloud service, for the typical faults in the hydraulic directional valve. The method, based on the Machine Learning Service (MLS) HUAWEI CLOUD, achieves accurate diagnosis of hydraulic valve faults by combining both the advantages of Principal Component Analysis (PCA) in dimensionality reduction and the eXtreme Gradient Boosting (XGBoost) algorithm. First, to obtain the principal component feature set of the pressure signal, PCA was utilized to reduce the dimension of the measured inlet and outlet pressure signals of the hydraulic directional valve. Second, a machine learning sample was constructed by replacing the original fault set with the principal component feature set. Third, the MLS was employed to create an XGBoost model to diagnose valve faults. Lastly, based on model evaluation indicators such as precision, the recall rate, and the *F*1 score, a test set was used to compare the XGBoost model with the Classification And Regression Trees (CART) model and the Random Forests (RFs) model, respectively. The research results indicate that the proposed method can effectively identify valve faults in the hydraulic directional valve and have higher fault diagnosis accuracy.

Keywords: hydraulic valve; fault diagnosis; principal component analysis (PCA); extreme gradient boosting (XGBoost); HUAWEI Cloud machine learning service (MLS)

1. Introduction

Hydraulic systems play an important role in a wide variety of industrial applications, such as robotics, manufacturing, aerospace, and engineering machinery. Monitoring the condition of hydraulic equipment can not only effectively improve productivity and reduce maintenance costs and downtime, but also improve the reliability and safety of this equipment in its application [1–3]. In particular, the hydraulic valve is the core control component of the hydraulic system, and it is widely used in numerous engineering applications to control the flow and pressure of fluids [4–6]. In the hydraulic system, a vibration analysis (VA) is the most popular and efficient condition monitoring technique for rotating systems including the hydraulic pump, electric motor, bearing, and more [7–16]. However, the working process of the valve core of the hydraulic valve is a reciprocating motion. These VA methods, which have been successfully applied in rotating machinery, will not be suitable for fault diagnosis and a condition monitoring signal analysis of non-rotating machinery, such as the hydraulic valve [17–19].

Many studies on fault diagnosis of the hydraulic valve have been conducted by theoretical approaches and test measurements, and certain research results have been obtained. Wu et al. [20] proposed a method for the mechanical fault diagnosis based on complex three-order cumulants. In the experiment regarding the fault diagnosis of the overflow valve, the results show that this method can improve the correction rate of diagnosis. Huang et al. [21] applied the theory of higher-order spectrum to the fault diagnosis of hydraulic valves. Li et al. [22] proposed a fault diagnosis method that involves choosing the fractal characteristic volume of a valve's displacement signal as a criterion to solve the nonlinear problems in the working process of autopilot hydraulic valves. Raduenz et al. [23] presented the development of a method for condition monitoring and online fault detection on proportional reversing valves. The effectiveness of the method to monitor and detect faults in valves with different sizes and constructive parameters was shown experimentally using five different proportional valves. Vianna et al. [24] presented a method to estimate degradation in a servo valve using an application of the Fading Extended Kalman Filter for system identification. Folmer et al. [25] also presented a data-driven fault detection system for valves, which uses historical process data obtained across company borders to detect faults by comparing standardized flow coefficients determined by DIN IEC 60534-2-1 in physical valve models. Moreover, many challenges emerge in the study of the condition monitoring and fault diagnosis of hydraulic valves. In particular, there are few research results for identifying hydraulic valve faults by pressure signals in the hydraulic system with condition monitoring on an Industrial Internet of Things (IIoT) platform.

Due to the availability of big data technology and data mining methods as well as the emergence of new IIoT platforms and machine learning algorithms, fault diagnosis for hydraulic valves based on big data for hydraulic system with condition monitoring is one of the focuses for this research [26–28]. Among them, Principal Component Analysis (PCA) is an effective method for dimensionality reduction in big data analysis. It is a multivariate statistical method, which compresses multiple linearly related variables into a few unrelated variables. PCA was first proposed by Pearson [29] in a study on optimal linear and plane fitting of spatial data. Fisher and Mackenzie [30] believed that PCA was more useful in the system response variance analysis than in system modeling, and they proposed a prototype of the Nonlinear Iterative Partial Least Squares (NIPALS) algorithm. Then, PCA was improved by Hoteling [31] and further developed into a common method widely used in data dimensionality reduction, fault diagnosis, and anomaly detection. For instance, Mohanty et al. [32] developed a new algorithm to identify bearing faults using empirical mode decomposition and principal component analysis (EMD-PCA) based on the average kurtosis technique. It was observed that this proposed combined approach effectively and adaptively identified inner ball faults. Stief et al. [33] proposed a sensor fusion approach to diagnose both electrical and mechanical faults in induction motors based on the combination of a two-stage Bayesian method and PCA. Caggiano [34] also proposed an advanced feature extraction methodology based on PCA. By introducing artificial neural networks to the PCA features, an accurate diagnosis of tool flank wear was achieved, with predicted values being very close to the measured tool wear values. Wang et al. [35] developed a variable selection algorithm based on PCA with multiple selection criteria, which can identify faults in wind turbines, determine the corresponding time and location where the fault occurs, and estimate its severity. Xiao et al. [36] also studied the application of PCA to fault diagnosis in Electro-Hydrostatic Actuators (EHAs). The experimental results demonstrated that PCA can effectively discriminate faults and their characteristics for EHAs, and could be used as an optional data fusion tool for the Prognostics and Health Management (PHM) of EHAs. Riba et al. [37] proposed a very fast, noninvasive, accurate, and easy-to-apply method to discriminate between paperboard samples produced from recovered and virgin fibers. For this method, FTIR spectroscopy was analyzed in combination with feature extraction methods such as PCA, PCA+ canonical variate analysis (CVA), extended canonical variate analysis (ECVA), and the *k* Nearest Neighbor algorithm (*k*NN) classifier. The experimental results proved that the proposed scheme allowed for the obtainment of a high classification accuracy with a very fast response.

In addition, the eXtreme Gradient Boosting (XGBoost) algorithm, proposed by Dr. Chen Tianqi in 2014, can automatically utilize the central processing unit (CPU) multi-threaded parallel computing and has the advantages of low computational complexity, fast running speed, and high accuracy, no matter whether the data scale is large or small [38,39]. At present, this method has been successfully applied in many fields, such as fault diagnosis, environmental prediction, and medical detection. Zhang et al. [40] designed an efficient machine learning method that combined random forests (RFs) with XGBoost and was used to establish the fault detection framework of data-driven wind turbines. The results indicated that the proposed approach was robust in various wind turbine models, including offshore ones, under different working conditions. Chakraborty and Elzarka [41] developed an XGBoost model with a dynamic threshold for early detection of faults in Heating Ventilation and Air Conditioning (HVAC) systems. Zhang et al. [42] applied the XGboost algorithm to the fault diagnosis of rolling bearings, and the results showed that the XGboost algorithm was superior to other tree algorithms in accuracy and time. Nguyen et al. [43] developed an XGBoost model to predict peak particle velocity (PPV). The results indicated that the developed XGBoost model, on both training and testing datasets, exhibited higher performance than the support vector machine (SVM), the Random Forests (RFs), and *k*NN models. Pan B et al. [44] applied the XGBoost algorithm to predict the concentration of PM2.5 per hour. Liu and Qiao [45] proposed a prediction method based on clustering and XGboost algorithms for the incidence of heart disease, which shows that the proposed method was feasible and effective. Fitriah et al. [46] proposed an algorithm combining PCA preprocessing with XGBoost classification to diagnose stroke patients in Indonesia, and the accuracy of diagnosis was increased by using fewer electrodes. PCA could reduce dimensionality and computation cost without decreasing classification accuracy. The XGBoost, as the scalable tree boosting classifier, can solve practical scalability problems with minimal resources.

Huawei launched the Machine Learning Service (MLS) in September 2017, which is a service that was launched on the IIoT platform for data mining and analysis by Huawei in September 2017 [47]. It has more than 300 algorithm function nodes, which can conveniently build visual workflow models to perform data processing, model training, evaluation, and prediction. In addition, Jupyter Notebook is integrated in MLS, and the algorithm functions can be extended by tools such as Python and R, in order to provide cloud customized services for the collection and analysis of massive data. Moreover, it can provide a cloud platform for the integration of technology, experience, and machine learning algorithms. At present, attempts are made to apply MLS in the fields of product recommendation, customer grouping, abnormality detection, predictive maintenance, and driving behavior analysis.

In summary, the existing fault diagnosis methods for hydraulic valves are not suitable for extracting fault features from pressure signals in hydraulic valve condition monitoring. It is very necessary to research a fault diagnosis method for hydraulic valves through a cloud service on the IIoT platform, where there is an inevitable demand. There will be a development trend for analyzing big data in hydraulic system condition monitoring in the future. In this paper, a novel fault diagnosis method is proposed, depending on a cloud service, for the typical faults in hydraulic directional valves. The method is based on the cloud service of MLS, using raw sensor data collected from inlet and outlet pressure signals in hydraulic valve condition monitoring, and it integrates both the advantages of the PCA descending dimension and the XGBoost classification.

The outline of the paper is as follows: Sections 2 and 3 summarize the PCA dimension reduction and the XGBoost algorithm principle. In Section 4, the hydraulic test bed is introduced, and the raw data acquisition scheme for condition monitoring is described based on the hydraulic system schematic diagram. In Section 5, the raw data for condition monitoring are analyzed, and inlet and outlet pressure signals of the hydraulic directional valve are selected as the sample. The PCA-XGBoost fault diagnosis model for hydraulic valves is built on an MLS cloud service platform, and, compared with the Principal Component Analysis and Classification And Regression Trees (PCA-CART) and the Principal Component Analysis and Random Forests (PCA-RFs) models, the test results indicate that the model is advanced. Section 6 concludes the proposed approach and shows future work regarding data analytics.

2. Principal Component Analysis-Based Data Dimensionality Reduction

2.1. Principle of PCA Dimensionality Reduction

PCA dimensionality reduction replaces the original dimension with a smaller number of unrelated dimensions. This occurs in order to map m-dimensional features to k-dimensional features ($k < m$). These unrelated dimensions are called principal components [48].

Suppose A is an $n \times m$ data matrix where each column represents a variable and each row represents a sample. The matrix can be decomposed into the sum of the outer products of m vectors, which is shown in the equation below.

$$A = t_1 p_1^{\mathrm{T}} + t_2 p_2^{\mathrm{T}} + \cdots + t_i p_i^{\mathrm{T}} + \cdots + t_m p_m^{\mathrm{T}} \tag{1}$$

where $t_i \in \mathrm{R}^n$ is defined as the column vector consisting of n observations of the i-th principal component t_i, which is called the score vector, $i = 1, 2, \cdots, m$. $p_i \in \mathrm{R}^m$ is called the load vector. Equation (1) can be further written in matrix form.

$$A = TP^{\mathrm{T}} \tag{2}$$

where $T = [t_1, t_2, \cdots, t_i, \cdots, t_m] \in \mathrm{R}^{n \times m}$ is called the score matrix, and $P = [p_1, p_2, \cdots, p_i, \cdots, p_m] \in \mathrm{R}^{m \times m}$ is called the load matrix.

If the score vectors are orthogonal to each other, then for any i and j, when $i \neq j$, $t_i^{\mathrm{T}} t_j = 0$ is satisfied. The load vectors are also orthogonal to each other, and the length of each load vector is 1. This is shown in the formulas below.

$$p_i^{\mathrm{T}} p_j = 0 i \neq j \tag{3}$$

$$p_i^{\mathrm{T}} p_j = 1 i = j \tag{4}$$

Multiply both sides of Equation (1) by p_i to get the following equation.

$$A p_i = t_1 p_1^{\mathrm{T}} p_i + t_2 p_2^{\mathrm{T}} p_i + \cdots + t_i p_i^{\mathrm{T}} p_i + \cdots + t_m p_m^{\mathrm{T}} p_i \tag{5}$$

Substitute Equations (3) and (4) into Equation (5) to get the equation shown below.

$$t_i = A p_i \tag{6}$$

As can be seen from Equation (6), each score vector is actually a projection of the data matrix in the direction of the corresponding load vector. The length of the vector t_i reflects the degree of coverage of the data matrix A in the p_i direction. The greater the length is, the greater the degree of coverage is. The score vectors are arranged from largest to smallest according to their length.

$$\|t_1\| > \|t_2\| > \cdots > \|t_m\| \tag{7}$$

Then the load vector p_1 represents the direction in which the data matrix A changes the most. p_2 is perpendicular to p_1 and represents the direction in which the data matrix A change is the second largest and p_m represents the direction in which the data matrix A changes the least.

Furthermore, through the principal component decomposition, the data matrix A can be transformed into the equation below.

$$A = t_1 p_1^{\mathrm{T}} + t_2 p_2^{\mathrm{T}} + \cdots + t_k p_k^{\mathrm{T}} + E \tag{8}$$

where E is the error matrix, representing the change of A on load vectors from p_{k+1} to p_m. In a practical application, the error matrix E can be ignored since k is much smaller than m, and the error matrix E is mainly caused by measurement noise. Therefore, the data matrix A can be approximately expressed as the following equation.

$$A \simeq t_1 p_1^{\mathrm{T}} + t_2 p_2^{\mathrm{T}} + \cdots + t_k p_k^{\mathrm{T}} \quad (9)$$

Thereby, the original dimension of the data matrix A can be reduced to the k dimension. In the process of PCA dimensionality reduction, eigenvalues and orthogonal normalized eigenvectors need to be solved. Principal components can be calculated by the Singular Value Decomposition (SVD) of a matrix.

2.2. Singular Value Decomposition

The principal component analysis of matrix A can be equivalent to the eigenvector analysis of covariance matrix $A^{\mathrm{T}}A$. The load vectors of the matrix A are the eigenvector of $A^{\mathrm{T}}A$. If the eigenvalues of $A^{\mathrm{T}}A$ are arranged as $\lambda_1 \geq \lambda_2 \geq \cdots \geq \lambda_m \geq 0$, the eigenvectors $p_1, p_2, \cdots, p_{\mathrm{m}}$, corresponding to the eigenvalues one by one, are the load vectors of the matrix A. The SVD of matrix A can be expressed by the equation below.

$$A = U\Sigma V^{\mathrm{T}} \quad (10)$$

In the equation,

$$U = [u_1, u_2, \cdots, u_n] \in \mathrm{R}^{n\times n} \quad (11)$$

$$V = [v_1, v_2, \cdots, v_m] \in \mathrm{R}^{m\times m} \quad (12)$$

$$\Sigma = \begin{bmatrix} \sigma_1 & 0 & \cdots & 0 \\ 0 & \sigma_2 & \cdots & 0 \\ 0 & 0 & \cdots & \sigma_m \\ \vdots & \vdots & & \vdots \\ 0 & 0 & \cdots & 0 \end{bmatrix} \in \mathrm{R}^{n\times m} \quad (13)$$

where $\sigma_1 > \sigma_2 > \cdots > \sigma_m$ are the singular values of the matrix A. The singular values of the data matrix A are actually the square roots of the eigenvalues of its covariance matrix $A^{\mathrm{T}}A$. Therefore, the following is true.

$$\begin{aligned} \sigma_1 &= \sqrt{\lambda_1} \\ \sigma_2 &= \sqrt{\lambda_2} \\ &\vdots \\ \sigma_m &= \sqrt{\lambda_m} \end{aligned} \quad (14)$$

If the columns in the matrices U and V are orthogonal to each other with a length of 1, then Equation (10) can be expressed as the formula below.

$$A = \sigma_1 u_1 v_1^{\mathrm{T}} + \sigma_2 u_2 v_2^{\mathrm{T}} + \cdots + \sigma_m u_m v_m^{\mathrm{T}} \quad (15)$$

If v_i is denoted as p_i and $\sigma_i u_i$ as t_i, Equation (15) is equivalent to Equation (1). $\sigma_i u_i$ is the i-th score vector of the data matrix A, and v_i is the load vector of the i-th principal component.

2.3. Determination of the Number of Principal Components

PCA is an analytical method to reduce the dimension by eliminating the information of independent variables with strict linear correlation or strong correlation. For m independent variables, up to m principal component vectors can be obtained. Usually, k principal components are used to replace m independent variables ($k < m$), and the information contained in them accounts for most of the information provided by the original m independent variables. In order to quantitatively describe the relative amount of information provided by principal components, the variance contribution rate δ_i of principal component vector $\boldsymbol{t}_i$ is defined by the equation below.

$$\delta_i = \frac{\lambda_i}{\sum\limits_{i=1}^{m} \lambda_i} \tag{16}$$

The cumulative contribution rate η_k of the first k principal components is defined as:

$$\eta_k = \frac{\sum\limits_{i=1}^{k} \lambda_i}{\sum\limits_{i=1}^{m} \lambda_i} \tag{17}$$

where λ_i is the variance of the principal component $\boldsymbol{t}_i$, and δ_i is the variance contribution rate of $\boldsymbol{t}_i$, which represents the contribution share of $\boldsymbol{t}_i$ to the total information contained in m variables. The cumulative contribution rate η_k of principal components is used to represent the proportion of the information contained in the first k principal components to the total information.

2.4. Main Steps of PCA

The steps of PCA based on Singular Value Decomposition (SVD) are as follows [49].

Input:

(1) data matrix $\boldsymbol{A} = \{x_1, x_2, \cdots, x_m\}$;

(2) dimension k of low-dimensional space.

Steps:

(1) Represent the sample data in the form of column vectors, and conduct zero centered for all samples: $\boldsymbol{x}_i \leftarrow \boldsymbol{x}_i - \frac{1}{m}\sum_{i=1}^{m} \boldsymbol{x}_i$;

(2) Calculate the covariance matrix $\boldsymbol{A}^{\mathrm{T}}\boldsymbol{A}$ of the sample;

(3) Conduct eigenvalue decomposition of the covariance matrix $\boldsymbol{A}^{\mathrm{T}}\boldsymbol{A}$;

(4) Determine the score vector $\boldsymbol{t}_1, \boldsymbol{t}_2, \cdots, \boldsymbol{t}_k$ corresponding to k eigenvalues.

Output:

(1) Score matrix $\boldsymbol{T} = [\boldsymbol{t}_1, \boldsymbol{t}_2, \cdots, \boldsymbol{t}_k]$.

3. Principles of the XGBoost Algorithm

XGBoost is an improved Gradient Boosting Decision Tree (GBDT) algorithm, and there is a big difference between them. GBDT uses only the first derivative in optimization, while XGBoost uses both the first and second derivatives. Moreover, XGBoost uses the tree model complexity as a regular term in the objective function to avoid overfitting [50].

3.1. Objective Function of the Model

XGBoost adds the regularization factor $\Omega(\boldsymbol{\theta})$ to represent the complexity of the tree based on the Gradient Boosting Decision Tree (GBDT) algorithm, and it defines the objective function of the optimization in the training model using the equation below.

$$Obj(\boldsymbol{\theta}) = L(\boldsymbol{\theta}) + \Omega(\boldsymbol{\theta}) \tag{18}$$

where $\boldsymbol{\theta}$ is the model parameter, $\Omega(\boldsymbol{\theta})$ is the regular term, which represents the complexity of the model, and $L(\boldsymbol{\theta})$ is the loss function, which represents the matching degree between the model and the training set.

For a given data set with n examples and m features, $D = \{(\mathbf{x}_i, y_i)\}(i = 1, 2, \ldots, n, \mathbf{x}_i \in R^m, y_i \in R)$; a tree ensemble model uses S additive functions $\boldsymbol{\theta} = \{f_1, f_2, \cdots f_s, \cdots, f_S\}$ to predict the output.

$$\hat{y}_i = \sum_{s=1}^{S} f_s(\mathbf{x}_i), f_s \in F \tag{19}$$

where $F = \left\{f(\mathbf{x}) = w_{q(\mathbf{x})}\right\}(w \in R^T, q : R^m \to T)$ is the space of regression trees (also known as CART). In this case, q represents the structure of each tree that maps an example to the corresponding leaf index. T is the number of leaves in the tree. Each f_s corresponds to an independent tree structure q and leaf weight w. Unlike decision trees, each regression tree contains a continuous score on each of the leaves, and we use w_i to represent the score on the i-th leaf. We will use the decision rules in the trees (given by q) to classify it into the leaves and calculate the final prediction by summing up the score in the corresponding leaves (given by w). To learn the set of functions used in the model, we minimize the following regularized objective.

$$Obj(\theta) = \sum_{i=1}^{n} l(y_i, \hat{y}_i) + \sum_{s=1}^{S} \Omega(f_s) \tag{20}$$

where $\sum_{i=1}^{n} l(y_i, \hat{y}_i)$ is a differentiable convex loss function that measures the difference between the prediction $\hat{y}_i$ and the target $\hat{y}_i$. The second term $\sum_{s=1}^{S} \Omega(f_s)$ penalizes the complexity of the trees.

3.2. Solution of Loss Function in the Objective Function

In the XGBoost model, the objective function (Equation (20)) is difficult to solve by using the traditional stochastic gradient descent algorithm. In addition, the additive training boosting method is needed to solve the value, whose specific learning and training process is shown below.

$$\begin{cases} \hat{y}_i^{(0)} = 0 \\ \hat{y}_i^{(1)} = f_1(\mathbf{x}_i) = \hat{y}_i^{(0)} + f_1(\mathbf{x}_i) \\ \hat{y}_i^{(2)} = f_1(\mathbf{x}_i) + f_2(\mathbf{x}_i) = \hat{y}_i^{(1)} + f_2(\mathbf{x}_i) \\ \cdots \\ \hat{y}_i^{(t)} = \sum_{s=1}^{t} f_s(\mathbf{x}_i) = \hat{y}_i^{(t-1)} + f_t(\mathbf{x}_i) \end{cases} \tag{21}$$

where $\hat{y}_i^{(t)}$ is the predicted value of the t-th round of the model, $\hat{y}_i^{(t-1)}$ is the predicted value of the $(t-1)$-th round, and $f_t(\mathbf{x}_i)$ is the prediction function added for the t-th round.

Substitute $\hat{y}_i^{(t)}$ in Equation (21) into Equation (20).

$$Obj^{(t)} = \sum_{i=1}^{n} l\left(y_i, \hat{y}_i^{(t-1)} + f_t(\mathbf{x}_i)\right) + \Omega(f_t) \tag{22}$$

For Equation (22), the purpose of iteration is to find the most appropriate $f_t(\mathbf{x}_i)$ to minimize the objective function.

The XGBoost algorithm performs second-order Taylor expansion to the objective function in the optimization process, which is explained via the formula below.

$$\begin{aligned} Obj^{(t)} &\simeq \sum_{i=1}^{n}\left[l\left(y_i, \hat{y}_i^{(t-1)}\right) + g_i f_t(\mathbf{x}_i) + \tfrac{1}{2} h_i f_t^2(\mathbf{x}_i)\right] + \Omega(f_t) + c \\ &= \sum_{i=1}^{n}\left[g_i f_t(\mathbf{x}_i) + \tfrac{1}{2} h_i f_t^2(\mathbf{x}_i)\right] + \Omega(f_t) + \left[\sum_{i=1}^{n} l\left(y_i, \hat{y}_i^{(t-1)}\right) + c\right] \end{aligned} \tag{23}$$

where g_i, h_i can be defined as:

$$\begin{cases} g_i = \partial_{\hat{y}_i^{(t-1)}} l\left(y_i, \hat{y}_i^{(t-1)}\right) \\ h_i = \partial^2_{\hat{y}_i^{(t-1)}} l\left(y_i, \hat{y}_i^{(t-1)}\right) \end{cases} \tag{24}$$

According to Equation (23), ignoring the influence of the constant value, the objective function optimized in step t can be simplified as:

$$Obj^{(t)} \simeq \sum_{i=1}^{n}\left[g_i f_t(\mathbf{x}_i) + \frac{1}{2}h_i f_t^2(\mathbf{x}_i)\right] + \Omega(f_t) \tag{25}$$

As can be seen from Equation (25), it is g_i, h_i that the objective optimization parameters depend on.

3.3. Complexity Calculation in the Objective Function

For a given tree ensemble model, the complexity of the model can be defined by the equation below.

$$\Omega(f_t) = \gamma T + \frac{1}{2}\lambda\sum_{j=1}^{T} {w_j}^2 \tag{26}$$

where γ and λ are both regularization factors. γ is the parameter used to control tree node splitting. When the cost function of a node after splitting is less than this value, it will not split. When it is greater than this value, it will split. λ is the regularization weight. T is the number of leaf nodes, and w_j is the weight of the leaf nodes.

Define $I_j = \{i|q(x_i) = j\}$ as the instance set of leaf j. We can substitute Equation (26) into the objective function Equation (25) as:

$$\begin{aligned} Obj^{(t)} &= \sum_{i=1}^{n}\left[g_i f_t(\mathbf{x}_i) + \tfrac{1}{2}h_i f_t^2(\mathbf{x}_i)\right] + \gamma T + \tfrac{1}{2}\lambda\sum_{j=1}^{T} w_j^2 \\ &= \sum_{i=1}^{n}\left[g_i w_{q(\mathbf{x}_i)} + \tfrac{1}{2}h_i w_{q(\mathbf{x}_i)}^2\right] + \gamma T + \tfrac{1}{2}\lambda\sum_{j=1}^{T} w_j^2 \\ &= \sum_{j=1}^{T}\left[\left(\Sigma_{i\in I_j} g_i\right)w_j + \tfrac{1}{2}\left(\Sigma_{i\in I_j} h_i + \lambda\right)w_j^2\right] + \gamma T \end{aligned} \tag{27}$$

If we define $G_j = \Sigma_{i\in I_j} g_i, H_j = \Sigma_{i\in I_j} h_i$, then Equation (27) can be abbreviated as:

$$Obj^{(t)} = \sum_{j=1}^{T}\left[G_j w_j + \frac{1}{2}\left(H_j + \lambda\right)w_j^2\right] + \gamma T \tag{28}$$

3.4. Optimization of the Objective Function

When a fixed structure of the tree is $q(\mathbf{x})\left(I_j = \{i|q(\mathbf{x}_i) = j\}\right)$, we can compute the optimal weight w_j^* of leaf j by using the equation below.

$$w_j^* = -\frac{\Sigma_{i\in I_j} g_i}{\Sigma_{i\in I_j} g_i + \lambda} = -\frac{G_j}{H_j + \lambda} \tag{29}$$

When Equation (29) is substituted into the objective function Equation (28), the optimal value of the objective function is found.

$$Obj^* = -\frac{1}{2}\sum_{j=1}^{T}\frac{G_j^2}{H_j + \lambda} + \gamma T \tag{30}$$

By optimizing the objective function, the optimal structure of the decision tree can be obtained. The value of the objective function can be understood as an index score of information gained, and, for the value of the function, the lower value is better.

A split finding algorithm is proposed in the XGBoost algorithm. This means that a splitting point is added to each leaf node. If a node is decomposed into two leaf nodes, then the score gained can be found using the equation below.

$$Gain = \frac{1}{2}\left[\frac{G_L^2}{H_L+\lambda} + \frac{G_R^2}{H_R+\lambda} - \frac{(G_L+G_R)^2}{H_L+H_R+\lambda}\right] - \gamma \quad (31)$$

where G_L is the sum of samples g_i distributed to the left cotyledon, G_R is the sum of samples g_i distributed to the right cotyledon, H_L is the sum of samples h_i distributed to the left cotyledon, and H_R is the sum of samples h_i distributed to the right cotyledon. In addition, in Equation (31), the first term in square brackets is the score of the left node, the second term is the score of the right node, and the third term is the score of the original node. Thus, select a feature as the reference quantity, and then scan from left to right with a certain step length in order to find out the gain of each splitting point. Take the point with the largest gain as the splitting point for the search, and it is not necessary to add a branch if the gain is less than γ.

Based on the principle of node splitting in the XGBoost, the model will continuously optimize itself, according to residuals during the iteration. Since the objective function of node splitting contains both an error term and a regularization term, the model has high precision.

4. Hydraulic Valve Failure Test

4.1. Introduction to the Experimental Platform

The experimental data sets for this study were derived from the UC Irvine Machine Learning Repository [51]. The data were collected from a hydraulic test bed that allowed a reversible change of the state or condition of various components, at the Mechatronics and Automation Technology Center of Saarbrucken University in Germany [52]. The hydraulic system consists of a primary working circuit (Figure 1a) and a secondary cooling-filtration circuit (Figure 1b), which are connected by the oil tank. In the working circuit with the main pump (electrical motor power of 3.3 kW), different load levels are cyclically repeated with the electro-hydraulic proportional valve (V11). It is possible that the typical cyclical operation and repeated load characteristic in an industrial application can be simulated by setting fixed working cycles with pre-defined load levels in the test. Meanwhile, the random load variations in mobile machines can be started by setting variable working cycles with pseudo-random load variations.

4.2. Data Acquisition System

The condition monitoring system for the hydraulic system is equipped with several sensors for measuring process values. Among them, there are six pressure sensors (PS1–PS6), two flow rate sensors (FS1, FS2), five temperature sensors (TS1–TS5), one motor power sensor (EPS1), and one vibration sensor (VS1) with standard industrial 4–20 mA current loop interfaces connected to a data acquisition system. In addition, sensors integrating EIA-232 and EIA-485 buses for oil particle contamination (CS and MCS) and oil parameter monitoring (COPS) are installed in the condition monitoring system. In total, the above 17 sensor signals are stored while the hydraulic system repeats pre-defined constant working cycles with changing conditions of hydraulic components to identify typical signal patterns. The sampling rate of the above sensors is set within the range of 1 Hz to 100 Hz, respectively, according to the different types of collected state signals. The specific sampling rate of each sensor is shown in Table 1.

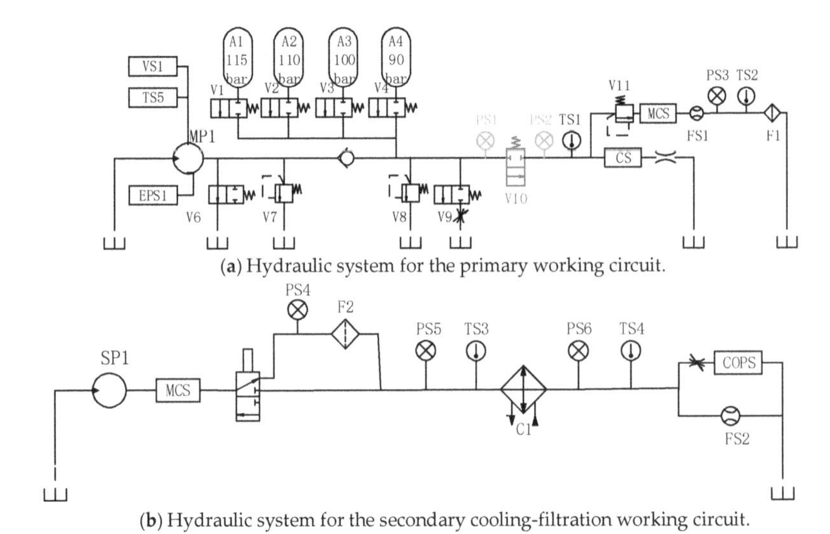

(**a**) Hydraulic system for the primary working circuit.

(**b**) Hydraulic system for the secondary cooling-filtration working circuit.

Figure 1. Schematic diagram of condition monitoring in hydraulic systems.

Table 1. Sampling rates of sensors.

Sensor	Physical Quantity	Unit	Sampling Rate	Attribute Information
PS1-6	Pressure	bar	100 Hz	Actual sensor
EPS1	Motor power	W	100 Hz	Actual sensor
FS1-2	Volume flow	L/min	10 Hz	Actual sensor
TS1-4	Temperature	°C	1 Hz	Actual sensor
VS1	Vibration	mm/s	1 Hz	Actual sensor
CE	Cooling efficiency	%	1 Hz	Virtual sensor
CP	Cooling power	kW	1 Hz	Virtual sensor
SE	Efficiency factor	%	1 Hz	Virtual sensor

The sensor data are connected and buffered on a Programmable Logic Controller (PLC) (Beckhoff CX5020) at run time and transferred to a computer by EtherCAT, where the data are stored for further analysis. It is possible to configure fault characterization measurements with a specifically developed tool such as LabVIEW and to, subsequently, perform them by using the PLC. Using this tool, measurements of the fault type, severity, and duration are taken, if necessary, to define different fault states, such as hydraulic pump internal leakage, hydraulic valve switching characteristic degradation, accumulator leakage, and cooler power degradation. Table 2 shows the components and respective parameters that are configurable to simulate fault scenarios. The experimental method has the advantages of not requiring damage to the mechanical structure of the hydraulic valve, repeatability, and reversibility of the fault state, and simple operation of the fault setting.

4.3. Hydraulic Valve Fault Setting and Data Acquisition

According to the primary working circuit in Figure 1a, the working cycle (duration 60 s) consists of different segments with transient and static load characteristics performed by the electro-hydraulic proportional pressure valve (V11) and the directional valve (V10) to simulate a typical machine operation. Under pre-defined load conditions, the current set points of the directional valve are 100%, 90%, 80%, and 73% of the nominal value. Thus, a simulation is conducted for the hydraulic valve (V10) in the fault states of normal, slight, medium, and severe valve operation deterioration, and raw sensor data are collected during characterization.

Table 2. Hydraulic test bed: components and their simulated fault conditions.

Hydraulic Component	Fault Conditions	Control Parameters	Possible Range
Valve (V10)	Switching characteristic degradation	Control current of V10	0 . . . 100% of nom. current.
Pump (MP1)	Internal leakage	Switchable bypass orifices V9	3 × 0.2 mm, 3 × 0.25 mm
Accumulators (A1–A4)	Gas leakage	Accumulators A1–A4 with different pre charge pressures	90, 100, 110, 115 bar
Cooler (C1)	Cooling power decrease	Fan duty cycle of C1	0 . . . 100% (0.6 . . . 2.2 kW)

The specific experimental process is under different fault conditions of the hydraulic valve (V10). The hydraulic system is cyclically operated for 2205 cycles with a pre-defined load on the electro-hydraulic servo valve (V11), and the running time of each cycle is 60 s. Among them, the number of cycles in which the hydraulic system is in an unstable state is 756 cycles, and, in a steady state, it is 1449 cycles. Therefore, in the steady state, the experimental data of 369 normal states, 360 slight fault states, 360 medium fault states, and 360 severe fault states are obtained, respectively, for a total of 1449 sets of data. Taking the partial pressure signals of the inlet (PS1) and outlet (PS2) measured by the hydraulic directional valve (V10) within the fixed working cycle with a duration of 60 s as an example, the changing rule of different fault states in the steady state is shown in Figure 2.

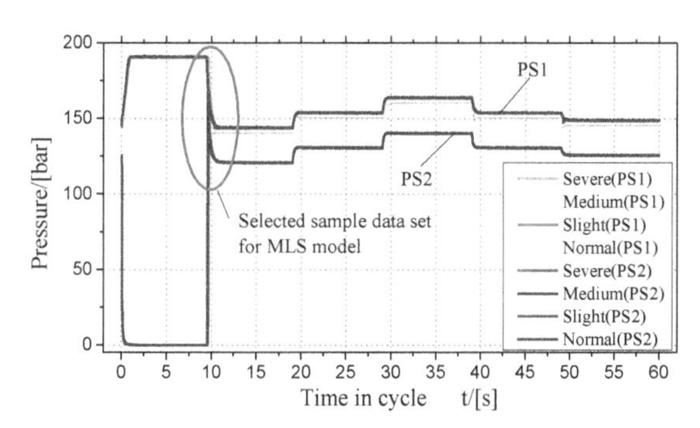

Figure 2. Fixed working cycle (measured by partial PS1 and PS2) with a pre-defined load.

5. Hydraulic Valve Fault Diagnosis Based on PCA and XGBoost

5.1. Acquisition of Sample Data for a Hydraulic Valve Fault Diagnosis

Figure 2 shows that, in each working cycle, the hydraulic directional valve (V10) is switched at about the 10th second, and the valve outlet pressure (PS2) rises rapidly when the valve inlet pressure (PS1) decreases significantly. It can be seen that the sensor data contain the fault characteristics of the hydraulic valve. As shown in Figure 3, based on the working principle and switching performance characteristics of the hydraulic valve, 100 data points between 9.3 s and 11.3 s in each cycle of PS1 (Figure 3a) and PS2 (Figure 3b) are, respectively, intercepted in the 1449 sets of data, which obtains two sample data sets of 1449 × 100 dimensions for the modeling analysis. These data sets contain the pressure changes before and after the hydraulic valve switching process.

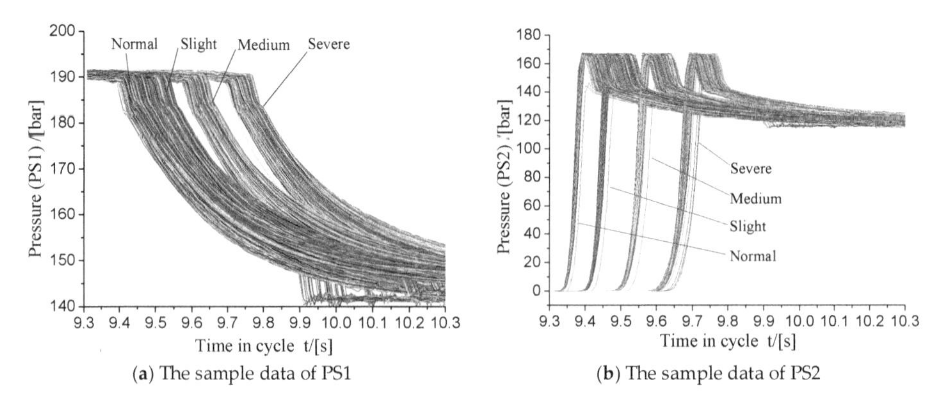

(**a**) The sample data of PS1

(**b**) The sample data of PS2

Figure 3. The sample data of the hydraulic valve fault diagnosis.

5.2. Dimensionality Reduction of a PCA-Based Training Set Sample

Dimensionality reduction of the training set sample data by the PCA method not only has the advantages of reducing the dimension of the training set and improving the speed of the model training, but also has the functions of eliminating the outliers of the signal and denoising the signal. Based on the dimensionality reduction principle and analysis steps of the PCA method described above, the variance contribution rates δ_i of the principal components of the PS1 and PS2 training set samples are calculated, according to Equation (16). The variance contribution rates δ_i of the first 1–18 principal elements are shown in Table 3.

Table 3. Variance contribution of partial principal components.

Number of Principal Component	1	2	3	4	5	6	7	8	9
Unit	%	%	%	%	%	%	%	%	%
PS1	90.36	4.96	3.02	0.68	0.24	0.18	0.12	0.07	0.06
PS2	66.72	20.90	11.19	0.46	0.21	0.17	0.12	0.05	0.03
Number of Principal Component	**10**	**11**	**12**	**13**	**14**	**15**	**16**	**17**	**18**
Unit	%	%	%	%	%	%	%	%	%
PS1	0.04	0.04	0.03	0.02	0.01	0.01	0.00	0.00	0.00
PS2	0.03	0.02	0.02	0.01	0.00	0.00	0.00	0.00	0.00

According to Equation (17), the relationship between the cumulative contribution rate η_k of the principal components of the PS1 and PS2 training set samples and the number of principal components is further plotted, as shown in Figure 4.

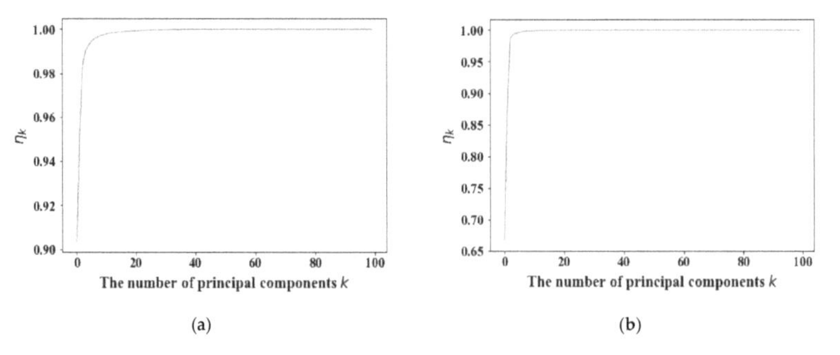

(a)

(b)

Figure 4. The cumulative contribution rate for the sample data. (**a**) The cumulative contribution rate η_k of PS1 training set samples; (**b**) The cumulative contribution rate η_k of PS2 training set samples.

However, there is no general method to select the optimal number of principal components to be retained. In order to retain the original information to the greatest extent, the variance contribution of principal components δ_i is set close to 0, and the cumulative contribution rate η_k is set close to 100%. According to Table 3 and Figure 4, from the variance contribution rates δ_i of the principal components, it can be known that, after PCA dimensionality reduction, the information of PS1 data is concentrated in the first 1–15 principal components, and the PS2 data information is concentrated in the first 1–13. Considering the balance of the model training data, the number of principal components k after dimensionality reduction to the PS1 and PS2 data sets is determined to be 15. The data set of 1449 × 100 dimensions can be compressed to 1449 × 15 dimensions. The data set after dimensionality reduction is applied to the modeling and learning process of the training sample data set, which is divided into training samples and test samples, according to a certain proportion, as displayed in Table 4.

Table 4. The proportion sets of sample data for the hydraulic valve (V10).

Fault Conditions for the Hydraulic Valve	Normal	Slight	Medium	Severe	Total
The total number of samples	369	360	360	360	1449
Number of training samples	257	252	238	264	1011
Number of test samples	112	108	122	96	438

5.3. Model Establishment Based on the XGBoost Algorithm

Huawei MLS integrates multiple algorithm nodes and can combine different nodes by dragging and connecting, and creating a corresponding visual workflow for data processing, model training, evaluation, and prediction, according to research tasks. At the same time, MLS integrates the function of the Jupyter notebook, which provides users with an interactive notebook as an integrated development environment for machine learning applications. The environment supports the writing of Python scripts and performs data analysis and model building by using the Spark native algorithm MLlib. Based on the workflow, a hydraulic valve fault diagnosis model combining PCA and the XGBoost algorithm is established in MLS. The specific process is shown in Figure 5.

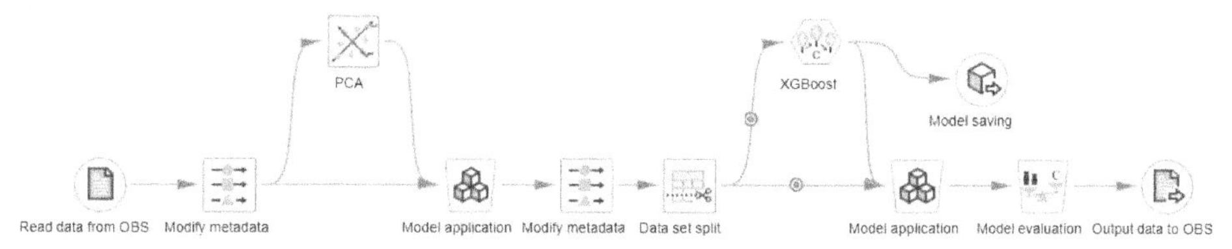

Figure 5. The model of principal component analysis (PCA) and eXtreme Gradient Boosting (XGBoost).

5.4. Model Evaluation

In the model shown in Figure 5, the training sample data are split into a training set and a test set by a "Data set split" module, and the split ratio is shown in Table 4. The test set is evaluated by the "Model Evaluation" module in the MLS. The models are quantitatively evaluated by using model evaluation indicators, such as confusion matrix, precision, recall rate, and an *F*1 score. The specific definitions of each indicator are as follows.

Taking the binary classification problem as an example, the sample data are divided into true positive (TP), false positive (FP), true negative (TN), and false negative (FN), according to the combination of its real category and machine learning prediction category. Then, TP + FP + TN + FN = Total number of samples.

(1) Precision

Precision indicates the proportion of the sample whose real category is positive in the sample with a positive prediction category. The calculation formula is shown below.

$$P = \frac{TP}{TP + FP} \quad (32)$$

(2) Recall Rate

The recall rate indicates that the proportion of the sample with a positive prediction category in the sample with a real positive category. The calculation formula is shown below.

$$R = \frac{TP}{TP + FN} \quad (33)$$

where the precision P and the recall rate R are a pair of contradictory indicators. In general, when the precision is high, the recall rate tends to be low. When the recall rate is high, the precision tends to be low.

(3) $F1$ Score

The $F1$ score takes the precision and recall rate into account and is their weighted harmonic mean. When the weights of precision and recall rate are the same, the harmonic mean obtained is called the $F1$ score, and the calculation formula is shown below.

$$F1score = 2 \times \frac{P \times R}{P + R} \quad (34)$$

(4) Confusion Matrix

The confusion matrix is used for evaluating the model when faced with a multi-classification problem, and the weight of each category is almost equal. Each column of the confusion matrix represents a prediction category, and the total number of data for each column represents the number of data predicted to be in the category. Each row represents the true attribution category of the data, and the total number of data for each row represents the number of data instances belonging to that category. For a confusion matrix, the larger the value on the diagonal is, the better the matrix. The smaller value of other locations are better.

The above-mentioned the Principal Component Analysis and eXtreme Gradient Boosting (PCA-XGBoost) model trained in the Huawei Cloud MLS is tested by the test set, and its specific indicators are shown in Table 5.

Table 5. Evaluation indexes of the model using test samples.

	Confusion Matrix				**Precision**	**Recall Rate**	***F1* Score**
Practical \ Predicted	**Normal**	**Slight**	**Medium**	**Severe**			
Normal	101	11	0	0	0.990	0.902	0.944
Slight	0	108	0	0	0.885	1.000	0.939
Medium	0	3	119	0	1.000	0.975	0.988
Severe	1	0	0	95	1.000	0.990	0.995

It can be seen from Table 5 that the diagonal value of the confusion matrix in the hydraulic valve fault diagnosis model is much larger than the value of the non-diagonal line. The precision and recall rate of the model are all above 88%, and the $F1$ score is higher than 93%. The above results show that the PCA-XGBoost model has high accuracy.

5.5. Comparison of Model Diagnosis Results

In the same environment of the Huawei Cloud MLS platform, the model constructed by the XGBoost algorithm was compared with the CART Tree classification model and the Random Forests (RFs) algorithm model. The comparison diagram is displayed in Figure 6.

The comparison is made based on the model evaluation indicators such as the precision, the recall rate, and the *F*1 score. Additionally, the comparison results are shown in Table 6.

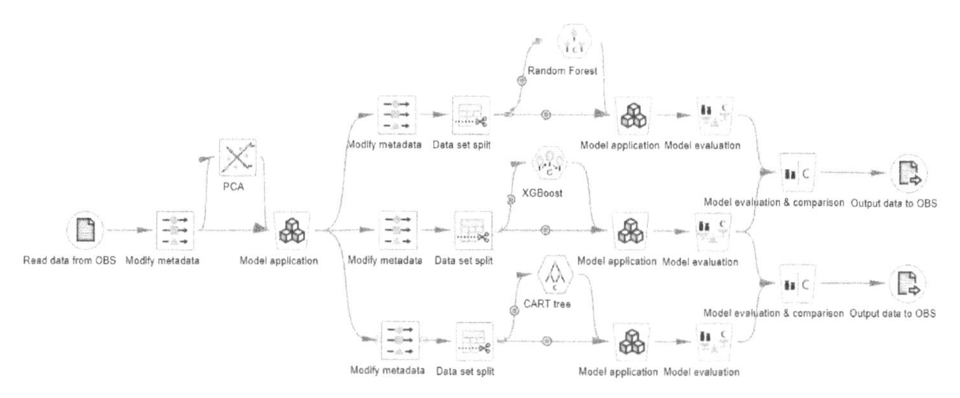

Figure 6. Model comparison analysis.

Table 6. Model evaluation comparison.

Model Algorithm	Fault Conditions for the Hydraulic Valve	Model Evaluation Index		
		Precision	Recall Rate	*F*1 Score
PCA-CART	Normal	0.878	0.902	0.890
	Slight	0.914	0.889	0.901
	Medium	0.907	0.959	0.932
	Severe	0.966	0.896	0.930
	Average value	0.916	0.911	0.913
PCA-RFs	Normal	0.916	0.875	0.895
	Slight	0.909	0.926	0.917
	Medium	0.929	0.959	0.944
	Severe	0.958	0.948	0.953
	Average value	0.928	0.927	0.927
PCA-XGBoost	Normal	0.990	0.902	0.944
	Slight	0.885	1.000	0.939
	Medium	1.000	0.975	0.988
	Severe	1.000	0.990	0.995
	Average value	0.969	0.967	0.966

As shown in Table 6, after the principal component dimensionality reduction of the data, the CART Tree, Random Forests, and XGBoost algorithms are, respectively, used to construct the fault diagnosis model of the hydraulic valve. Afterward, the models are tested through the test set. The test results indicate that the average precision of the XGBoost model is 96.9%0.969, the average recall rate is 96.7%, and the average *F*1 score is96.6%. The values of the evaluation indicators are higher than those of the CART Tree and Random Forests models, which can not only prove the superiority of the algorithm, but also demonstrate the effectiveness of this algorithm for hydraulic valve fault diagnosis.

6. Conclusions

This paper mainly studies the fault diagnosis of hydraulic valves. Based on the status monitoring data of the measured inlet and outlet pressure signals of the hydraulic valve, PCA was adopted to reduce the dimensions of the data, and the XGBoost algorithm was used to construct a machine learning model for hydraulic valve fault diagnosis. By testing the evaluation indexes of the machine learning model, the effectiveness and superiority of the above method are proved. The main conclusions are as follows.

(1) In this study, the pressure signals of the hydraulic valve are utilized as the sample data for fault diagnosis in order to realize accurate diagnosis and classification of hydraulic valve faults. Then, a novel fault diagnosis method for hydraulic valves based on the variation characteristics of pressure signals is proposed.

(2) PCA dimensionality reduction for the original data set of feature vectors can not only significantly reduce the dimension of the feature vector, but also remove redundant information in the original data set. The principal component feature set after dimensionality reduction is used to train the XGBoost machine learning, in order to construct the fault diagnosis model for the hydraulic valve. The test results indicate that the precision mean of the model is 96.9%, the recall rate mean is 96.7%, and the $F1$ score mean is 96.6% on the test set. Compared with the decision tree and random forest models, the constructed model has higher accuracy.

(3) This research builds a fault diagnosis model for the hydraulic valve in the visual workflow of HUAWEI Cloud MLS, and carries out data processing, model training, evaluation, and prediction. In this way, hydraulic valve fault diagnosis, machine learning algorithms, and HUAWEI cloud are organically combined together, which can provide a theoretical basis and practical guidance for the remote fault diagnosis of hydraulic components and the predictive maintenance of hydraulic systems.

Author Contributions: W.J. and A.J. conceived and designed the method. S.Z. and H.N. analyzed the data. Y.L. and Y.Z. wrote the paper.

Funding: This research was funded by The National Natural Science Foundation of China grant number 51875498, 51475405, 51805214; This research was funded by Key Program of Hebei Natural Science Foundation grant number E2018203339; This research was funded by Innovation Foundation for Graduate Students of Hebei Province grant number CXZZBS2018045; This research was funded by China Postdoctoral Science Foundation grant number 2019M651722; This research was funded by Young Problems in the Special Project of Basic Research of Yanshan University grant number 15LGB005. And The APC was funded by 51875498.

Conflicts of Interest: The authors declare no conflict of interest.

References

1. Schneider, T.; Helwig, N.; Schütze, A. Automatic feature extraction and selection for classification of cyclical time series data. *TM-Tech. Mess.* **2017**, *84*, 198–206. [CrossRef]
2. Goharrizi, A.Y.; Sepehri, N. Application of fast Fourier and wavelet transforms towards actuator leakage diagnosis: A comparative study. *Int. J. Fluid Power* **2013**, *14*, 39–51. [CrossRef]
3. Watton, J. *Modelling, Monitoring and Diagnostic Techniques for Fluid Power Systems*; Springer Science & Business Media: New York, NY, USA, 2007.
4. Qian, J.Y.; Chen, M.R.; Liu, X.L.; Jin, Z.J. A numerical investigation of the flow of nanofluids through a micro Tesla valve. *J. Zhejiang Univ. Sci. A* **2019**, *20*, 50–60. [CrossRef]
5. Qian, J.Y.; Gao, Z.X.; Liu, B.Z.; Jin, Z.J. Parametric study on fluid dynamics of pilot-control angle globe valve. *ASME J. Fluids Eng.* **2018**, *140*, 111103. [CrossRef]
6. Zhang, J.; Xia, S.; Ye, S.; Xu, B.; Song, W.; Zhu, S.; Xiang, J. Experimental investigation on the noise reduction of an axial piston pump using free-layer damping material treatment. *Appl. Acoust.* **2018**, *139*, 1–7. [CrossRef]
7. Ye, S.; Zhang, J.; Xu, B.; Zhu, S.; Xiang, J.; Tang, H. Theoretical investigation of the contributions of the excitation forces to the vibration of an axial piston pump. *Mech. Syst. Signal Process.* **2019**, *129*, 201–217. [CrossRef]
8. Wang, C.; Hu, B.; Zhu, Y.; Wang, X.; Luo, C.; Cheng, L. Numerical study on the gas-water two-phase flow in the self-priming process of self-priming centrifugal pump. *Processes* **2019**, *7*, 330. [CrossRef]

9. Wang, C.; Chen, X.; Qiu, N.; Zhu, Y.; Shi, W. Numerical and experimental study on the pressure fluctuation, vibration, and noise of multistage pump with radial diffuser. *J. Braz. Soc. Mech. Sci. Eng.* **2018**, *40*, 481. [CrossRef]
10. Zheng, H.; Wang, R.; Xu, W.; Wang, Y.; Zhu, W. Combining a HMM with a genetic algorithm for the fault diagnosis of photovoltaic inverters. *J. Power Electron.* **2017**, *17*, 1014–1026.
11. Xu, X.; Wang, W.; Zou, N.; Chen, L.; Cui, X. A comparative study of sensor fault diagnosis methods based on observer for ECAS system. *Mech. Syst. Signal Process.* **2017**, *87*, 169–183. [CrossRef]
12. Sun, H.; Yuan, S.; Luo, Y. Cyclic spectral analysis of vibration signals for centrifugal pump fault characterization. *IEEE Sens. J.* **2018**, *18*, 2925–2933. [CrossRef]
13. Tang, S.; Gu, J.; Tang, K.; Zou, R.; Sun, X.; Uddin, S. A Fault-signal-based generalizing remaining useful life prognostics method for wheel hub bearings. *Appl. Sci.* **2019**, *9*, 1080. [CrossRef]
14. Mao, Y.; Liu, G.; Zhao, W.; Ji, J. Vibration prediction in fault-tolerant flux-switching permanent-magnet machine under healthy and faulty conditions. *IET Electr. Power Appl.* **2017**, *11*, 19–28. [CrossRef]
15. Chen, T.; Chen, L.; Xu, X.; Cai, Y.; Jiang, H.; Sun, X. Passive fault-tolerant path following control of autonomous distributed drive electric vehicle considering steering system fault. *Mech. Syst. Signal Process.* **2019**, *123*, 298–315. [CrossRef]
16. Zhou, H.; Liu, G.; Zhao, W.; Yu, X.; Gao, M. dynamic performance improvement of five-phase permanent-magnet motor with short-circuit fault. *IEEE Trans. Ind. Electron.* **2018**, *65*, 145–155. [CrossRef]
17. Schneider, T.; Helwig, N.; Schütze, A. Industrial condition monitoring with smart sensors using automated feature extraction and selection. *Meas. Sci. Technol.* **2018**, *29*, 094002. [CrossRef]
18. Zhu, Y.; Tang, S.; Quan, L.; Jiang, W.; Zhou, L. Extraction method for signal effective component based on extreme-point symmetric mode decomposition and Kullback-Leibler divergence. *J. Braz. Soc. Mech. Sci. Eng.* **2019**, *41*, 100. [CrossRef]
19. Zhu, Y.; Qian, P.; Tang, S.; Jiang, W.; Li, W.; Zhao, J. Amplitude-frequency characteristics analysis for vertical vibration of hydraulic AGC system under nonlinear action. *AIP Adv.* **2019**, *9*, 035019. [CrossRef]
20. Wu, W.; Yang, S.; Zhou, T. Application of complex three-order cumulants to fault diagnosis of hydraulic valve. *J. Tianjin Univ.* **2013**, *46*, 590–595.
21. Gao, Y.; Huang, Y. Application of AR bi-spectrum in fault diagnosis of reducing valve. *Mach. Des. Manuf.* **2011**, *11*, 70–72.
22. Li, T.; Fan, M.; Huang, Q.; Li, Z. Research on fault diagnosis method for auto pilot hydraulic valve based on fractal theory. *China Meas. Test* **2012**, *38*, 1–5.
23. Raduenz, H.; Mendoza, Y.E.A.; Ferronatto, D.; Souza, F.J.; da Cunha Bastos, P.P.; Soares, J.M.C.; De Negri, V.J. Online fault detection system for proportional hydraulic valves. *J. Braz. Soc. Mech. Sci. Eng.* **2018**, *40*, 331. [CrossRef]
24. Vianna, W.O.L.; de Souza Ribeiro, L.G.; Yoneyama, T. Electro hydraulic servovalve health monitoring using fading extended Kalman filter. In Proceedings of the 2015 IEEE Conference on Prognostics and Health Management (PHM), Austin, TX, USA, 2015, 22–25 June 2015; IEEE: Piscataway, NJ, USA, 2015; pp. 1–6.
25. Folmer, J.; Schrüfer, C.; Fuchs, J.; Vermum, C.; Vogel-Heuser, B. Data-driven valve diagnosis to increase the overall equipment effectiveness in process industry. In Proceedings of the 2016 IEEE 14th International Conference on Industrial Informatics (INDIN), Poitiers, France, 18–21 July 2016; IEEE: Piscataway, NJ, USA, 2016; pp. 1082–1087.
26. Lei, Y.; Jia, F.; Kong, D.; Lin, J.; Xing, S. Opportunities and challenges of machinery intelligent fault diagnosis in big data era. *Chin. J. Mech. Eng.* **2018**, *54*, 94–104. [CrossRef]
27. Pei, H.; Hu, C.; Si, X.; Zhang, J.; Pang, Z.; Zhang, P. Review of machine learning based remaining useful life prediction methods for equipment. *J. Mech. Eng.* **2019**, *55*, 1–13.
28. Zhu, Y.; Jiang, W.; Kong, X.; Quan, L.; Zhang, Y. A chaos wolf optimization algorithm with self-adaptive variable step-size. *AIP Adv.* **2017**, *7*, 105024. [CrossRef]
29. Pearson, K. On lines and planes of closest fit to systems of points in space. *Philos. Mag. A* **1901**, *6*, 559–572. [CrossRef]
30. Fisher, R.A.; Mackenzie, W.A. Studies in crop variation. II. The manurial response of different potato varieties. *J. Agric. Sci.* **1923**, *13*, 311–320. [CrossRef]
31. Hotelling, H. Analysis of a complex of statistical variables into principal components. *J. Educ. Psychol.* **1933**, *24*, 417–441. [CrossRef]

32. Mohanty, S.; Gupta, K.K.; Raju, K.S. Adaptive fault identification of bearing using empirical mode decomposition–principal component analysis-based average kurtosis technique. *IET Sci. Meas. Technol.* **2017**, *11*, 30–40. [CrossRef]
33. Stief, A.; Ottewill, J.; Baranowski, J.; Orkisz, M. A PCA and two-stage bayesian sensor fusion approach for diagnosing electrical and mechanical faults in induction motors. *IEEE Trans. Ind. Electron.* **2019**, *66*, 9510–9520. [CrossRef]
34. Caggiano, A. Tool wear prediction in Ti-6Al-4V machining through multiple sensor monitoring and PCA features pattern recognition. *Sensors* **2018**, *18*, 823. [CrossRef] [PubMed]
35. Wang, Y.; Ma, X.; Qian, P. Wind turbine fault detection and identification through PCA-based optimal variable selection. *IEEE Trans. Sustain. Energy* **2018**, *9*, 1627–1635. [CrossRef]
36. Xiao, X.; Zhao, S.; Chen, K.; Zhang, M.; Liu, L. Application of principal component analysis in fault diagnosis of electro-hydrostatic actuators. *Missiles Space Veh.* **2019**, *366*, 98–104.
37. Riba, J.R.; Canals, T.; Cantero, R. Recovered paperboard samples identification by means of mid-infrared sensors. *IEEE Sensors J.* **2013**, *13*, 2763–2770. [CrossRef]
38. Chen, T.; Guestrin, C. Xgboost: A scalable tree boosting system. In Proceedings of the 22nd Acm Sigkdd International Conference on Knowledge Discovery and Data Mining ACM, San Francisco, CA, USA, 13–17 August 2016; pp. 785–794.
39. Nielsen, D. Tree Boosting with XGBoost-Why Does XGBoost Win "Every" Machine Learning Competition? Master's Thesis, Norwegian University of Science and Technology, Trondheim, Noreg, 2016.
40. Zhang, D.; Qian, L.; Mao, B.; Huang, C.; Huang, B.; Si, Y. A data-driven design for fault detection of wind turbines using random forests and XGBoost. *IEEE Access* **2018**, *6*, 21020–21031. [CrossRef]
41. Chakraborty, D.; Elzarka, H. Early detection of faults in HVAC systems using an XGBoost model with a dynamic threshold. *Energy Build.* **2019**, *185*, 326–344. [CrossRef]
42. Zhang, R.; Li, B.; Jiao, B. *Application of XGBoost Algorithm in Bearing Fault Diagnosis. IOP Conference Series: Materials Science and Engineering*; IOP Publishing: Bristol, UK, 2019; Volume 490, p. 072062.
43. Nguyen, H.; Bui, X.N.; Bui, H.B.; Cuong, D.T. Developing an XGboost model to predict blast-induced peak particle velocity in an open-pit mine: A case study. *Acta Geophys.* **2019**, *67*, 477–490. [CrossRef]
44. Pan, B. *Application of XGBoost algorithm in hourly PM2. 5 concentration prediction. IOP Conference Series: Earth and Environmental Science*; IOP Publishing: Bristol, UK, 2018; Volume 113, p. 012127.
45. Liu, Y.; Qiao, M. Heart disease prediction based on clustering and XGboost. *Comput. Syst. Appl.* **2019**, *28*, 228–232.
46. Fitriah, N.; Wijaya, S.K.; Fanany, M.I.; Badri, C.; Rezal, M. EEG channels reduction using PCA to increase XGBoost's accuracy for stroke detection. In Proceedings of the AIP Conference, 10–11 July 2017; AIP Publishing: Melville, NY, USA, 2017; Volume 1862, p. 030128.
47. Zhao, W.; Dong, L. *Machine Learning*; Posts & Telecom Press: Beijing, China, 2018.
48. Zhang, J. *Multivariable Statistical Process Control*; Chemical Industry Press: Beijing, China, 2000.
49. Wang, G. *Principal Component Analysis and Partial Least Square Method*; Tsinghua University Press: Beijing, China, 2012.
50. Wang, X. *A Research on CTR Prediction Based on Ensemble of RF, XGBoost and FFM*; Zhejiang University: Hangzhou, China, 2018.
51. Available online: http://archive.ics.uci.edu/ml/datasets/Conditionmonit-oringofhydraulicsystems (accessed on 26 April 2018).
52. Helwig, N.; Pignanelli, E.; Schütze, A. Condition monitoring of a complex hydraulic system using multivariate statistics. In Proceedings of the 2015 IEEE International Instrumentation and Measurement Technology Conference (I2MTC), Pisa, Italy, 11–14 May 2015; IEEE: Piscataway, NJ, USA, 2015; pp. 210–215.

Article

The Simulation of Vortex Structures Induced by Different Local Vibrations at the Wall in a Flat-Plate Laminar Boundary Layer

Weidong Cao *, Zhixiang Jia and Qiqi Zhang

Research Institute of Fluid Engineering Equipment Technology, Jiangsu University, Zhenjiang 212013, China
* Correspondence: cwd@ujs.edu.cn; Tel.: +86-139-5281-6468

Received: 26 June 2019; Accepted: 21 August 2019; Published: 23 August 2019

Abstract: The compact finite difference scheme on non-uniform meshes and the Fourier spectral hybrid method are used to directly simulate the evolution of vortex structures in a laminar boundary layer over a flat plate. To this end, two initial local vibration disturbances, namely, the positive–negative and the negative–positive models, at the wall were adopted. The numerical results show that the maximum amplitudes of vortex structures experience a process of linear growth and nonlinear rapid growth. The vertical disturbance velocity and mean flow shear and the derivative term of the stream-wise disturbance velocity and the span-wise disturbance velocity, are important factors for vortex structure development; the high- and low-speed stripe and the stream-wise vortex are consistent with structures seen in full turbulence. The maximum amplitude of the negative–positive model grows more quickly than that of the negative–positive model, and the detailed vortex structures are different for the two models. The mean flow profiles both become plump, which leads to the instability of the laminar boundary layer. The way in which the disturbance is generated with different local vibrations influences the dynamics of vortex structures in a laminar boundary layer.

Keywords: vortex structures; boundary layer; direct numerical simulation; amplitude; vortices

---

1. Introduction

At the beginning of a transition in shear flow, a large number of vortex structures are formed. The formation mechanism and dynamic characteristics of vortex structures are still key issues in fluid mechanics research.

Lee presented direct comparisons of experimental results on transitions in wall-bound flows obtained by flow visualizations, hot-film measurement, and Particle Image Velocimetry (PIV), along with a brief mention of relevant theoretical progresses, based on a critical review of about 120 selected publications. Despite the somewhat different initial disturbance conditions used in the experiments, the flow structures were found to be practically the same [1]. Sharma presented a new theory of coherent structures in wall turbulence. The theory is the first to predict packets of hairpin vortices and other structures in turbulence and their dynamics, based on an analysis of the Navier–Stokes equations, under an assumption of a turbulent mean profile [2]. Wedin studied finite-amplitude coherent structures with a reflection symmetry in the span-wise direction of a parallel boundary layer flow; some states computed displayed a span-wise spacing between streaks of the same length scale as turbulence flow structures observed in experiments [3]. Wu presented a mathematical theory to describe the nonlinear dynamics of coherent structures. The formulation was based on a triple decomposition of the instantaneous flow into a mean field, coherent fluctuations, and small-scale turbulence, but with the mean-flow distortion induced by nonlinear interactions of coherent fluctuations being treated as part of the organized motion [4]. Mcmullan implemented large eddy simulations of the plane mixing layer for the purpose of reducing the stream-wise vortex structure that may exist in these flows. Both an initially laminar and initially turbulent mixing layer were considered in this study. The initially laminar flow originated from Blasius profiles with a white noise fluctuation environment, whilst the initially turbulent flow had an inflow condition obtained from an inflow turbulence generation method. Flow visualization images demonstrated that both mixing layers contained organized turbulent coherent structures, and that the structures contained rows of stream-wise vortices distributed across the span of the mixing layer [5]. Wall studied three spatially extended traveling wave exact coherent states, together with one span-wise localized state for channel flow. Two of the extended flows were derived by the homotopy method from solutions. Both these flows were asymmetric with respect to the channel center plane, and featured streaky structures in stream-wise velocity flanked by staggered vortical structures; one of these flows featured two streak/vortex systems per span-wise wavelength [6]. Kang studied the direct numerical simulation data of a wave packet in laminar turbulent transition in a Blasius boundary layer. The decomposition of this wave packet into a set of modes could be achieved in a wide variety of ways [7]. Shinneeb experimentally studied the turbulent wake generated by a vertical sharp-edged flat plate suspended in a shallow channel flow with a gap near the bed. Two different gap heights were studied which were compared with the no-gap flow case. The Reynolds number based on the water depth was 45,000. Extensive measurements of the flow field in the vertical and horizontal planes were made using a PIV system. The large vortices were exposed by analyzing the PIV velocity fields using the proper orthogonal decomposition method [8]. Lemarechal experimentally investigated the laminar turbulent transition of a Blasius boundary layer-like flow at the Institute of Aerodynamics and Gas Dynamics, University of Stuttgart. The late stage of controlled transition with K-type breakdown was investigated with the temperature-sensitive paint (TSP) method on the flat-plate surface. The test conditions enabled the TSP method to resolve the complete transition process temporally and spatially. Therefore, it was possible to detect the coherent structures occurring in the late stage of laminar–turbulent transition from the visualizations on the flat-plate surface, namely, Λ and Ω vortices [9]. The effects of isolated, cylindrical roughness elements on laminar–turbulent transition in a flat-plate boundary layer were investigated in a laminar water channel. Most predictions by global linear stability theory could be confirmed, but additional observations in the physical flow demonstrated that not all features could be captured adequately by global linear stability theory [10].

There are many kinds of small disturbance models for boundary layer transition, such as combination Tollmier–Schlichting(T–S) waves, multi-eddy structures which satisfy continuity equation, slot jets, and so on. The initial disturbance sources of other physical models except slot jets are still not clear. In this paper, vortex structures induced by local instantaneous small wall vibration combinations similar to the real physical mechanism are adopted, and the internal mechanisms of the influence of different disturbance combinations on the flow stability are analyzed.

To perform approximate simulations of the wall local forced vibration, an initial disturbance was set in the form of local single-period micro-vibration at the bottom of the wall of $y = 0$. However, the influence of physical deformation on the mesh was neglected since it is outside of the scope of this paper. Two disturbance models were implemented, namely, the positive–negative (P–N) model and the negative–positive (N–P) model. The P–N model is defined as follows: the normal disturbance velocity at mesh points in the circle region, $\sqrt{(x-10.5)^2 + z^2} < 2$ is supposed to be $v\prime = 0.02\sin(2\pi \cdot t/15)$, where $0 \le t \le 15$, and that in the circle region, $\sqrt{(x-14.7)^2 + z^2} < 2$, is supposed to be $v\prime = -0.02\sin(2\pi \cdot t/15)$, where $0 \le t \le 15$. Similarly, the N–P model is defined as follows: the normal disturbance velocity at mesh points in the circle region, $\sqrt{(x-10.5)^2 + z^2} < 2$, is supposed to be $v\prime = -0.02\sin(2\pi \cdot t/15)$, where $0 \le t \le 15$, and that in the circle region, $\sqrt{(x-14.7)^2 + z^2} < 2$, is supposed to be $v\prime = 0.02\sin(2\pi \cdot t/15)$, where $0 \le t \le 15$. The amplitude of the small disturbance velocity is commonly chosen to be about 1% of the maximum mean basic flow velocity. A relatively small amplitude of the disturbance velocity can lead to a relatively long time of disturbance evolution. The computational domain and initial disturbance location are shown in Figure 1.

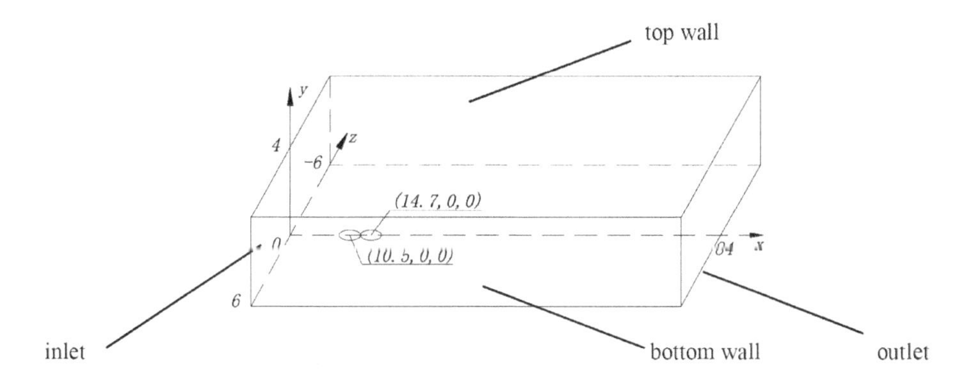

Figure 1. Distribution of initial disturbance at the wall.

2. Governing Equations and Numerical Method

2.1. Governing Equations

The governing equations are the non-dimensional incompressible Navier–Stokes equations and the continuity equation.

$$\frac{\partial \vec{u\prime}}{\partial t} + (\vec{u}_0 \cdot \nabla)\vec{u\prime} + (\vec{u\prime} \cdot \nabla)\vec{u}_0 + (\vec{u\prime} \cdot \nabla)\vec{u\prime} = -\nabla p\prime + \frac{1}{Re}\nabla^2 \vec{u\prime} \tag{1}$$

$$\nabla \cdot \vec{u\prime} = 0, \tag{2}$$

where ∇ is the gradient operator, ∇^2 is the Laplacian, R_e is the Reynolds number, $\vec{u}_0 = (u_0, v_0)$ and p_0 are the velocity and the pressure of Blasius solutions, and $\vec{u}' = (u\prime, v\prime, w\prime)$ and p' are the disturbance velocity and pressure of vortex structures. In this paper, $R_e = U_\infty \cdot \delta / v = 2000$, where U_∞ is the free stream velocity of Blasius basic flow, δ is the upstream thickness of the boundary layer in Blasius basic flow, and v is the kinematics viscosity. The velocity of Blasius basic flow $\vec{u}_0 = (u_0, v_0)$ can be obtained through the Falkner–Skan equation, $f''' + ff'' = 0$.

2.2. Numerical Methods

The direct numerical simulations of Equations (1) and (2) were implemented as follows: a third-order mixed explicit-implicit scheme was employed for the time discretization, and the space discretization combined the higher-accuracy compact finite differences of non-uniform meshes with the Fourier spectral expansion. The nonlinear terms were approximated by a fifth-order upwind compact difference scheme for non-uniform meshes. The treat of pressure terms was approximated by a third-order center finite difference scheme with five points. The viscous terms were approximated by a fifth-order compact difference scheme for non-uniform meshes. Detailed numerical methods and verifications of simulation accuracy were given in Reference [11].

The computational time step was 0.01. Owing to the limitation of computational capacity, the range of directions x, y, z was limited to 84, 4, and 12, respectively. The number of Fourier modes was 16, which implies that the number of collocation points in the z-direction was 32, and the numbers of mesh points in the x- and y-directions were 240 and 150, respectively. Uniform and non-uniform meshes were applied in the x- and y-directions, respectively. The node coordinate $y(k)$ in the y-direction can be expressed by Equation (3) below, which is used to refine the mesh in the near-wall region.

$$\begin{aligned} y(\mathrm{k}) &= 4[1 - \tanh\tfrac{300-2k}{149}]/\tanh 2 - c_y(150-k)/150 \\ c_y &= 4(1 - \tanh\tfrac{300}{149})/\tanh 2 \end{aligned} \tag{3}$$

2.3. Boundary Layer

Boundary layer conditions were as follows:

Inflow boundary conditions, $x = 0$, $\vec{u}' = 0$, and $\partial p\prime/\partial x = 0$;
Outflow boundary conditions, $x = 84$, non-reflecting boundary condition, $\partial p\prime/\partial x = 0$;
Boundary conditions at the top wall, $y = 4$, $\partial \vec{u}'/\partial y = 0$, $p\prime = 0$;
Boundary condition at the bottom wall, $y = 0$, $\partial p\prime/\partial y = 0$, if $0 \le t \le 15$, $v\prime$ is shown above, $t > 15$, $\vec{u}' = 0$.

To verify the independence of results on mesh and time-step sizes, the numbers of mesh points in the x- and y-directions were raised to 480 and 200, respectively, and the computational time step was reduced to 0.005. The maximum absolute values of the velocity of the vortex structures $u\prime$, $v\prime$, and $w\prime$ with two kinds of meshes and time steps were compared, as shown in Figure 2. Lines represent the simulation results of the grid and time step used in this paper, and circles represent the simulation results of the raised grid and reduced time step, $t < 25$; there is almost no difference in the comparison of simulation results. Because the simulation efficiency was too low when the number of grids was increased and the time step was reduced, a total of more than one million grids were used in this paper, and the time step was set as 0.01.

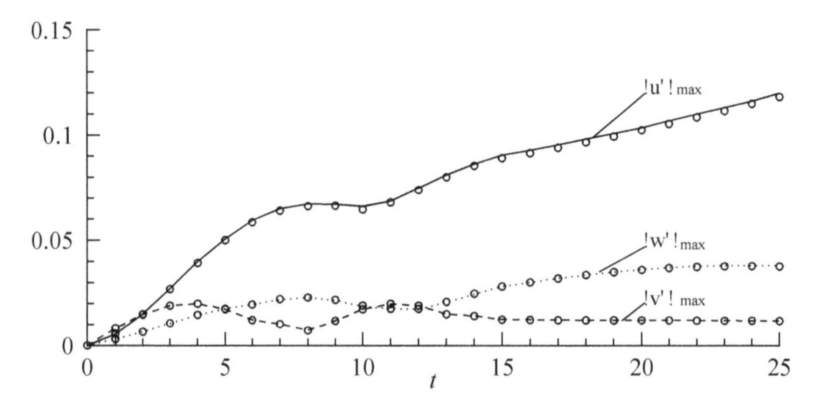

Figure 2. Mesh and time-step independence verification.

3. Numerical Results and Analyses

3.1. Disturbance Amplitude of Vortex Structures

Figure 2 shows the evolution of the maximum disturbance amplitude (A) of the vortex structures as they originated from the P–N model and N–P model and propagated downstream in the boundary layer. The maximum disturbance amplitude (A) is defined as

$$A = \sqrt{|u\prime|^2_{\max} + |v\prime|^2_{\max} + |w\prime|^2_{\max}}. \tag{4}$$

At $t = 15$, when the wall local forced vibrations stop, the maximum amplitude of vortex structures derived by the P–N and N–P models was almost 0.1, which is obviously much greater than the initial forced vibration amplitude of 0.02. For $t < 65$, the maximum amplitude of vortex structures derived by the N–P model showed almost no difference compared to that of the P–N model; in this range, the maximum amplitude of the two models gradually rose from 0 to about 0.35. The maximum amplitude of the N–P model was almost unchanged from $t = 65$ to $t = 80$, and showed rapid growth for $t > 80$.

On the other hand, however, the maximum amplitude of v using the N–P model obviously showed a rapid growth for $t > 65$.

It can also be seen from Figure 3 that the largest contribution to the maximum amplitude was that of $|u\prime|_{max}$, with $u\prime$ being the stream-wise velocity, while the second and third contributions were those of $|w\prime|_{max}$ and $|v\prime|_{max}$, respectively. In addition, as $|u\prime|_{max} > 0.6$, both $|w\prime|_{max}$ and $|v\prime|_{max}$ increased gradually.

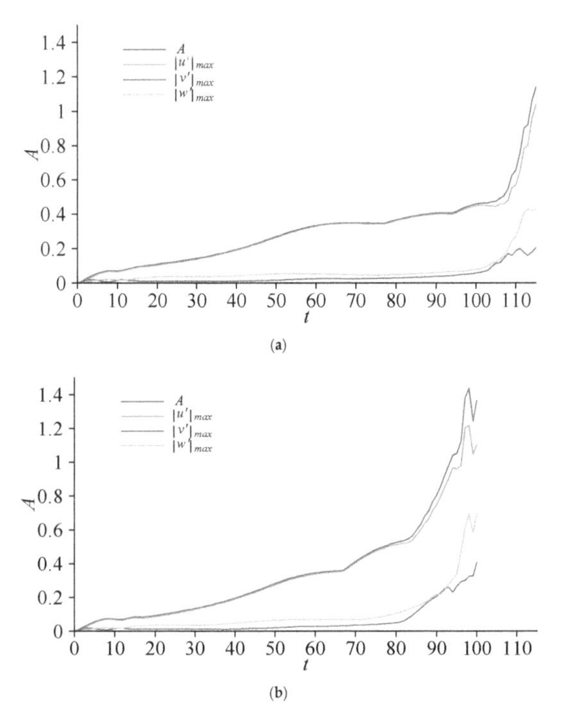

Figure 3. Maximum disturbance amplitude: (**a**) positive–negative (P–N) model; (**b**) negative–positive (N–P) model.

In order to study the reasons for the rapid growth of velocity disturbance in the stream-wise direction, the governing equation for the velocity and pressure disturbances of vortex structures in the x-direction was written in the following form:

$$\begin{aligned}\frac{\partial u\prime}{\partial t} &= \left(-\frac{\partial p\prime}{\partial x}\right) + \left(-u\prime\frac{\partial u\prime}{\partial x} - U_0\frac{\partial u\prime}{\partial x} - u\prime\frac{\partial U_0}{\partial x}\right) + \left(-v\prime\frac{\partial u\prime}{\partial y} - V_0\frac{\partial u\prime}{\partial y} - v\prime\frac{\partial U_0}{\partial y}\right) \\ &+ \left(-w\prime\frac{\partial u\prime}{\partial z}\right) + \frac{1}{\mathrm{Re}}\left(\frac{\partial^2 u\prime}{\partial x^2} + \frac{\partial^2 u\prime}{\partial y^2} + \frac{\partial^2 u\prime}{\partial z^2}\right)\end{aligned} \tag{5}$$

where $u\prime, v\prime, w\prime$ are the components of the velocity of the vortex structures, $p\prime$ is the pressure disturbance, and u_0 and v_0 are the stream-wise and vertical velocity components of the Blasius solution. The various terms appearing in Equation (5) can be grouped in the following form, where b, c, and d are all nonlinear terms, $c = c1 + c2 + c3$, and e is the viscous term:

$$\begin{cases} a = u\prime \times 0.1 \\ b = (-u\prime \frac{\partial u\prime}{\partial x} - u_0 \frac{\partial u\prime}{\partial x} - u\prime \frac{\partial u_0}{\partial x}) \\ c = (-v\prime \frac{\partial u\prime}{\partial y} - v_0 \frac{\partial u\prime}{\partial y} - v\prime \frac{\partial u_0}{\partial y}) \\ c1 = -v\prime \frac{\partial u\prime}{\partial y}, c2 = -v_0 \frac{\partial u\prime}{\partial y}, c3 = -v\prime \frac{\partial u_0}{\partial y} \\ d = (-w\prime \frac{\partial u\prime}{\partial z}) \\ e = \frac{1}{\mathrm{Re}} (\frac{\partial^2 u\prime}{\partial x^2} + \frac{\partial^2 u\prime}{\partial y^2} + \frac{\partial^2 u\prime}{\partial z^2}) \end{cases}. \tag{6}$$

The evolution of grouped terms in time is shown in Figure 4. In particular, Figure 4a shows the evolution trends of a, b, c, d, and e at the position of maximum stream-wise velocity disturbance with $u\prime > 0$ for the P–N model; the characteristics of a were similar to those shown in Figure 3a. Compared with the viscous term e, the nonlinear terms including b, c, and d contributed to the acceleration of the growth of $u\prime$, while term b contributed a little. For $t < 55$, it can be seen that c was the most significant term for the disturbance amplitude growth.

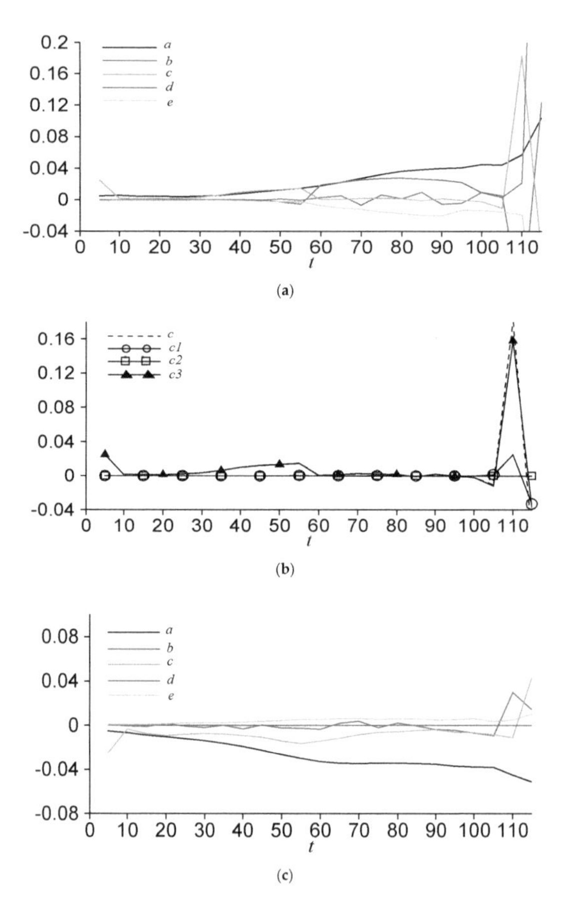

Figure 4. *Cont.*

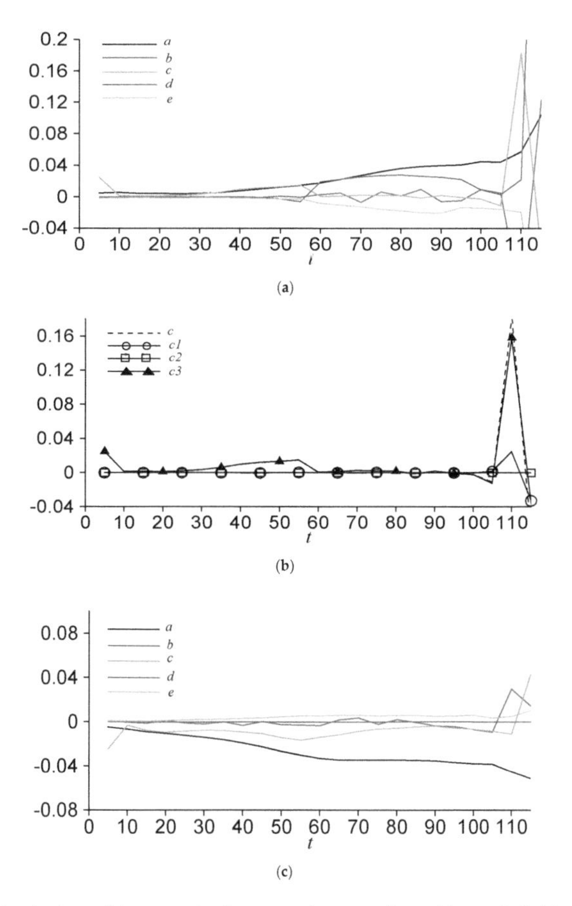

Figure 4. Evolution in time of the terms in the governing equations: (**a**) $u\prime > 0$, P–N model; (**b**) $u\prime > 0$, P–N model; (**c**) $u\prime < 0$, P–N model; (**d**) $u\prime < 0$, P–N model; (**e**) $u\prime > 0$, N–P model; (**f**) $u\prime < 0$, N–P model.

Figure 4b shows the evolution trends of c, $c1$, $c2$, and $c3$, whereby c was almost equal to $c3$, as $c3$ is the product of $-v\prime$ and the mean shear rate of boundary layer $\partial u_0/\partial y$. Since $\partial u_0/\partial y > 0$, vertical velocity disturbance at the position of maximum stream-wise velocity disturbance was of course negative. However, for $t > 55$, term d becomes important for the disturbance amplitude growth; the span-wise disturbance velocity $w\prime$ and the derivative term of the stream-wise disturbance velocity $\partial u\prime/\partial z$ increase significantly.

Figure 4c shows the evolution trends of a, b, c, d, and e at the position of maximum stream-wise disturbance velocity with $u\prime < 0$ for the P–N model. $|u\prime|_{max}$ was less than that in Figure 3a, which means that the strength of the stream-wise high-speed disturbance velocity was greater than that of the stream-wise low-speed disturbance velocity. Nonlinear terms including b, c, and d can promote further reduction of the stream-wise disturbance velocity, while viscous term e can prevent it. It can be seen that c was the most significant term for the disturbance amplitude growth, while term d seemed unimportant for $|u\prime|_{max}$ with $u\prime < 0$. Figure 4d shows the evolution trends of c, $c1$, $c2$, and $c3$, whereby $c3$ was almost equal to c, and, due to $c3 < 0$, vertical disturbance velocity at the position of maximum stream-wise disturbance velocity was of course positive.

Figure 4e shows the evolution trends of a, b, c, d, and e at the position of maximum stream-wise disturbance velocity with $u\prime > 0$ for the N–P model. For $t < 50$, it can be seen that c was the most significant term for the disturbance amplitude growth. Although $|e|$ in Figure 4e was greater than that in Figure 4a, term d in Figure 4e rose very fast for $t > 50$, which was the main reason for the growth of the vortex structures in the N–P model. It can be inferred that terms $w\prime$ and $\partial u\prime/\partial z$ of the vortex structures of the N–P model were relatively greater than those of the P–N model.

Compared to Figure 4c, the stream-wise low-speed disturbance velocity strength of the N–P model shown in Figure 4f was greater than that of the P–N model. Term c was the key factor for the growth of low-speed disturbance velocity. In contrast, terms c and d were the only important factors for the growth of high-speed disturbance velocity.

3.2. Stripe of Vortex Structures

Figure 5a provides the velocity vector of the vortex structures for the P–N model at $t = 15$, when the local micro-vibration at the wall stops. Because the disturbance at the wall was symmetrical along the plane $z = 0$, only the results of $0 < z < 6$ are shown. It can be seen that the initial form of the vortex structures was complex, whereby the vortex structures mainly concentrated near $z = 0$, $x = 15$, and showed complex three-dimensional (3D) vortices in space. The core region of disturbance velocity was at the position of $x > 15$ due to the influence of basic flow in the laminar boundary layer.

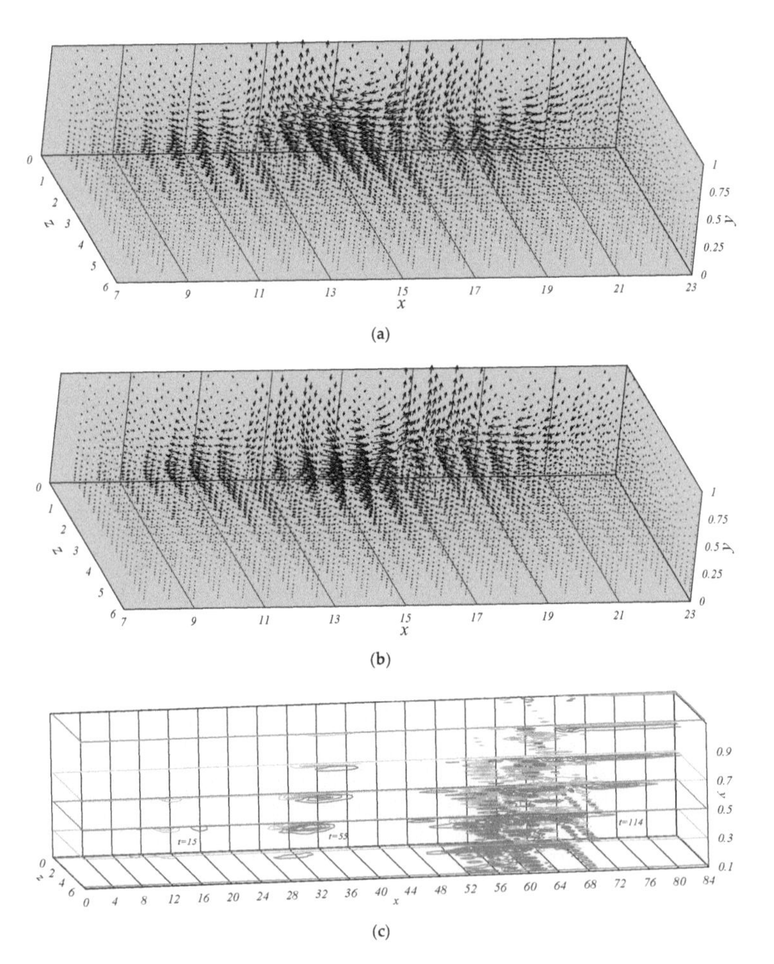

Figure 5. *Cont.*

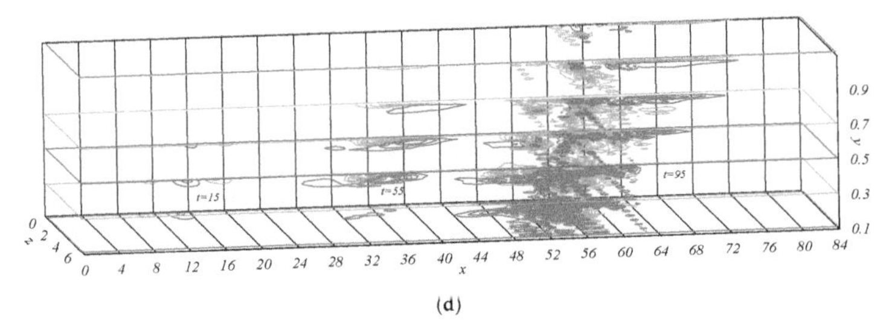

(d)

Figure 5. Vectors and contours of vortex structures: (**a**) velocity vectors at t = 15, P–N model; (**b**) velocity vectors at t = 15, N–P model; (**c**) stream-wise high- and low-speed disturbance velocity, contour increment of 0.03, P–N model; (**d**) stream-wise high- and low-speed disturbance velocity, contour increment of 0.03, N–P model.

Figure 5b provides the velocity vector of the vortex structures at t = 15 of the N–P model. The contrast between Figure 4a,b shows the opposite nature of the vortices; however, in fact, there were differences in the amplitudes of the velocities. For example, at the position of x = 18.2, y = 0.309, z = 0, the velocities of the P–N model and N–P model are compared in Table 1.

Table 1. Velocity comparison. P–N—positive–negative; N–P—negative–positive.

Model	u'	v'	w'
P–N	$2.260294610168348 \times 10^{-2}$	$-5.310678689829363 \times 10^{-3}$	$5.266260498348791 \times 10^{-18}$
N–P	$-6.111656282953886 \times 10^{-2}$	$7.780919736954207 \times 10^{-3}$	$-2.068746728425687 \times 10^{-18}$

Apart from the different spatial evolution in time, the stream-wise high-speed and low-speed disturbance velocities of different vortex structures exhibited different properties. The vortex structure amplitude of the P–N model increased to about 1.0 at t = 114, and that of the N–P model increased to about 1.0 at t = 95. Figure 5c provides the stream-wise disturbance velocity distributions of high-speed and low-speed fluids of the P–N model at t = 15, 55, and 114, at y = 0.1, 0.3, 0.5, 0.7, 0.9, and 1.1. Because the vortex disturbances were symmetrically distributed along plane z = 0, only the contour between planes of z = 0 and z = 6 is displayed; the contour increment is 0.03, and red lines represent the disturbance velocity of the high-speed fluid, $u\prime > 0$. Green lines represent the disturbance velocity of the low-speed fluid, $u\prime < 0$. At t = 15, high-speed and low-speed fluids mainly distributed in a local area near x =16–20, y = 0.3, z = 0, and the intensity of the low-speed fluid was greater than that of the high-speed fluid, with the former being below the latter. With the evolution of the vortex structure, at t = 55, the intensity and area of both the high-speed and low-speed fluids increased, but it seems that the area of the high-speed fluid was larger than that of the low-speed fluid. The low-speed fluid concentrated near the plane of z = 0, and the high-speed fluid existed in relatively large areas. At t = 144, the vortex structure was further inclined due to the shear action of the basic flow in the laminar boundary layer. At the same time, it can be seen that the high-speed fluid mainly concentrated near the wall, which may have led to an increase in the friction shear force at the wall region. The area occupied by the high-speed fluid was larger than that occupied by the low-speed fluid. The spatial range of the high-speed fluid increased more than that of the low-speed fluid in all directions.

Figure 5d provides the stream-wise disturbance velocity distributions of the high-speed and low-speed fluids of the N–P model at t = 15, 55, and 114, at y = 0.1, 0.3, 0.5, 0.7, 0.9, and 1.1. Compared with Figure 4c, at t = 15, although the high-speed and low-speed fluids mainly distributed in a small area near x = 16–20, y = 0.3, z = 0, the intensity of the high-speed flow was slightly greater than that of the low-speed flow, while the high-speed fluid was lower than the low-speed fluid. The low-speed fluid main distributed at the location near y = 0.5 and x = 20, and there was no high-speed fluid at the same position in Figure 5c. At $t = 55$, the high-speed fluid exceeded the low-speed fluid, and the elongation range of the high-speed fluid in the stream-wise direction was larger than that in Figure 4c. At $t = 95$, the amplitude of vortex structures rose rapidly to nearly 1.0, and the scales of the high-speed fluid and low-speed fluid were smaller than those in Figure 5c in the stream-wise direction; however, there were mainly high-speed stripes near the wall, and the characteristics of the high-speed fluid occupied a larger area than the low-speed fluid, similar to Figure 5c. Results of the vortex structure stripe agree with the results of Wall et al. [12].

The rough center positions of the high-speed and low-speed fluids in the plane of $y = 0.3$ of the P–N and N–P models are listed in Table 2. The forward speed in the stream-wise direction of the P–N model was approximately $0.45U_\infty$, and that of the N–P model was approximately $0.525U_\infty$.

Table 2. Center position evolution with time.

Model		x	
P–N	16 ($t = 15$)	34 ($t = 55$)	60 ($t = 114$)
N–P	16 ($t = 15$)	38 ($t = 55$)	58 ($t = 95$)

3.3. Mean Flow Profile and Neutral Curve

The black dashed lines in Figure 6a,b represent the velocity u_0 of Blasius basic flow at $x = 64$ for the P–N model and at $x = 56$ for the N–P model. The red solid lines in Figure 6a,b represent the mean value of the stream-wise disturbance velocity within a local region adding to Blasius basic flow; the local region was $52 < x < 76$, $-6 < z < 6$ for the P–N model and $44 < x < 68$, $-6 < z < 6$ for the N–P model. As can be seen, the mean velocity profile deformations in the area of the vortex structures could easily be discerned. On one hand, because of the presence of stream-wise disturbance velocity, the shear stress close to the wall increased. On the other hand, under the conditions of different initial disturbances at the walls in the laminar boundary flow, the mean velocity profiles all became plump after a period of evolution. Although vortex structures were only in their initial stages, the velocity profiles had a tendency to evolve into the turbulent mean velocity profiles. Furthermore, the mean velocity profiles were plumper at $y < 0.3$, which indicates that the stream-wise high-speed disturbance velocity mainly distributed near the wall region.

The green solid lines in Figure 6a,b represent the mean value of the stream-wise disturbance velocity within a local scope adding to Blasius basic flow; the local region was $52 < x < 76$, $-3 < z < 3$ for the P–N model and $44 < x < 68$, $-3 < z < 3$ for the N–P model. The black solid lines in Figure 6a,b represent the mean value of the stream-wise disturbance velocity within a local region adding to Blasius basic flow; the local region was $52 < x < 76$, $-1.5 < z < 1.5$ for the P–N model and $44 < x < 68$, $-1.5 < z < 1.5$ for the N–P model. It can be seen from Figure 6a,b that the mean velocity profiles had inflection points in the local scope of $-1.5 < z < 1.5$; the flow stability in this scope would, thus, be altered. According to the theory of linear stability, the solution of a three-dimensional T–S wave is obtained by solving the eigenvalue problem of Orr–Sommerfeld equations.

$$\vec{u}\prime = a_0\left[\vec{u}(y))\right]e^{i(\alpha x+\beta z-\omega t)} + c.c, \tag{7}$$

where $c \cdot c$ is the conjugate complex, $\alpha = \alpha_r + i\alpha_i$ is the stream-wise wave number, α_r is the real part, α_i is the imaginary part, β is the span-wise wave number, ω is the frequency, a_0 is the initial amplitude, $\vec{u}(y) = \{u(y), v(y), w(y)\}$ is the eigenvalue velocity, and $\alpha_i = 0$ represents the points on the neutral curve. For each span-wise wave number β, the corresponding unstable T–S wave with maximum local growth rate $-\alpha_i$ exists.

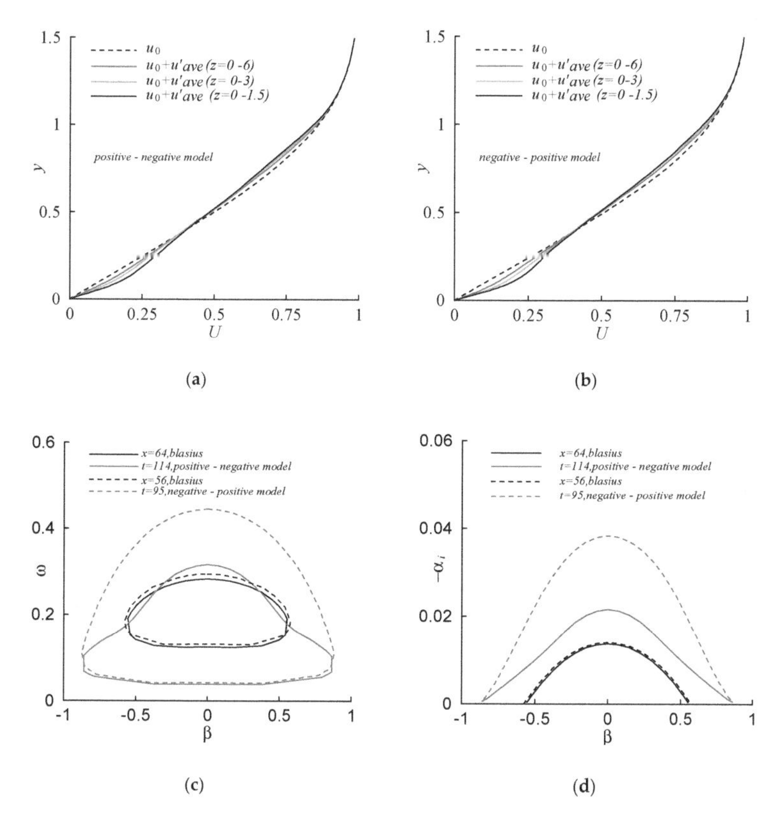

Figure 6. Mean stream-wise velocity and flow stability: (**a**) t = 114; (**b**) t = 95; (**c**) neutral curve; (**d**) growth rate.

T–S waves inside the neutral curve are unstable. The black solid line in Figure 6c is the neutral curve at x = 64, and the black dotted line is the neutral curve at x = 56 in the Blasius boundary layer. The frequency of the neutral curve at x = 64 was less than that at x = 56. The red solid line in Figure 6c is the neutral curve based on the mean value of the stream-wise disturbance velocity added to the Blasius basic flow within the local region $52 < x < 76$, $-1.5 < z < 1.5$ of the P–N model. The range of the neutral curve (red solid line) was obviously larger than that of the black solid line. The red dashed line in Figure 6c is the neutral curve based on the mean value of the stream-wise disturbance velocity added to the Blasius basic flow within the local region $44 < x < 68$, $-1.5 < z < 1.5$ of the N–P model. The range of the neutral curve (red dashed line) was obviously larger than that of the black dashed line. Due to the existence of the vortex structures, the neutral curve range of the N–P mode was relatively the largest.

The black solid and dashed lines in Figure 6d correspond to the maximum local growth rates $-\alpha_i$ at x = 64 and x = 56 in the Blasius boundary layer. The red solid line in Figure 6d is the maximum local growth rate $-\alpha_i$ based on the mean value of the stream-wise disturbance velocity added to the Blasius basic flow within the local region $52 < x < 76$, $-1.5 < z < 1.5$ of the P–N model. The amplitude of the maximum local growth rate $-\alpha_i$ of the red solid line was obviously larger than that of the black solid line. The red dashed line in Figure 6d is the maximum local growth rate $-\alpha_i$ based on the mean value of the stream-wise disturbance velocity added to the Blasius basic flow within the local region $44 < x < 68$, $-1.5 < z < 1.5$ of the N–P model. Also, the amplitude of the maximum local growth rate $-\alpha_i$ of the red dashed line was larger than that of the black dashed line. The amplitude of the maximum local growth rate $-\alpha_i$ of the N–P model was relatively the largest. Growth rates of the T–S waves and profile characteristics of the mean flow had the ability to promote each other. The self-sustaining structures in the logarithmic region of the boundary agree with the results obtained by Yang [11].

3.4. Stream-Wise Vortices

Figure 7a shows the distribution of stream-wise vortices ϖ_x at different times of t = 15, 55, and 114 of the P–N model. Figure 7b shows the distribution of stream-wise vortices ϖ_x at different times of t = 15, 55, and 114 of the N–P model. Contours in three section planes at each time are given, and the contour increment was 0.1; the red color denotes $\varpi_x > 0$, while the green color denotes $\varpi_x < 0$.

$$\varpi_x = \partial w\prime/\partial y - \partial v\prime/\partial z. \tag{8}$$

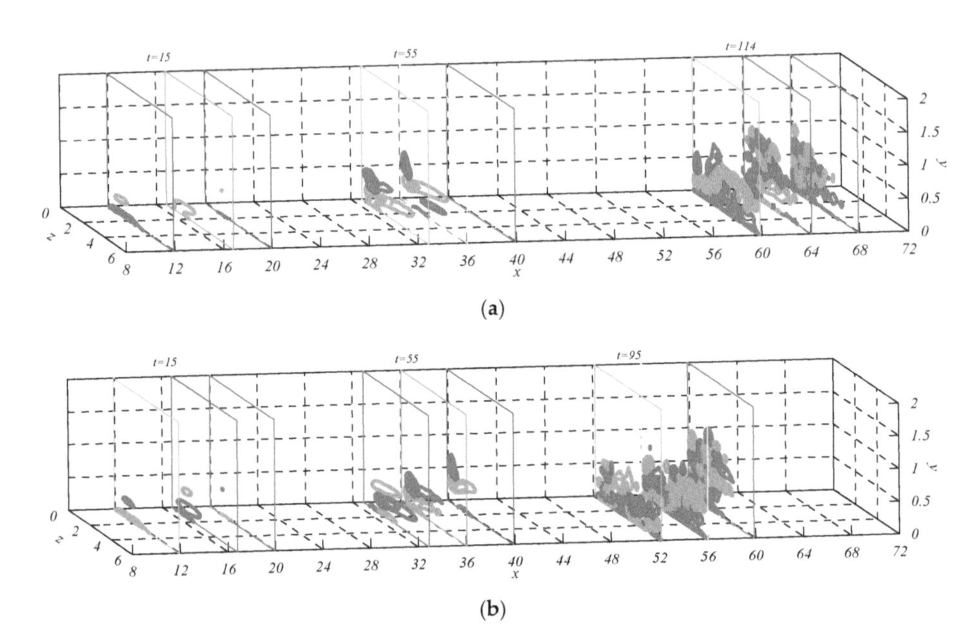

Figure 7. Stream-wise vortices: (**a**) P–N model; (**b**) N–P model.

The intensity of stream-wise vortices became stronger and stronger with the evolution of vortex structures. Furthermore, different scales of stream-wise vortices came into being, the centers of vortices went up, and the influencing areas gradually diffused, followed by strong ejection and sweeping at the centers of these vortices. In addition, the sizes of stream-wise vortices increased continuously. According to Biot–Savart theory, the vortex tube is lengthened in the stream-wise direction, which amplifies the vortex intensity. At last, both stream-wise and span-wise velocities grow quickly, followed by dramatic ejection and sweeping, the formation of strong shear layers locally, and the flow becoming unstable. However, the main affected area is in the vicinity of the wall, which is the region near the wall where hydrodynamic instability occurs actively. Results of the vortex distribution agree with the research results of Shinneeb et al. [12].

4. Conclusions

A direct numerical simulation of the nonlinear evolution of vortex structures in a laminar boundary layer induced by different local micro-vibrations on the wall was carried out. The characteristics of vortex structures were studied in detail, and the conclusions are summarized below.

With the evolution of time, the amplitude of vortex structures undergoes a process of linear increase and nonlinear rapid increase, and the amplitude of the stream-wise velocity disturbance is the relative maximum. The growth rate of vortex structures in the P–N model is slightly less than that in the N–P model, and the growth of the P–N model vortex structure is most affected by the vertical velocity and boundary layer shear. The growth of vortex structures in the N–P model is greatly influenced by the derivative term of the stream-wise disturbance velocity and span-wise disturbance velocity. In the core region of vortex structures, the inflection point of the velocity profile exists, which leads to the expansion of the range of neutral curves, as well as the increased growth rate of T–S waves. The change in mean flow profile further induces or promotes the growth or formation of vortex structures. These results will help to understand and further study how coherent structures in turbulence come into being and evolve.

Author Contributions: W.C. designed the scheme and wrote the program, organized paper; Z.J. helped to analyze the data; Q.Z. contributed some figures.

Funding: This work was supported by the National Key R&D Program of China (2018YFC0810506) and the Key R&D Program of Zhenjiang (SH2017049, BK20161472).

Conflicts of Interest: The author(s) declared no potential conflicts of interest with respect to the research, authorship, and/or publication of this article.

References

1. Lee, C.B.; Wu, J.Z. Transition in wall-bounded flows. *Appl. Mech. Rev.* **2008**, *61*, 030802. [CrossRef]
2. Sharma, A.S.; Mckeon, B.J. On coherent structure in wall turbulence. *J. Fluid Mech.* **2013**, *728*, 196–238. [CrossRef]
3. Wedin, H.; Bottaro, A.; Hanifi, A.; Zampogna, G. Unstable flow structures in the Blasius boundary layer. *Eur. Physi. J. E* **2014**, *37*, 34. [CrossRef] [PubMed]
4. Wu, X.S.; Zhuang, X.L. Nonlinear dynamics of large-scale coherent structures in turbulent free shear layers. *J. Fluid Mech.* **2015**, *787*, 396–439. [CrossRef]
5. Mcmullan, W.A.; Garretty, S.J. Streamwise vortices in plane mixing layers originating from laminar or turbulent initial conditions. In Proceedings of the 53rd AIAA Aerospace Sciences Meeting, Kissimmee, FL, USA, 5 January 2015.
6. Wall, D.P.; Nagata, M. Exact coherent states in channel flow. *J. Fluid Mech.* **2016**, *788*, 444–468. [CrossRef]
7. Kang, K.L.; Yeo, K.S. Hybrid POD-FFT analysis of nonlinear evolving coherent structures of DNS wave packet in laminar-turbulent transition. *Phys. Fluids* **2017**, *29*, 084105. [CrossRef]
8. Shinneeb, A.M.; Balachandar, R.; Zouhri, K. Effect of gap flow on the shallow wake of a sharp-edged bluff body-Coherent structures. *Phys. Fluids* **2018**, *30*, 065107. [CrossRef]
9. Lemarechal, J.; Klein, C.; Henne, U.; Puckert, D.K.; Rist, U. Detection of Lambda and Omega vortices with the temperature-sensitive paint method in the late stage of controlled laminar-turbulent transition. *Exp. Fluids* **2019**, *60*, 91. [CrossRef]
10. Puckert, D.K.; Rist, U. Experiments on critical Reynolds number and global instability in roughness-induced laminar-turbulent transition. *J. Fluid Mech.* **2018**, *844*, 878–904. [CrossRef]
11. Lu, C.G.; Cao, W.D.; Qian, J.H. A study on numerical method of Navier–Stokes equations and nonlinear-evolution of the coherent structure in a laminar boundary layer. *J. Hydrodyn. Ser. B* **2006**, *18*, 110–116. [CrossRef]
12. Yang, Q.; Willis, A.P.; Hwang, Y. Exact coherent states of attached eddies in channel flow. *J. Fluid Mech.* **2019**, *862*, 1029–1059. [CrossRef]

Article

Near-Wall Flow Characteristics of a Centrifugal Impeller with Low Specific Speed

Weidong Cao [1,2,*], Zhixiang Jia [1,2] and Qiqi Zhang [1,2]

1 Research Institute of Fluid Engineering Equipment Technology, Jiangsu University, Zhenjiang 212013, China
2 China National Research Center of Pumps, Jiangsu University, Zhenjiang 212013, China
* Correspondence: cwd@ujs.edu.cn; Tel.: +86 1395-281-6468

Received: 14 July 2019; Accepted: 1 August 2019; Published: 5 August 2019

Abstract: In order to study the near-wall region flow characteristics in a low-specific-speed centrifugal impeller, based on ANSYS-CFX 15.0 software, Reynolds averaged Navier-Stokes (RANS) methods and renormalization group (RNG) k-ε turbulence model were used to simulate the whole flow field of a low specific speed centrifugal pump with five blades under different flow rates. Simulation results of external characteristics of the pump were in good agreement with experimental results. Profiles were set on the pressure side and suction side of impeller blades at the distances of 0.5 mm and 2 mm, respectively, to study the distributions of flow characteristics near the wall region of five groups of blades. The results show that the near-wall region flow characteristics of five groups of blades were similar, but the static pressure, relative velocity, cross flow velocity, and turbulent kinetic energy of profiles on the pressure side were quite different to those on the suction sides, and these characteristics also changed with the alternation of flow rate. As the flow rate was 13 m^3/h or 20 m^3/h, within the radius range of 40 to 50 mm, there was an extent of negative relative velocity of the profiles on the pressure side, and a counter-current happened not on the suction side, but on the pressure side in the low specific speed centrifugal impeller. The flow characteristics of profiles at the distances of 0.5 mm and 2 mm also showed a small difference.

Keywords: low specific speed; centrifugal pump; impeller; near-wall region; flow characteristics

1. Introduction

Low-specific-speed centrifugal pumps have the characteristics of small flow rate, large head, and low efficiency, which are widely used in aerospace, military industry, electric power, water conservancy, chemical industry, and other important fields. After decades of efforts, with the continuous improvement of experimental means and the development of numerical calculation methods for the internal flow of turbo-machinery, more understanding and consensus have been gained on the internal flow characteristics and more achievements have been achieved in hydraulic design optimization methods. Micro internal flow structure controlling has been the method for improving the efficiency of centrifugal pumps with low specific speed.

Karrasik, a well-known pump expert in the United States, pointed out that since flow loss depends on the boundary layer, the efficiency of low-specific-speed centrifugal pumps can be improved by properly controlling the boundary layer of the flow channel [1]. Young-Do et al. studied the internal flow and performance of a centrifugal pump impeller with very low specific speed through an external characteristic test and particle image velocimetry (PIV) internal flow field test. The results show that there is a large recirculation at the outlet of a semi-open impeller, which significantly reduces the absolute tangential velocity and thus reduces the pump's head [2]. Westra et al. simulated the internal flow field of a low-specific-speed centrifugal pump, where Spalart and Allmaras' turbulence model was selected and the near-wall mesh scale was dense, and found that the maximum error of the relative velocity obtained by PIV was 6%. From the relative velocity distribution, it can be seen that, from the pressure side of the blade to the suction side, when the velocity gradually increases to a certain maximum, the relative velocity on the suction surface of the blade decelerates, and the jet-wake area decreases with the increasing of flow rate. With the spatial change of the flow separation point, the secondary flow structure and intensity also change continuously [3]. Pedersen et al. used PIV and laser Doppler vibrometer to measure the flow inside the impeller of a centrifugal pump. It was found that there were two channels in the impeller under the condition of a small flow rate: the flow in one channel was controlled by the impeller rotation, which was similar to that under the design flow rate, while in the other channel, there was a relatively static stall and the entrance area of the blade was blocked by a cluster, with edthe remaining part of the flow passage being occupied by some vortices [4]. Shao et al. used PIV technology to measure the velocity of the internal flow field of a centrifugal pump impeller with large blade outlet angle, there exists a low energy fluid wake region with relatively low velocity near the suction side of the blade, where there exists a jet region with relatively high velocity while near the pressure side. The flow structure is especially obvious under small flow rates. These flow structures are the result of the interaction between the fluid viscosity, boundary layer, and secondary flow [5]. Tsujita and co-workers' numerical simulation and experiments show that, if the boundary layer thickness of the suction side is larger than that of pressure side at the inflection of the streamline in the impeller inlet area of the centrifugal pump, it will be helpful to restrain the formation of vortices in the internal channel to reduce hydraulic loss [6]. Cui et al.'s numerical simulation of ultra-high speed and low specific speed centrifugal pumps with long and short blades shows that the reflux zone mainly concentrates on the inlet of the long blade on the suction side, the middle of the long blade on the pressure side, and the outlet of the short blade on the suction side; these refluxes have a great influence on the internal and external characteristics of pumps [7]. Limbach et al. simulated the cavitation flow of a low-specific-speed centrifugal pump under different flow rates and different surface roughness conditions, and for non-cavitating flow, the measured and calculated head are in good agreement. According to the measurement, for rough walls, net positive suction head rises slightly higher [8]. Cao et al. studied the influence of the impeller eccentricity on the performance of a low-specific-speed centrifugal pump, and found that with the impeller eccentricity increasing, the head and efficiency become lower and the area of low pressure at the suction side of the short blade inlet becomes gradually smaller; however, the area of low pressure at the suction side of the long blade inlet becomes larger, and the pressure at the impeller inlet and tongue become larger too [9]. Zhang et al. simulated the start–stop process of three kinds of fluids in a low specific speed centrifugal pump, and the effects of viscosity on transient performance, head-flow curve and internal flow structure in impeller and volute were studied. Results show that, the liquid with higher viscosity than water may reduce the operation reliability of low specific speed centrifugal pump during start-up period [10]. Chen et al. explored internal flow and its unsteady characteristics in a low-specific-speed centrifugal pump. There were vortices in various sizes and numbers in flow channels of the impeller under different flow conditions. A high velocity zone was found in two adjacent channels near the tongue. However, the zone disappeared gradually with the increase of flow rate [11]. Dong et. al. studied the effect of the front streamline wrapping angles variation of a low-specific-speed centrifugal impeller on energy performance, the pressure pulsation, interior and exterior noise characteristics, where the front sweep angle variation was found to have an insignificant influence on centrifugal pump performance characteristics; however, it influences fluid hydrodynamics around the volute tongue [12]. The flow field in a low specific speed impeller was reconstructed by Zhang et. al. [13], the root-mean-square error for pressure prediction was 0.84% and the velocity prediction error was within 0.5 m/s, its computation time for the flow field prediction was less than 1/240 of the computational fluid dynamics. Proper orthogonal decomposition (POD) base modal analysis was carried out on the sample set, and the base modal characteristic of the flow field and its energy distribution were analyzed.

The optimization design of low specific speed pumps and the study of large-scale flow characteristics have been carried out, but there are few studies on the flow characteristics near the blade wall. For this paper, ANSYS CFX 15.0 software used to simulate the unsteady flow of a low specific speed pump by analyzing the near-wall region flow on blades near the tongue and far from the tongue such that the distribution of viscous stress, relative velocity, and static pressure was obtained.

2. Geometric Model and Computational Grid

A low specific speed pump was designed with the following parameters: flow rate $Q = 13\ m^3/h$, head $H = 26$ m, rotational speed $n = 2880$ r/min, impeller inlet diameter $D_1 = 44$ mm, outlet diameter $D_2 = 146$ mm, outlet width $b_2 = 6$ mm, blade outlet angle $\beta_2 = 40°$, blade wrap angle $\phi = 120°$, blade thickness from inlet to outlet was 2 mm to 5 mm, blade number $z = 5$. The width of the volute was $B_3 = 18$ mm and the diameter of the volute outlet is $D_4 = 40$ mm. The geometric model is shown in Figure 1.

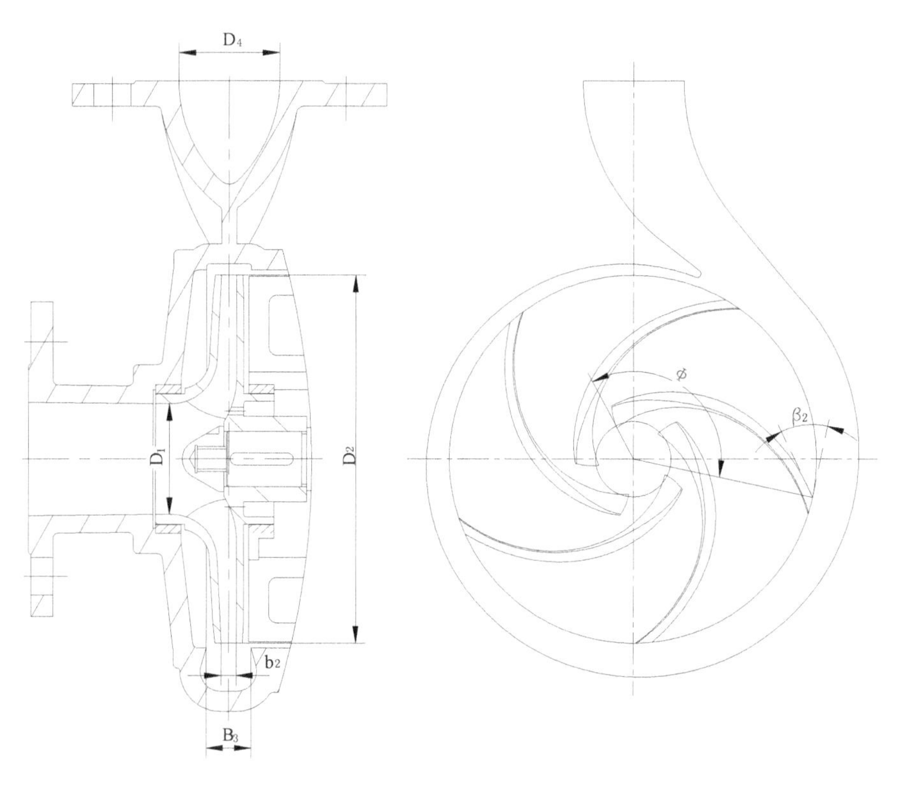

Figure 1. Geometric model.

2.1. Simulation Method

The pump was modeled using Creo 2.0 software. Figure 2 is the main water body model of its computational area, including the inlet section, impeller, volute shell, outlet extension section, etc. The leakage loss of seal ring and balance hole, friction loss of the disc at the front and rear cover plate of the impeller were taken into account when establishing the model. During the numerical simulation of turbulence flow in the low-specific-speed centrifugal pump, the Reynolds averaged Navier-Stokes (RANS) method was used, and the renormalization group (RNG) k-ε turbulence model considering the effects of separated flow and eddy flow was selected [14]. The turbulence model governing equations are as follows:

$$\frac{\partial}{\partial t}(\rho k) + \frac{\partial}{\partial x_i}(\rho k u_i) = \frac{\partial}{\partial x_j}[(\alpha_\varepsilon k \mu_{eff} \frac{\mu_t}{\sigma_k}) \frac{\partial k}{\partial x_j}] + G_k + G_b - \rho\varepsilon - Y_M + S_k \quad (1)$$

$$\frac{\partial}{\partial t}(\rho\varepsilon) + \frac{\partial}{\partial x_i}(\rho\varepsilon u_i) = \frac{\partial}{\partial x_j}[(\alpha_\varepsilon k \mu_{eff} \frac{\mu_t}{\sigma_k}) \frac{\partial \varepsilon}{\partial x_j}] + C_{1\varepsilon}\frac{\varepsilon}{k}(G_k + C_{3\varepsilon}G_b) - C_{2\varepsilon}\rho\frac{\varepsilon^2}{k} + S_\varepsilon \quad (2)$$

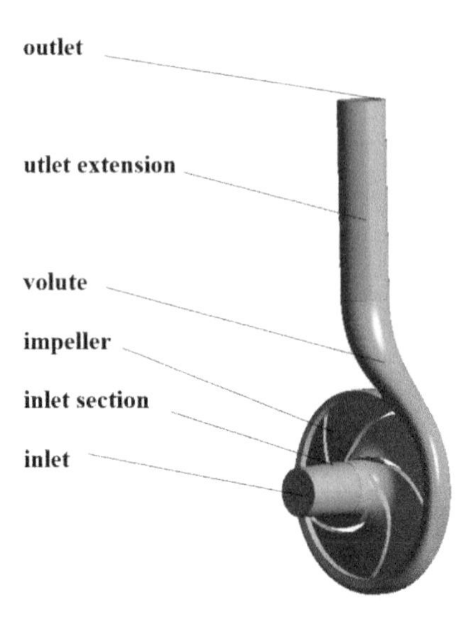

Figure 2. Water body model.

Compared with standard k-ε turbulence model, the rotation effect on α_k and α_ε has been considered in the above formula, and the other terms have the same meaning as the standard k-ε turbulence model.

The boundary condition at the outlet was chosen as the pump's flow rate, while the boundary condition at the inlet was chosen as the total pressure, and the reference pressure was set be 1.0 atm at the inlet. The walls formed by the impeller were defined as the rotating boundary and its rotating speed was 2880 r/min, the other walls were defined as the non-slip boundary, and the wall roughness was set to be 0.025 mm uniformly. Simulations were carried out using the commercial software ANSYS-CFX, the governing equations were discretized using the finite volume method, the pressure term was solved using the central difference scheme, the velocity term was solved using the second-order upwind difference scheme, the turbulent kinetic energy term and the turbulent energy dissipation rate term were solved using the second-order upwind difference scheme, and the near-wall flow was approximated using the standard wall function method. The convergence accuracy was set to 10^{-5}. In this paper, unstructured tetrahedral meshes were used to divide the computational water domain. The grid independence of numerical simulation was studied under a design flow rate, and it was found that when the number of grids was more than 3.9 million, the fluctuation of the head and hydraulic efficiency was small, and the relative fluctuation range was less than 1%; therefore, it can be considered that the number of grids had no effect on the calculation results.

2.2. Experimental Method and Result

The test bench mainly included a turbine flow-meter, valve, pressure transmitter, speed sensor, data acquisition instrument, etc. The valve at the outlet was used to control the flow rate, the pressure sensor was used to measure the inlet and outlet pressure of the pump, and the power of the pump was also measured and calculated. The head (H) and efficiency (η) comparison between the simulation and experiment is shown in Figure 3. It can be found that the simulation value was slightly higher than the experimental value, the maximum relative error of head was about 4%, and the maximum relative error of efficiency was about 2%. Because of the complexity of the internal flow in the low-specific-speed centrifugal pump, the existing turbulence models could not accurately adapt to simulate the complex effects of the surface curvature, Coriolis force, and centrifugal force. There were also some manufacturing errors in the model pump, and there were inevitable errors in the head and efficiency obtained from experiments and simulation; however, the trends under different flow rates was close.

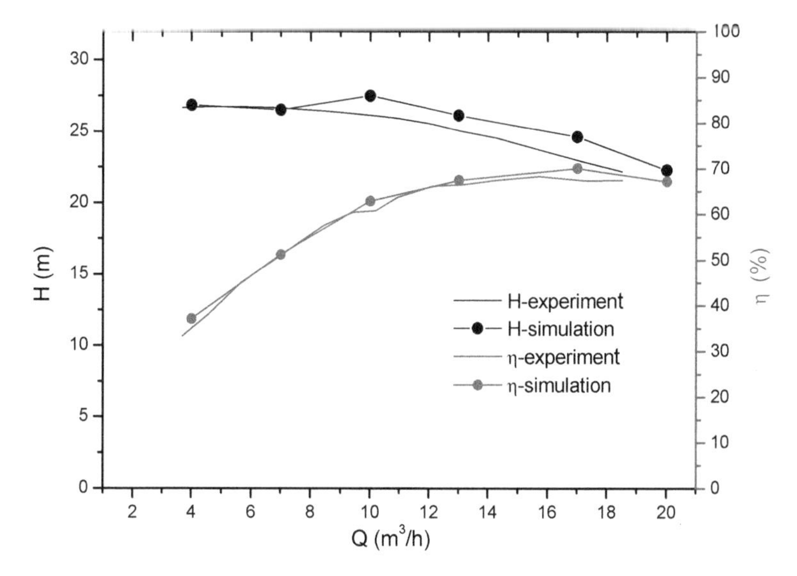

Figure 3. Comparison of simulation and experimental results.

3. Internal Flow Characteristic Analysis

In order to study the near-wall flow characteristics of the centrifugal pump's impeller with a low specific speed, the blades were numbered I, II, III, IV, and V; also, the flow channels between the blades were numbered 1, 2, 3, 4, and 5, respectively, as shown in Figure 4. The profiles of impeller blade I in the middle section with the distance of 0.5 mm and 2 mm are marked as lpI-0.5 and lpI-2 on the pressure side, and lsI-0.5 and lsI-2 on the suction side, respectively. The profiles of the impeller blade 2 in middle section are marked as lpII-0.5, lpII-2, lsI-0.5, and lsII-2 in accordance with the distance of 0.5 mm and 2 mm, as appropriate. Similar marking methods were used for the near-wall profiles of blades III, IV, and V.

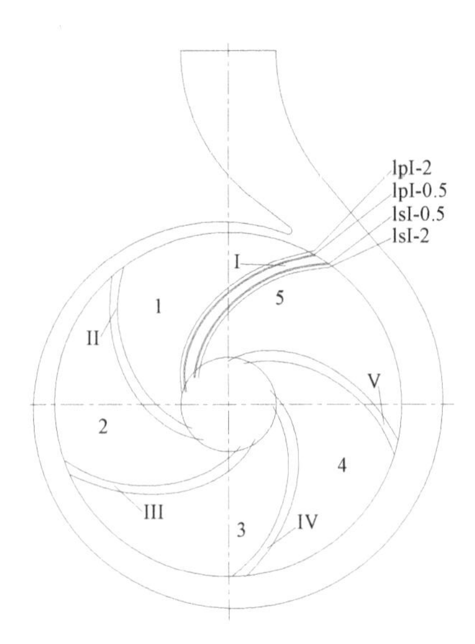

Figure 4. Near-wall profiles of blades.

3.1. *Whole Static Pressure and Relative Velocity*

From the static pressure distribution shown in Figure 5, it can be seen that, the static pressure increased gradually from the impeller inlet to the volute outlet, and the pressure distribution characteristics in the flow channels of 1, 2, 3, 4, and 5 were generally similar. The static pressure in the flow channels of Q = 7 m^3/h was almost the same as that of Q = 13 m^3/h. When increasing the flow rate to 20 m^3/h, the static pressure in the flow channels decreased obviously. The area of low pressure in channel 5 was the largest. The static pressure in the volute of Q = 7 m^3/h was larger than that of Q = 13 m^3/h and Q = 20 m^3/h, and the relative pressure at the wall of the volute was larger than that at the outlet of the impeller.

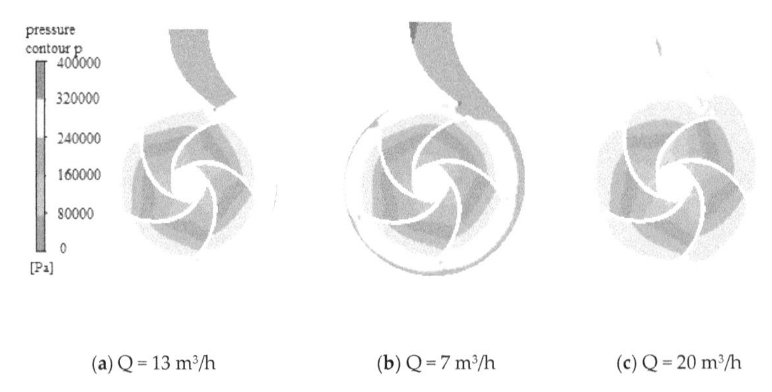

(**a**) Q = 13 m^3/h (**b**) Q = 7 m^3/h (**c**) Q = 20 m^3/h

Figure 5. Static pressure distributions under different flow rates.

From the relative velocity distributions shown in Figure 6, it can be seen that the relative velocity increased with the increasing of flow rate. For Q = 7 m^3/h, the relative velocity in the impeller blade channels and diffusive sections of volute shell was small, and the relative velocity from the tongue to the throat of the volute shell was large. For Q = 13 m^3/h, the relative velocity inside the impeller increased obviously and the relative velocity near the pressure sides of blades was obviously less than that near the suction sides of blades. When the flow rate increased to Q = 20 m^3/h, the relative velocity increased obviously, both inside the impeller and inside the volute; however, there was a local low-speed zone on the suction sides of blades.

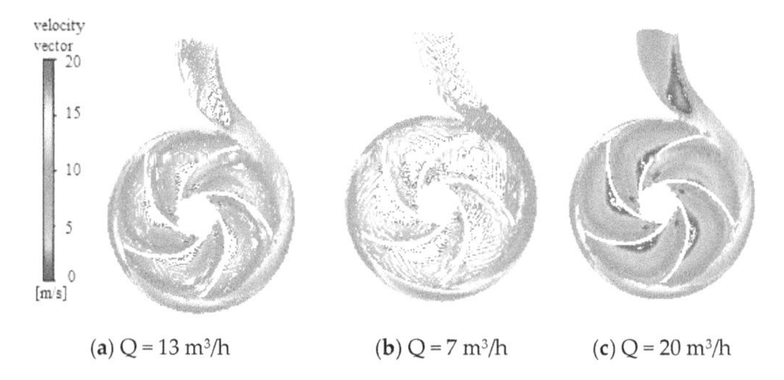

(**a**) Q = 13 m^3/h (**b**) Q = 7 m^3/h (**c**) Q = 20 m^3/h

Figure 6. Relative velocity distributions under different flow rates.

3.2. Pressure Distribution Near the Wall Region in the Impeller

Figure 7 shows the static pressure distributions of profiles at the distance of 0.5 mm from the blades with Q = 13 m^3/h. The pressure of the profiles at the distance of 0.5 mm from blade I is shown with a black solid line on the pressure side and with a black dotted line on the suction side, and the other pressure of profiles from blades II, III, IV, and V are shown with their respective colorful lines. The pressure distributions of near-wall profiles on pressure sides, as well as on the suction sides of five blades, were almost the same, and the pressure on the pressure sides was greater than that on the suction sides. The pressure on the pressure side at the outlet of blade I was slightly higher than that of other blades due to the affection of the tongue separation.

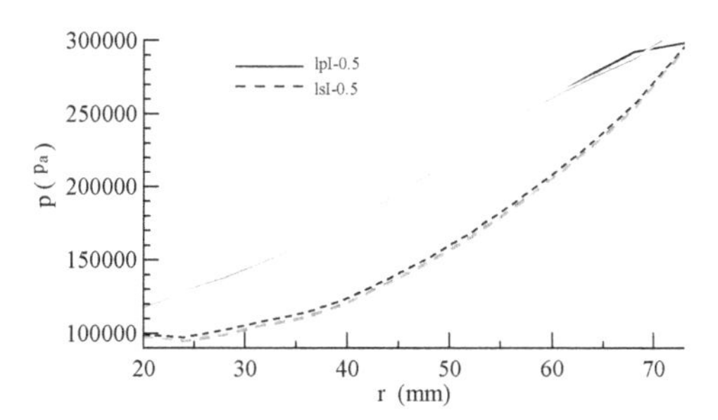

Figure 7. Static pressure distributions of profiles at the distance of 0.5 mm from the blade with Q = 13 m^3/h.

Figure 8 shows the static pressure distributions of the profiles near the wall under different flow rates on blade I, where the smaller the flow rate was, the greater the pressure on the pressure sides and the suction sides of the blades was. The pressure of the profiles at a distance of 2 mm on the pressure side was greater than that at a distance of 0.5 mm on the pressure side, but the pressure of the profiles at a distance of 2 mm on the suction side was less than that at a distance of 0.5 mm on the suction side. For Q = 7 m^3/h, 13 m^3/h, and 20 m^3/h, the pressure difference on the pressure sides between the radii of 20 mm to 60 mm was very little, and the pressure difference on the pressure sides between the radii of 60 mm to 73 mm became obvious. For Q = 7 m^3/h and 13 m^3/h, the pressure on the pressure sides of the near wall kept increasing. For Q = 20 m^3/h, the pressure even decreased slightly with increasing radius. The pressure on the suction sides near the wall generally decreased with the increase of flow rates, especially when the flow rate increased to 20 m^3/h, where the lowest pressure appeared at the radius of 25 mm. Compared with the pressure on the suction side at Q = 7 m^3/h, the pressure decreased to about 40,000 Pa, and the pressures at distances of 0.5 mm and 2 mm on the suction side interlaced, which shows the complexity of the flow.

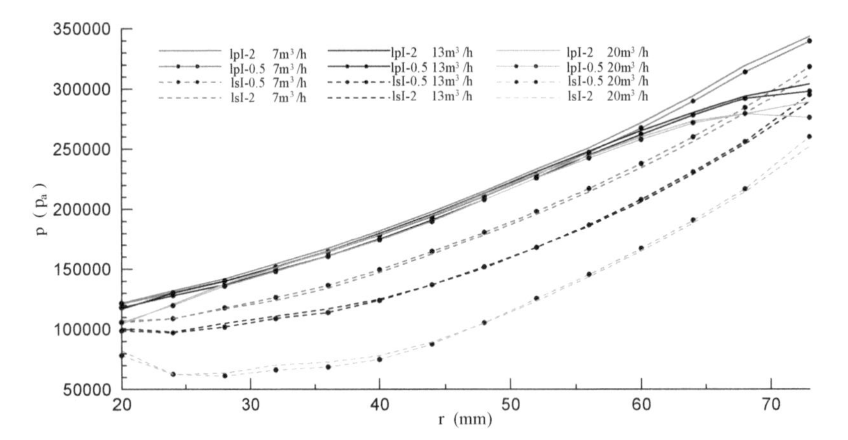

Figure 8. Static pressure distributions of profiles near wall region at different flow rates.

3.3. *Relative Velocity Distribution Near the Wall of Blade*

In order to analyze the near wall flow on the blade surfaces, the velocity vectors of the near-wall flow were projected onto the profiles of lpI-0.5, lpI-2, lsI-0.5, and lsI-2. The relative velocity was recorded as W and $W = (u, v) \cdot (\cos\alpha, \cos\beta)$, where (u, v) is the velocity vector near the wall and $(\cos\alpha, \cos\beta)$ the directional derivative of the profiles. The direction from the blade inlet to the outlet was positive, and if there was a negative value, then there was countercurrent near the wall.

Figure 9 shows the relative velocity distributions of profiles at the distance of 0.5 mm from blades with Q = 13 m^3/h, where the relative velocity of profiles at the distance of 0.5 mm from blade I is shown with a black solid line on the pressure side and with a black dotted line on the suction side, and the other relative velocity of profiles of blades II, III, IV, and V are shown with their respective colorful lines. The relative velocity distributions of the five blades were almost the same. The relative velocity on the pressure side was greater than that on the suction side in the range of radii from 20 mm to 68 mm, because the near-wall flow at the radius of 73 mm on the pressure side was the inner flow of the volute, the near-wall flow at the radius of 73 mm on the suction side was still in the blade channel 5, and the relative velocity on the pressure side was greater than that on the suction side. The relative velocity on the pressure side and suction side of blade I was slightly lower than that of other blades. Within the radius of 40 mm to 50 mm, the relative velocity on the pressure side appeared negative, which suggests the existence of a countercurrent. There was positive flow, deceleration flow, and acceleration flow, in turn, on the pressure sides. There was a maximum relative velocity in the range of radii from 30 mm to 40 mm in the inlet area on the suction side, and the relative velocity suddenly decreased in the range of radii from 40 mm to 50 mm.

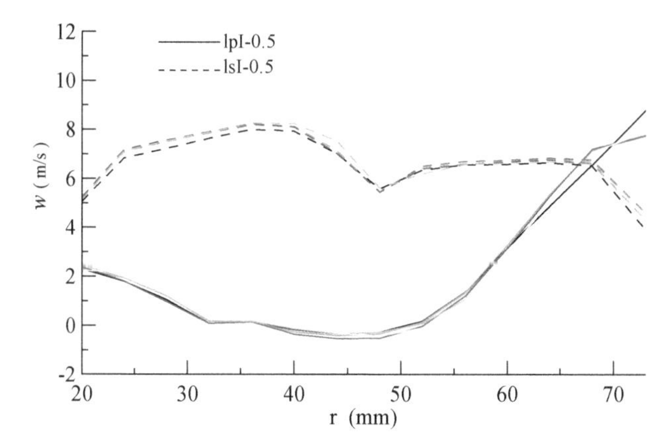

Figure 9. Relative velocity distributions of profiles at the distance of 0.5 mm with Q = 13 m^3/h.

Figure 10 shows the relative velocity distributions of profiles under different flow rates of blade I. Except at the outlet on the blade pressure side, the relative velocity on the suction side was larger than that on the pressure side in most areas. For Q = 7 m^3/h, the relative velocity on the pressure side was positive as a whole. In the range of radii from 20 mm to 35 mm, the relative velocity of profiles lpI-0.5 and lpI-2 on the pressure side had a certain difference. When the radius was greater than 35 mm, the velocity difference was very small. In the range of radii from 20 mm to 32 mm, the relative velocity on the suction side was obviously greater than that on the pressure side, and for the radius of 32 mm, the relative velocity on the pressure side dropped suddenly to below that on the suction side and then rose slowly.

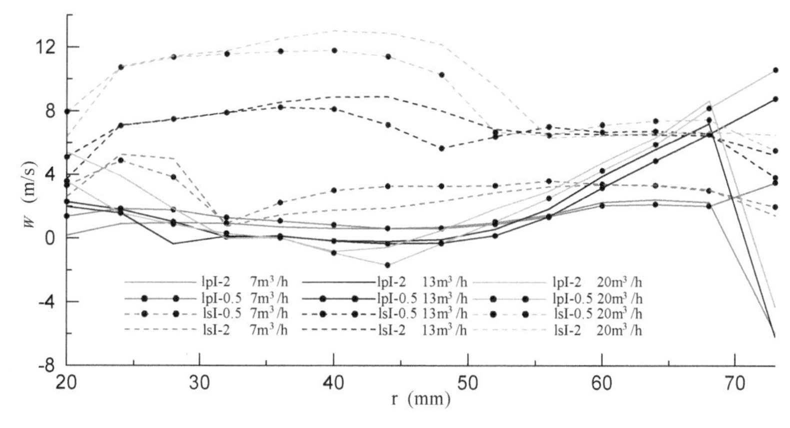

Figure 10. Relative velocity distributions of profiles with different flow rates.

For Q = 13 m^3/h, the relative velocity on the suction side and the pressure side showed an obvious difference, and the relative velocity on the suction side was generally positive. In the range of radii from 35 mm to 55 mm, the relative velocity of profiles lpI-0.5 and lpI-2 showed some differences. In the range of radii from 40 mm to 50 mm, there was an extent of negative relative velocity of the profiles on the pressure side where the countercurrent zone existed. When the radius was greater than 55 mm, the relative velocity on the pressure side rose rapidly, and in the area near the outlet of the blade, the relative velocity amplitude on the pressure side rose rapidly to the same order as that on the suction side.

The difference between the relative velocity on the suction side and the pressure side when Q = 20 m^3/h was larger than that when Q = 13 m^3/h, and the maximum relative velocity on the suction side was close to 13 m/s when Q = 20 m^3/h. Within the range of radii from 30 mm to 55 mm, there was a certain difference between the relative velocity of profiles lpI-0.5 and lpI-2 on the suction side. Within the range of radii from 40 mm to 50 mm, there was an extent of negative relative velocity of the profiles on the pressure side, and when the radius was greater than 55 mm, the relative velocity on the pressure side rose rapidly. In the region near the blade outlet, the relative velocity amplitude on the pressure side rose rapidly to even larger than that on the suction side. It was also found by Wang et al. that there is a clockwise relative flow vortex in the flow channel of a low specific speed impeller with counterclockwise rotation, thus a positive relative velocity forms on the suction surface of blades, and a negative relative velocity may form on the pressure surface of the blades, and the center of the vortex is about one third of the position from the inlet to the outlet of the blades [13].

3.4. Cross Flow Velocity Distribution

The flow near a blade's surface is three-dimensional, and the main flow should be along the derivative direction of the profiles mentioned above. If the cross-flow velocity W_z in the z-direction appears with a large value, it will interfere with the main flow greatly, and even large vortex fields and viscous loss form. Figure 11 shows the cross-flow velocity distributions of profiles near blade I under different flow rates. Compared with Figure 10, the amplitude of cross-flow velocity W_z was significantly smaller than the relative velocity W. The W_z of profiles on the suction side was generally negative, showing a tendency to flow to the back cover plate. The W_z of profiles on the pressure side was generally small and fluctuated near zero value. The amplitude of W_z on the suction side for Q = 7 m^3/h and 20 m^3/h was larger than that for Q = 13 m^3/h. Under three flow rates, when W_z of profile lpI-2 is compared with W_z of profile lpI-0.5, and W_z of profile lsI-2 is compared with W_z of profile lsI-0.5, it can be found that the cross-flows were different.

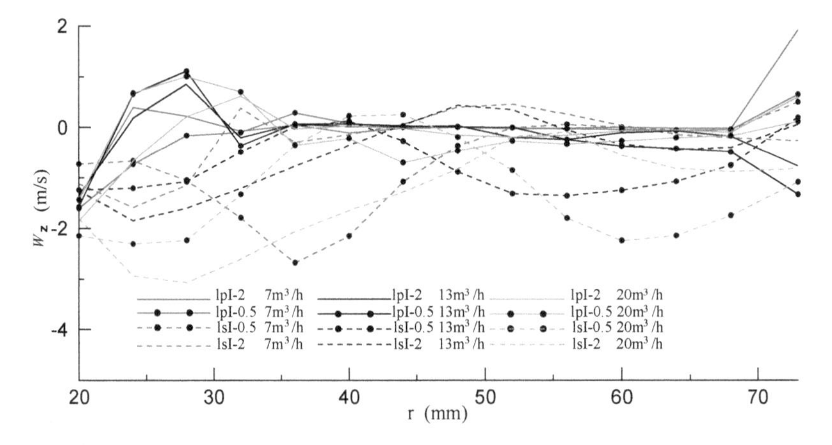

Figure 11. Cross flow velocity distributions of profiles near blade I.

3.5. Shear Stress Distribution

Shear stress is caused by viscous friction on the wall, which represents the spatial variation rate of the relative velocity near the wall. It is also the main cause of viscous dissipation and power loss. The study of the distribution of shear stress provides a guiding value for further optimization design. Figure 12 shows the distribution of the shear stress on the impeller walls for Q = 13 m^3/h. Figure 12a shows that the shear stress on the suction sides of blades was greater than that on the pressure sides, which is consistent with the results that the relative velocity on suction sides was greater than that on the pressure side, as shown in Figures 8 and 9, where relatively large shear stress at the outlet of the pressure surface existed locally. Figure 12b,c shows that the distribution of shear stress around different blades along the circumferential direction was generally similar, and the location of a larger shear stress distribution around the front and rear cover plates was generally consistent, which indirectly indicated that the three-dimensional characteristics of flow in the low specific speed impeller were not obvious. However, the shear stress on the front cover was slightly larger than that on the back cover.

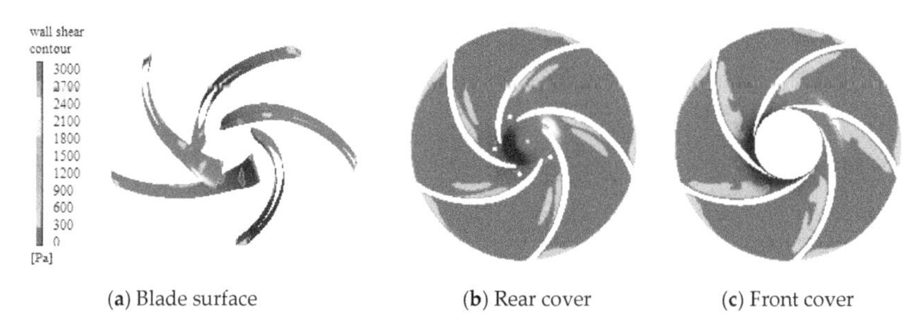

Figure 12. Shear stress distribution on the impeller wall, Q = 13 m^3/h.

3.6. Turbulent Kinetic Energy

The turbulent kinetic energy is related to the velocity fluctuation and turbulent dissipation, and it is a measure of the intensity of turbulence. Usually, the turbulent kinetic energy is estimated from the formula, $k = 3(\overline{u} \cdot I)^2/2$, where $\overline{u}$ is the average velocity and I the intensity of turbulence. The larger the average velocity and intensity of turbulence are, the larger the turbulent kinetic energy is.

Figure 13 shows the turbulent kinetic energy distribution of profiles on blade I. In the range of radii from 25 mm to 30 mm, the amplitude of the turbulent kinetic energy was small, especially at a relatively low level on the pressure side. Affected by the rotor–static transition, the turbulent kinetic energy rose sharply at the outlet of the blade on pressure side, and the larger the flow rate was, the greater the turbulent kinetic energy on the pressure side was. The turbulent kinetic energy on the suction side was generally larger than that on the pressure side, and the larger the flow rate was, the larger the turbulent kinetic energy on the suction side was. The closer the profile was to the blade surface on the pressure side, the smaller the turbulent kinetic energy was; on the contrary, the closer the profile was to the blade surface on the suction side, the greater the turbulent kinetic energy was.

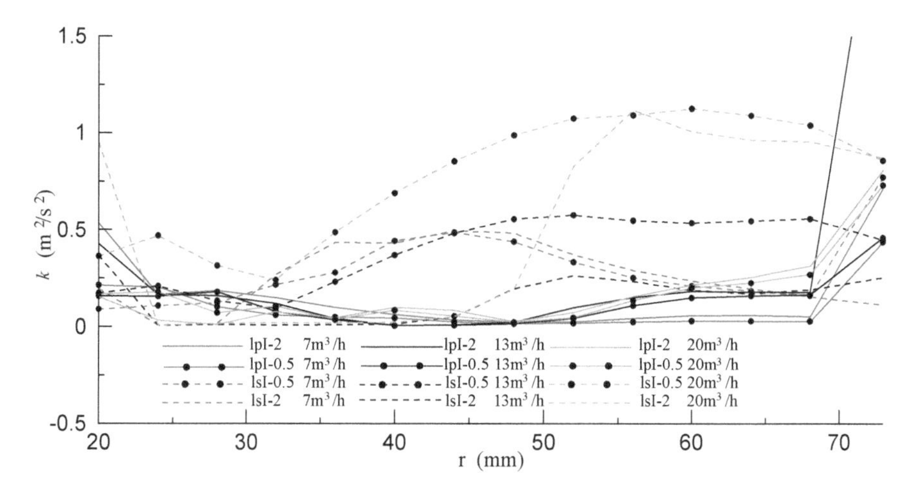

Figure 13. Turbulent kinetic energy distributions.

4. Conclusions

In this paper, some near-wall region flow characteristics of impeller blades, and front and rear cover plates of a low-specific-speed centrifugal pump were studied by means of numerical simulation of the internal flow and an experiment of external characteristics. The main results are as follows:

(1) Under different flow rates from 4 m^3/h to 20 m^3/h, the head and efficiency from the simulations were close to the experimental results, and the simulation results were credible.

(2) Under the design flow rate, the pressure distribution of profiles on the pressure side of different blades was almost the same, as well as that on the suction side, and the static pressure on the pressure side was greater than that on the suction side. The pressure on the blade near the tongue was greatly affected by the separator flow. The pressure of profiles at the distance of 2 mm from the blade pressure side was greater than that of profiles at the distance of 0.5 mm from the blade pressure side, but the pressure of profiles at the distance of 2 mm from the blade suction side was less than that of profiles at the distance of 0.5 mm from the blade suction side. With the increasing of the flow rate, the pressure on the pressure side did not decrease significantly, but the pressure on the suction side decreased more obviously.

(3) Under the design flow rate, the relative velocity distribution of profiles on the pressure side of different blades was almost the same, as well as that on the suction side; it started from the forward flow near the impeller inlet, decelerated or even formed counter-currents, and accelerated again. The relative velocity achieved its maximum in a certain range of the inlet area on the suction side, and then suddenly decreased, but there was no countercurrent in general. With an increasing flow rate, the relative velocity difference between the suction side and the pressure side increased obviously, and there was a countercurrent on the pressure side within the range of radii from 40 mm to 50 mm when the flow rate was 13 m^3/h or 20 m^3/h.

(4) Under three flow rates of 7 m^3/h, 13 m^3/h, and 20 m^3/h, the cross-flow velocity was relatively different. The greater the flow rate was, the greater the turbulent kinetic energy on the pressure side was. The turbulent kinetic energy on the suction sides was larger than that on the pressure side.

Author Contributions: W.C. designed the pump, simulated and analyzed data, organized paper; Z.J. contributed meshes construction and simulation assistance; Q.Z. contributed experimental process and experimental results analysis.

Funding: This work was supported by the National Key R&D Program of China (2018YFC0810506) and the Key R&D Program of Zhenjiang (SH2017049, BK20161472).

Conflicts of Interest: The author(s) declared no potential conflicts of interest with respect to the research, authorship, and/or publication of this article.

References

1. Karraisk, I.J. Future centrifugal pumps. *Pump Technol.* **1978**, *14*, 52–56.
2. Choi, Y.D.; Kurokawa, J.; Matsui, J. Performance and internal flow characteristics of a very low specific speed centrifugal pump. *J. Fluids Eng.* **2005**, *128*, 341–349. [CrossRef]
3. Westra, R.W.; Broersma, L.; Andel, K.; Kruyt, N.P. PIV measurements and CFD computations of secondary flow in a centrifugal pump impeller. *J. Fluids Eng.* **2010**, *132*. [CrossRef]
4. Pedersen, N.; Larsen, P.S.; Jacobsen, C.B. Flow in a centrifugal pump impeller at design and off-design conditions part I: Particle image velocimetry (PIV) and laser doppler velocimetry (LDV) measurements. *J. Fluid Eng.* **2003**, *125*, 61–72. [CrossRef]
5. Shao, J.; Liu, S.H.; Zhang, G.Y. PIV measurement and analysis of internal flow in impeller of centrifugal pump with large outlet angle. *Pump Technol.* **2010**, *45*, 1–7.
6. Tsujita, H.; Migita, K. Investigation for secondary flow and loss generation mechanisms within centrifugal impeller by using rotating curved duct: 2nd report, influence of inlet pitchwise velocity distribution. *Trans. Jpn. Soc. Mech. Eng. Ser. B* **2010**, *76*, 814–822. [CrossRef]
7. Cui, B.L.; Lin, Y.G.; Jin, Y.Z. Numerical simulation of flow in centrifugal pump with complex impeller. *J. Therm. Sci.* **2011**, *20*, 47–52. [CrossRef]

8. Limbach, P.; Skoda, R. Numerical and experimental analysis of cavitating flow in a low specific speed centrifugal pump with different surface roughness. *J. Fluids Eng.* **2017**, *139*. [CrossRef]
9. Cao, W.D.; Yao, L.J.; Liu, B.; Zhang, Y. The influence of impeller eccentricity on centrifugal pump. *Adv. Mech. Eng.* **2017**, *9*. [CrossRef]
10. Zhang, Y.L.; Zhu, Z.C.; Li, W.G.; Xiao, J.J. Effects of viscosity on transient behavior of a low specific speed centrifugal pump in starting and stopping periods. *Int. J. Fluid Mech. Res.* **2018**, *45*, 1–20. [CrossRef]
11. Chen, J.; Wang, Y.; Liu, H.L.; Shao, C.; Zhang, X. Internal flow and analysis of its unsteady characteristics in centrifugal pump with ultra-low specific-speed. *J. Drain. Irrig. Mach. Eng.* **2018**, *36*, 377–383.
12. Dong, L.; Zhao, Y.Q.; Liu, H.L.; Dai, C. The effect of front streamline wrapping angle variation in a super low specific speed centrifugal pump. *Proc. Inst. Mech. Eng. Part. C: J. Mech. Eng. Sci.* **2018**, *232*. [CrossRef]
13. Zhang, R.H.; Chen, X.B.; Guo, G.Q.; Li, R. Reconstruction and modal analysis for flow field of low Specific speed centrifugal pump impeller. *Trans. Chin. Soc. Agric. Mach.* **2018**, *49*, 143–149.
14. Wang, C.; Zhang, Y.X.; Li, Z.W.; Xu, A.; Xu, C.; Shi, Z. Pressure fluctuation-vortex interaction in an ultra-low specific-speed centrifugal pump. *J. Low Freq. Noise Vib. Act. Control.* **2019**, *38*, 527–543. [CrossRef]

Article

The Impact of Erythrocytes Injury on Blood Flow in Bionic Arteriole with Stenosis Segment

Donghai Li [1,†], Guiling Li [2,†], Yuanyuan Chen [3], Jia Man [4], Qingyu Wu [5], Mingkui Zhang [5], Haosheng Chen [6,*] and Yu Zhang [1,*]

1. School of Medicine, Tsinghua University, Beijing 100084, China; donghai_li10@163.com
2. School of Pharmaceutical Sciences, Tsinghua University, Beijing 100084, China; lgl@mail.tsinghua.edu.cn
3. School of Mechanical Engineering and Automation, Beijing Advanced Innovation Center for Biomedical Engineering, Beihang University, Beijing 100191, China; chenyuanyuan0526@sina.com
4. School of Mechanical Engineering, Shandong University; Key Laboratory of High-efficiency and Clean Mechanical Manufacture at Shandong University, Ministry of Education, Shandong 250061, China; mj@sdu.edu.cn
5. First Hospital of Tsinghua University, School of Medicine, Tsinghua University, Beijing 100084, China; wuqingyu@mail.tsinghua.edu.cn (Q.W.); zhangmingkui@mail.tsinghua.edu.cn (M.Z.)
6. School of Mechanical Engineering, State Key Laboratory of Tribology, Tsinghua University, Beijing 100084, China

* Correspondence: chenhs@tsinghua.edu.cn (H.C.); yuzhang2014@tsinghua.edu.cn (Y.Z.)

† The two authors contributed equally to this work.

Received: 18 May 2019; Accepted: 11 June 2019; Published: 14 June 2019

Abstract: Ventricular assist device (VAD) implantation is an effective treatment for patients with end-stage heart failure. However, patients who undergo long-term application of VADs experience a series of VAD-related adverse effects including pump thrombosis, which is induced by rotate impeller-caused blood cell injury and hemolysis. Blood cell trauma-related flow patterns are the key mechanism for understanding thrombus formation. In this study, we established a new method to evaluate the blood cell damage and investigate the real-time characteristics of blood flow patterns in vitro using rheometer and bionic microfluidic devices. The variation of plasma free hemoglobin (PFH) and lactic dehydrogenase (LDH) in the rheometer test showed that high shear stress was the main factor causing erythrocyte membrane injury, while the long-term exposure of high shear stress further aggravated this trauma. Following this rheometer test, the damaged erythrocytes were collected and injected into a bionic microfluidic device. The captured images of bionic microfluidic device tests showed that with the increase of shear stress suffered by the erythrocyte, the migration rate of damaged erythrocyte in bionic microchannel significantly decreased and, meanwhile, aggregation of erythrocyte was clearly observed. Our results indicate that mechanical shear stress caused by erythrocyte injury leads to thrombus formulation and adhesion in arterioles.

Keywords: erythrocyte injury; flow pattern; bionic microfluidic device; thrombosis

1. Introduction

The concept of mechanical circulatory support (MCS) was developed rapidly to resolve the shortage of donor hearts for heart transplantation and the sustained increase of heart failure (HF) mobility and mortality. MCSs with ventricular assist devices (VADs) were originally designed as a temporary therapy for bridge-to-transplant (BTT). Nevertheless, VAD implantation is an efficacious treatment as a destination therapy for patients with end-stage HF, which is refractory to current medical therapy and can effectively improve the long-time survival rate to approximately 81% and 70% at 1 year and 2 years, respectively [1,2]. The therapeutic benefits of VAD implantation are that it also improves the functional capacity and quality of life. However, patients who undergo long-term application of VADs experience a series of VAD-related adverse events, including right heart failure, pump thrombosis, gastrointestinal bleeding, driveline infection, stroke, and aortic insufficiency [2–10].

Among these VAD-related complications, thrombosis is concerning as a multifactorial complication which is related to other adverse events, such as hemolysis, stroke and bleeding, and can induce rapid clinical deterioration [11]. The mechanism of thrombosis has been studied for decades, and the induced factors of thrombosis lead to a decreased or turbulent flow pattern in VADs and suboptimal anticoagulation. Furthermore, hemolysis and von Willebrand deficiency, which are related to mechanical damage caused by rotate impellers, can also induce thrombosis. It is difficult to visualize the thrombosis in vivo. Therefore, various in vitro test methods and devices have been applied to monitor and evaluate blood clotting and platelet function to avoid thrombosis, such as assays for bleeding time, activated clotting time (ACT), activated partial thromboplastin time (APTT), thromboelastography, and platelet aggregometry. However, all of these parameters are tested under static or irrelevant flow conditions, which fail to incorporate the true flow pattern and status in blood vessels. Furthermore, the test devices are also quite different from actual vessel structures, especially the structure of stenosis arteries segments. Therefore, there is an urgent demand for a new method to investigate the real-time thrombus formation process and evaluate the relationship with the assessment of the blood cell damage in the region of VAD application.

Microfluidic devices, on which the blood flow pattern in simulated small vessels can be observed directly, have recently been used to study erythrocyte movement and blood flow in small vessels in vitro [12]. Based on bionic microfluidic devices, atherosclerotic confinement of vessels and blood flow patterns (including flow rate and physiological and pathological shear stress) in microscale arterioles can be simulated for basic combined research on whole blood and extracellular matrix surface coating or treatment [13–15]. However, the flow patterns were controlled by injecting flow rate and pressure, but were too coarse to maintain consistent inner shear stress force and flow rate in a device with several microscale channels. Furthermore, traces of experiment blood were difficult to collect and measure for the evaluation of the blood cell damage. In our previous research, the relations among erythrocyte morphology, membrane damage, and the concentration of specific plasma proteins had been revealed and confirmed, and the changes of plasma free hemoglobin (PFH) and lactic dehydrogenase (LDH) concentrations in plasma can be measured and analyzed to evaluate the degree of erythrocyte rupture and the degree of erythrocyte membrane injury, respectively [16]. Here, we collected fresh whole blood from sheep and exposed it to a fixed shear stress environment generated by a rheometer for blood cell damage assessment. Then, the damaged blood was collected and injected into a bionic microfluidic device, which was designed to mimic arterioles with several continuous stenosis segments, to investigate and evaluate the effects of red cell damage at different degrees on the flow pattern inside the microfluidic channel.

2. Materials and Methods

2.1. Blood Collection

Fresh total blood was collected from the jugular veins of 3 healthy adult male sheep (aged about 4 years old, with a body weight of around 80 kg) for fixed shear stress and microfluidic channel tests. A large bore needle (16G) was applied to collect 400 mL blood into a 500 mL sterile PVC bag prefilled with 56 mL of citrate phosphate dextrose adenine (CPDA-1) anticoagulant; the procedure was performed according to American Society for Testing Materials (ASTM) Standard Practices F1830, and the aseptic technique was followed strictly. The fresh sheep blood was stored and transported with the temperature between 2 and 8 °C, and the time interval from collection to experiment was within 8 h. Before the experiment started, the collected sheep blood was kept quiescent in room temperature and rewarmed slowly to 24 °C while maintaining the temperature of 37 °C with a water bath during the experiment process. According to ASTM Standard Practices F1841, the hematocrit of blood applied in the tests was adjusted to 30% by a phosphate-buffered saline.

2.2. *Hemolysis Evaluation under Fixed Shear Stress*

A rheometer (MCR 310, Physica, Anton Paar GmbH, Graz, Austria) and coaxial cylinder testing mold CC27/E were applied to provide a well-defined test environment of fixed shear stress (Figure 1). The rotor and container were immersed in 75% ethanol for 5 min for sterilization and rinsed thoroughly with normal saline (NS) at the beginning of the experiment. NS was also used to rinse the residual blood at each test interval. Pre-warmed blood (6 mL) were slowly injected into containers for each testing. The gap between a rotor and container was set as 0.5 mm and the temperature of container was maintained at 37 °C. The shear stress was set as 0, 25, 50, 75 and 100 Pa, with the exposure time of 5, 10 and 15 min. At the end of the tests, two 1.5 mL blood samples were collected from the bottom of the containers. Each blood sample was centrifuged twice at 3000 rpm for 5 min at a temperature of 4 °C. Supernatant was collected to measure the concentration of PFH and LDH, while the curves showing the changes of PFH and LDH concentrations over time at each shear stress gradient were also drawn.

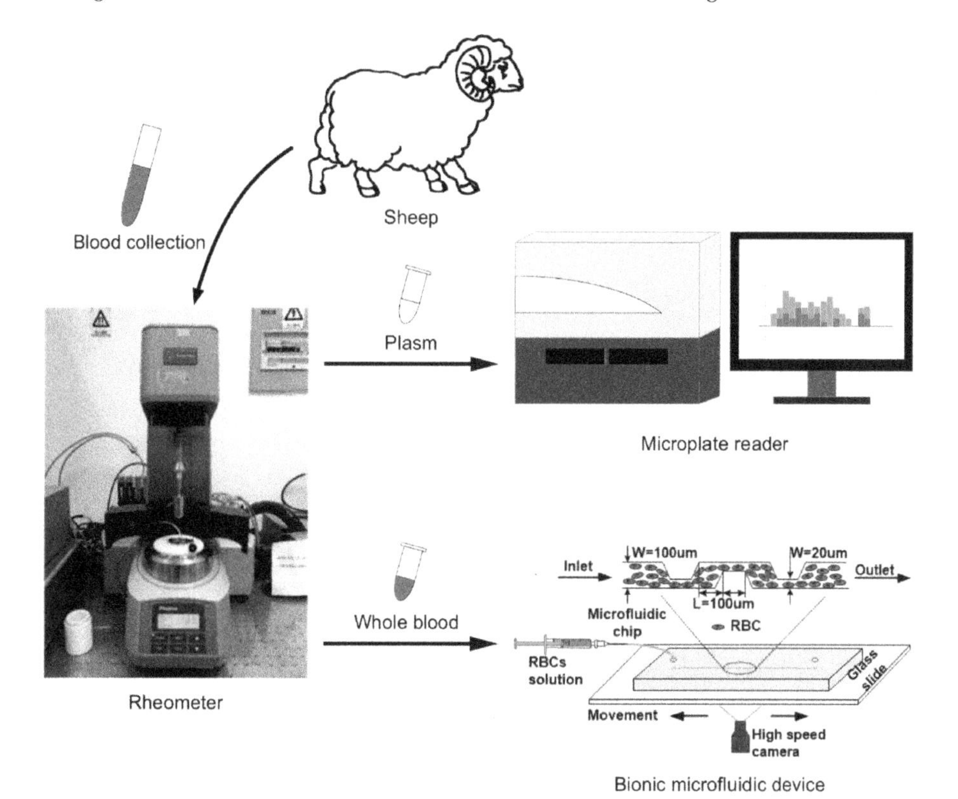

Figure 1. Schematics of the impact of erythrocytes injury caused by mechanical shear stress force on flow patterns in bionic arteriole with stenosis segment.

2.3. *Bionic Microfluidic Device Design and Visualization of Injected Erythrocyte Trajectories*

The bionic microfluidic channel was designed using AutoCAD software and fabricated through standard lithographic method [17]. The microfluidic device was composed of polydimethylsiloxane (PDMS) and a glass plate, which is one layer with the height of 20 µm, and the detailed channel parameters are shown in Figure 1. After exposed to stable shear stress on a rheometer platform for

10 min, the whole blood sample was collected and injected into the inlet hole of the microchannel with the flow rate of 0.12 mL/min. The visualization of injected flowing erythrocyte trajectories in the microchannel were captured using a high-speed camera (M710, Phantom Co., New Jersey, USA) under the bright field provided by an inverted microscope (Leica DME6000 B, Leica Co., Wetzlar, Germany). The video analysis was processed using ImageJ software.

2.4. Statistical Analysis

The statistical analyses are performed with the one-way analysis of variance (ANOVA) using the prism software (GraphPad, San Diego, CA, USA). The data are reported as mean ± standard error, and p-values of * $p < 0.05$, ** $p < 0.01$, *** $p < 0.001$ were considered statistically significant.

3. Results

3.1. Variation of PFH Concentration with Gradient Change of Shear Stress and Exposure Time

PFH is a classic label which is first applied to evaluate the red blood cell damage caused by VADs. The world-recognized CF-VADs testing process (American Society of Testing Materials Standard Practices F1830, ASTM F1841) and evaluation index, including normalized index of hemolysis (NIH) and modified index of hemolysis (MIH), were established on the basis of PFH concentration measurement. According to the curve of PFH concentration variation, shown in Figure 2a, when the blood was exposed in a mechanical shear stress environment, within 5 min, the concentrations of PFH release under 25 to 100 Pa rose synchronously without notable differences. When the exposure time increased up to 10 min, the concentration of PFH rose rapidly in the group with 75 and 100 Pa shear stress. In particular, extremely high PFH releases were detected in the 75 and 100 Pa groups when the exposure time reached 15 min. Meanwhile, as shown in Figure 2b, in a low shear stress environment (25 Pa and 50 Pa), there were no significant changes of PFH concentration with the increasing of exposure time. However, in the high shear stress condition, especially under 100 Pa, PFH release rose extremely high with the extension of exposure time. Therefore, long exposure time under high shear stress environment would induce more damage to erythrocytes.

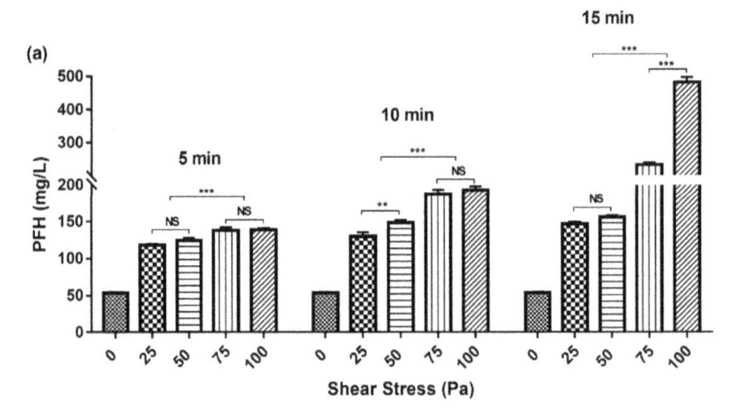

Figure 2. *Cont.*

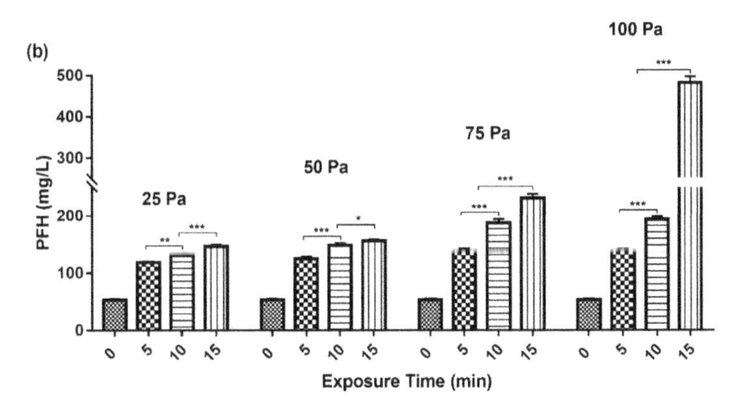

Figure 2. The changes of plasma plasma free hemoglobin (PFH) concentration under different shear stress and exposure times. (**a**) With the increase of shear stress, the concentration of PFH released into plasma showed a slow rising trend within 5 min and 10 min, but with the extension of exposure time up to 15 min, the concentration of PFH showed a rapid increase trend, especially in the shear stress of 100 Pa. (**b**) With the increase of exposure time, the concentration of PFH released into plasma showed a slow rise trend at 25 and 50 Pa, but the shear stress up to 75 and 100 Pa, the concentration of PFH showed a rapid increase trend, especially in the exposure time 15 min.

3.2. LDH Release with Gradient Change of Shear Stress and Exposure Time

Compared to PFH, LDH is a novel and more sensitive protein label to assess the trauma of the erythrocyte membrane. Differently to PFH release based on erythrocyte rupture, injuries to the red blood cell membrane will induce an increase of LDH concentration in plasma. As the results show in Figure 3a, there was no significant difference in LDH release between the three groups with the exposure times of 5, 10, and 15 min in a low shear stress environment within 50 Pa. The increase of LDH concentration was smooth and maintained a low level. However, when blood was exposed to a high shear stress condition (especially at 100 Pa), the concentration of LDH rose rapidly with the increase of exposure time. Meanwhile, as the curve shows in Figure 3b, LDH release was insensitive to the prolonging of exposure time at low shear stress conditions in 25 Pa and 50 Pa. However, the concentration of plasma LDH increased dramatically in a high shear stress environment, and the longer the exposure was, the greater the influence became. Therefore, high shear stress force was the main factor causing erythrocyte membrane injury, and the long exposure time could aggravate this trauma.

3.3. Flow Patterns Monitoring with Bionic Microfluidic Devices

After being exposed to a 25, 50, 75, and 100 Pa shear stress environment for 10 min, blood samples were collected for the bionic microfluidic device test. The initial flow rate of blood sample injected into microfluidic devices was maintained as 0.12 mL/min by a micropump. As shown in Figure 4, compared to the morphological characteristics of red blood cells (RBCs) exposed to 25 and 50 Pa shear stress force, more poikilocytes (characterized by sphericity), swelling, and multiple spines were found in the 75 and 100 Pa groups. The mean migration rate of erythrocytes at the posterior region of stenosis in the bionic microfluidic channel, which were subjected to 0, 25, 50, 75, and 100 Pa for 10 min, were 467.55 ± 74.16, 444.39 ± 41.82, 354.43 ± 46.93, 250.91 ± 34.67, and 152.87 ± 27.33 μm/s, respectively. With the increase of shear stress force applied, the migration rate of damaged erythrocyte in the bionic microchannel declined significantly. However, the aggregation of RBCs rose notably with the increase to an exposed shear stress environment, which was synchronous with the degree of erythrocyte damage assessed previously. Moreover, thrombus was captured in the microchannel of the 75 and 100 Pa groups. The migration of thrombus and surrounding flow rate of 75 and 100 Pa groups were slower than the other two groups. These results indicate that slow flow patterns can also lead to thrombus deposition and adhesion in arterioles.

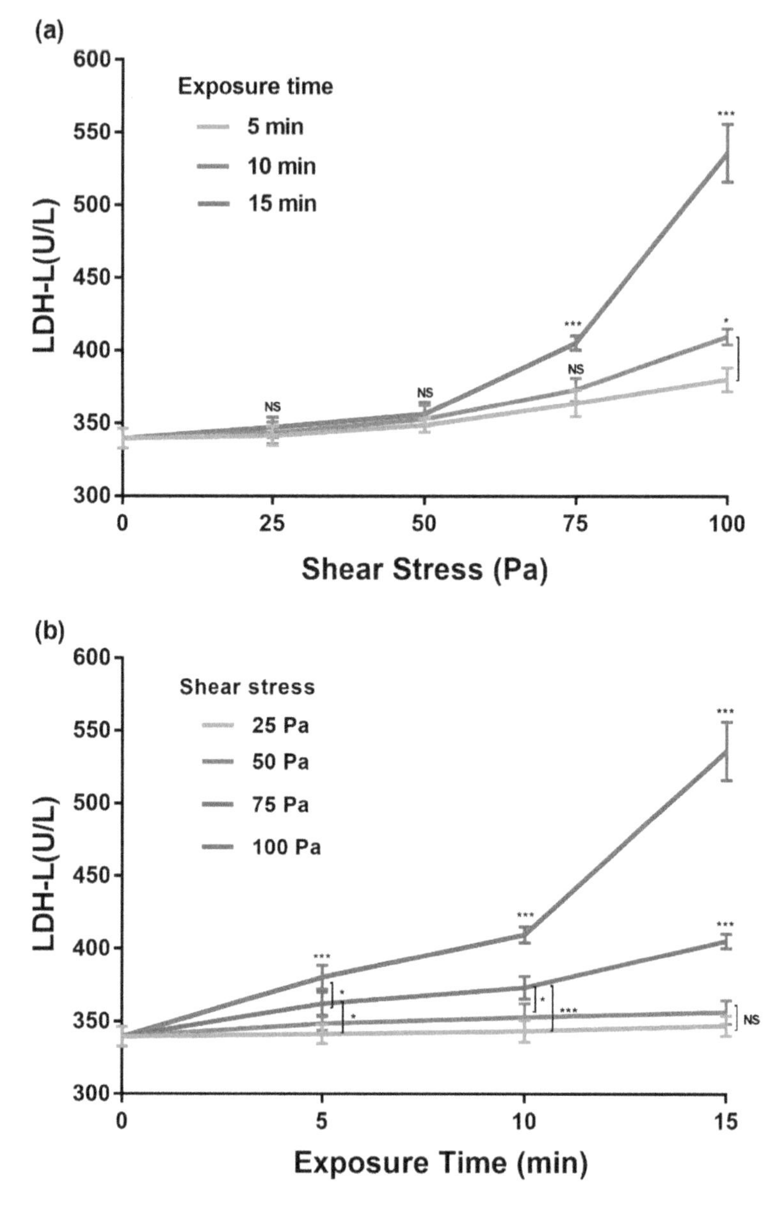

Figure 3. The changes of plasma lactic dehydrogenase (LDH) concentration under different shear stress and exposure times. (**a**) With the increase of shear stress, the concentration of plasma LDH showed an increase trend, especially above 50 Pa, the LDH release presented a rapid raise; (**b**) With the extension of the exposure time, the concentration of plasma LDH in 25 and 50 Pa shear stress environments rose slowly, but in an 100 Pa environment, showed a rapid increase.

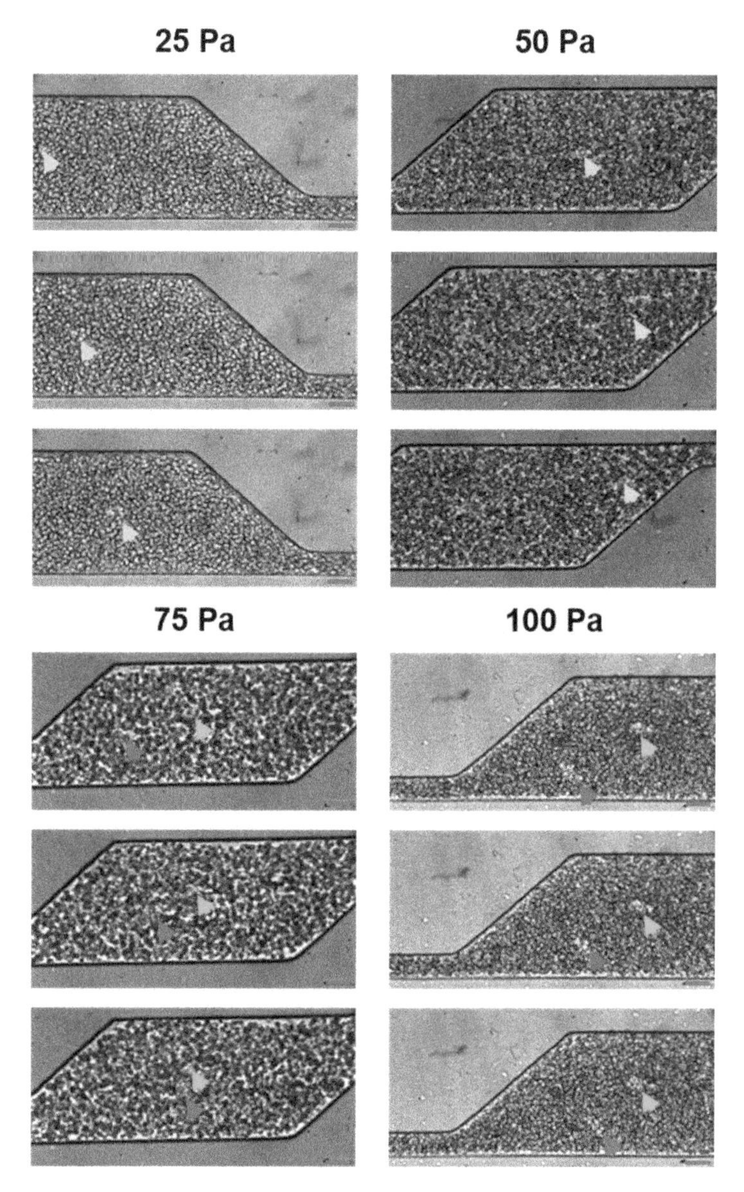

Figure 4. Migration rate and aggregation of damaged erythrocyte in microfluidic devices. All the blood samples were subjected to each shear stress environment for 10 min. The initial flow rate injected into devices was 0.12 mL/min. With the increase of shear stress, the blood cell migration rate declined, but the aggregation increased. Yellow arrow: erythrocyte; Blue arrow: adhesion thrombus; Red arrow: moving thrombus (scale bar: 20 μm).

4. Discussion

Erythrocyte damage caused by high shear stress generated by the rotation of the inner impeller is the main problem of VADs that cannot be ignored, and has been demonstrated to induce damage of the erythrocyte membrane, increase cellular fragility, reduce deformability, enhance aggregation and, finally, increase the risk of intervascular hemolysis and thrombosis [18]. The variation of PFH and LDH in the rheometer tests showed that high shear stress was the main factor causing red blood cell membrane injury. This result is consistent with our previous observations, and an effective interpretation of the influence of shear stress force on hemolysis [19]. Meanwhile, with the increase of exposure time in a fixed shear stress environment, the release of PFH and LDH rose gradually. This is consistent with the phenomenon reported by Shimono, that a lot of erythrocytes were suddenly ruptured in the long-term in vitro hemolysis tests [20]. The result indicated the cumulative damage of erythrocyte membrane increased with the extension of exposure time and, finally, aggravated the occurrence of hemolysis.

Increased blood viscosity, effects on the vessel wall, platelet activation, and thrombin generation, and the effects of fibrinogen and fibrin are established risk factors for thrombosis [21]. Blood consists of a large number of blood cells suspended in plasma. The total of white cells and platelets occupy less than 1% of blood volume, while RBCs are the dominant component of the blood, accounting for 40%–45% of blood volume [18]. Growing evidence suggests erythrocytes contribute to blood thrombosis because erythrocytes mediate blood rheology, interact with fibrinogen and fibrin, interact with other cells, and also support thrombin generation [22]. The rheological properties of blood are primarily governed by the concentration of RBCs, size, sharpness, deformability, intrinsic viscoelastic properties and fibrinogen-binding ability [23–25]. Therefore, the investigation of blood flow rheology and thrombus generation process directly in arterioles and its relationship with blood cell trauma under high shear stress environments is the key mechanism for research of thrombus formation. However, it is difficult to identify the mechanisms of thrombosis in clinical and epidemiologic studies, due to inaccurate identification of causation. Moreover, previous in vivo studies based on animal models, it was also difficult to interpret the effects of erythrocyte on thrombosis, due to the preparation and evaluation of the injured erythrocyte and the limitations of current observation techniques. Thus, microfluidic devices are applied to observe the dynamic movement patterns of erythrocyte. Previous microfluidic device experiments have confirmed that thrombosis can be induced by blood contact with exogenous materials and local flow conditions [26–28]. These studies depended on the parallel plate channel structure on microfluidic devices to provide similar arteriolar flow conditions in live vasculature. It is impossible to control the effect of specific shear stress on RBCs in the microchannel. Previous studies have shown that exposure to shear stress environments would induce erythrocyte membrane injury, resulting in swelling, irregular deformation, and reduce deformability. This phenomenon can also be observed directly in this research. All of these, especially when the stenosis of the blood vessel affects the blood fluid rheology, will aggravate the erythrocyte aggregation and eventually accelerate thrombosis. In this project, the erythrocyte damage in a fixed shear stress environment has been evaluated. The variations of flow patterns in bionic microchannels reflected accurately the effect of erythrocyte injury on aggregation ability and the flow state in small vessels. The captured images of bionic microfluidic device tests show that there was no significant difference for the migration rate at 0 and 25 Pa ($p = 0.52 > 0.05$). However, significant differences were found between other groups ($p < 0.01$). The results indicate that with an increase of shear stress suffered by the erythrocyte, the migration rate of damaged erythrocyte in bionic microchannels significantly decreased. Meanwhile the aggregation of erythrocyte was clearly observed in 75 and 100 Pa groups. The results indicate that mechanical shear stress caused by erythrocyte injury, which enhanced aggregation ability of erythrocytes and increased blood viscosity, resulted in decreased blood rheological performance, eventually leading to thrombus formulation and adhesion in arterioles.

5. Conclusions

In this study, we used rheometer and bionic microfluidic devices to investigate and evaluate the effect of injury erythrocytes on flow patterns in bionic arterioles with stenosis segment. The variation of PFH concentration under different shear stress and exposure time in rheometers showed that the long exposure time in high shear stress environments would induce more damage of erythrocytes.

Meanwhile, LDH release with gradient changes under different shear stress and exposure times revealed that high shear stress was the main factor causing erythrocyte membrane injury, and a long exposure time could aggravate this trauma. Following this rheometer test, the damaged erythrocytes were collected and injected into a bionic microfluidic device which was designed to mimic arterioles with several continuous stenosis segments, to investigate and evaluate the flow pattern modification by red cell damage at different degrees. The results show that with the increase of shear stress, the migration rate of damaged erythrocyte declined and the aggregation of erythrocyte was clearly observed in bionic microchannels, which was consistent with the degree of erythrocyte damage assessed through the rheometer test. Our results indicate that mechanical shear stress was caused by erythrocyte injury, which enhanced aggregation ability of erythrocytes and increased blood viscosity and resulted in decreased blood rheological performance, eventually leading to thrombus formulation and adhesion in arterioles.

Author Contributions: D.L. and G.L. developed the concept of the article and wrote the outline of the paper draft; Y.Z. supported work in funding and reviewed the paper; H.C. reviewed the paper; Y.C. and J.M. analyzed the data; Q.W. and M.Z. redacted the paper. All authors of this article provided substantive comments.

Funding: This research was funded by Tsinghua University Initiative Scientific Research Program, grant number 20141081265.

Acknowledgments: Guiling Li acknowledges the support from Center for Life Sciences at Tsinghua and Peking Universities. Jia Man acknowledges the support from Key Laboratory of High-efficiency and Clean Mechanical Manufacture at Shandong University, Ministry of Education.

Conflicts of Interest: The authors declare no conflict of interest.

References

1. Slaughter, M.S.; Rogers, J.G.; Milano, C.A.; Russell, S.D.; Conte, J.V.; Feldman, D.; Sun, B.; Tatooles, A.J.; Delgado, R.M., 3rd; Long, J.W.; et al. Advanced heart failure treated with continuous-flow left ventricular assist device. *N. Engl. J. Med.* **2009**, *361*, 2241–2251. [CrossRef] [PubMed]
2. Kirklin, J.K.; Pagani, F.D.; Kormos, R.L.; Stevenson, L.W.; Blume, E.D.; Myers, S.L.; Miller, M.A.; Baldwin, J.T.; Young, J.B.; Naftel, D.C. Eighth annual INTERMACS report: Special focus on framing the impact of adverse events. *J. Heart Lung Transplant.* **2017**, *36*, 1080–1086. [CrossRef] [PubMed]
3. Kapur, N.K.; Esposito, M.L.; Bader, Y.; Morine, K.J.; Kiernan, M.S.; Pham, D.T.; Burkhoff, D. Mechanical Circulatory Support Devices for Acute Right Ventricular Failure. *Circulation* **2017**, *136*, 314–326. [CrossRef] [PubMed]
4. Kirklin, J.K.; Naftel, D.C.; Pagani, F.D.; Kormos, R.L.; Myers, S.; Acker, M.A.; Rogers, J.; Slaughter, M.S.; Stevenson, L.W. Pump thrombosis in the Thoratec HeartMate II device: An update analysis of the INTERMACS Registry. *J. Heart Lung Transpl.* **2015**, *34*, 1515–1526. [CrossRef] [PubMed]
5. Najjar, S.S.; Slaughter, M.S.; Pagani, F.D.; Starling, R.C.; McGee, E.C.; Eckman, P.; Tatooles, A.J.; Moazami, N.; Kormos, R.L.; Hathaway, D.R.; et al. An analysis of pump thrombus events in patients in the HeartWare ADVANCE bridge to transplant and continued access protocol trial. *J. Heart Lung Transplant.* **2014**, *33*, 23–34. [CrossRef] [PubMed]
6. Goldstein, D.J.; Aaronson, K.D.; Tatooles, A.J.; Silvestry, S.C.; Jeevanandam, V.; Gordon, R.; Hathaway, D.R.; Najarian, K.B.; Slaughter, M.S.; Investigators, A. Gastrointestinal bleeding in recipients of the HeartWare Ventricular Assist System. *JACC Heart Fail.* **2015**, *3*, 303–313. [CrossRef] [PubMed]
7. Joy, P.S.; Kumar, G.; Guddati, A.K.; Bhama, J.K.; Cadaret, L.M. Risk Factors and Outcomes of Gastrointestinal Bleeding in Left Ventricular Assist Device Recipients. *Am. J. Cardiol.* **2016**, *117*, 240–244. [CrossRef] [PubMed]
8. Willey, J.Z.; Gavalas, M.V.; Trinh, P.N.; Yuzefpolskaya, M.; Reshad Garan, A.; Levin, A.P.; Takeda, K.; Takayama, H.; Fried, J.; Naka, Y.; et al. Outcomes after stroke complicating left ventricular assist device. *J. Heart Lung Transplant.* **2016**, *35*, 1003–1009. [CrossRef]
9. Cowger, J.; Rao, V.; Massey, T.; Sun, B.; May-Newman, K.; Jorde, U.; Estep, J.D. Comprehensive review and suggested strategies for the detection and management of aortic insufficiency in patients with a continuous-flow left ventricular assist device. *J. Heart Lung Transpl.* **2015**, *34*, 149–157. [CrossRef]

10. Jorde, U.P.; Uriel, N.; Nahumi, N.; Bejar, D.; Gonzalez-Costello, J.; Thomas, S.S.; Han, J.; Morrison, K.A.; Jones, S.; Kodali, S.; et al. Prevalence, Significance, and Management of Aortic Insufficiency in Continuous Flow Left Ventricular Assist Device Recipients. *Circ Heart Fail.* **2014**, *7*, 310–319. [CrossRef]
11. Uriel, N.; Han, J.; Morrison, K.A.; Nahumi, N.; Yuzefpolskaya, M.; Garan, A.R.; Duong, J.; Colombo, P.C.; Takayama, H.; Thomas, S.; et al. Device thrombosis in HeartMate II continuous-flow Left ventricular assist devices: A multifactorial phenomenon. *J. Heart Lung Transplant.* **2014**, *33*, 51–59. [CrossRef] [PubMed]
12. Jain, A.; Graveline, A.; Waterhouse, A.; Vernet, A.; Flaumenhaft, R.; Ingber, D.E. A shear gradient-activated microfluidic device for automated monitoring of whole blood haemostasis and platelet function. *Nat. Commun.* **2016**, *7*, 10176. [CrossRef] [PubMed]
13. Westein, E.; van der Meer, A.D.; Kuijpers, M.J.; Frimat, J.P.; van den Berg, A.; Heemskerk, J.W. Atherosclerotic geometries exacerbate pathological thrombus formation poststenosis in a von Willebrand factor-dependent manner. *Proc. Natl. Acad. Sci. USA* **2013**, *110*, 1357–1362. [CrossRef] [PubMed]
14. Li, M.; Hotaling, N.A.; Ku, D.N.; Forest, C.R. Microfluidic thrombosis under multiple shear rates and antiplatelet therapy doses. *PLoS ONE* **2014**, *9*, e82493. [CrossRef]
15. Para, A.N.; Ku, D.N. A low-volume, single pass in-vitro system of high shear thrombosis in a stenosis. *Thromb. Res.* **2013**, *131*, 418–424. [CrossRef] [PubMed]
16. Li, D.; Wu, Q.; Liu, S.; Chen, Y.; Chen, H.; Ruan, Y.; Zhang, Y. Lactic Dehydrogenase in the In Vitro Evaluation of Hemolytic Properties of Ventricular Assist Device. *Artif. Organs* **2017**, *41*, E274–E284. [CrossRef] [PubMed]
17. McDonald, J.C.; Duffy, D.C.; Anderson, J.R.; Chiu, D.T.; Wu, H.K.; Schueller, O.J.A.; Whitesides, G.M. Fabrication of microfluidic systems in poly(dimethylsiloxane). *Electrophoresis* **2000**, *21*, 27–40. [CrossRef]
18. Baskurt, O.u.K. *Handbook of Hemorheology and Hemodynamics*; IOS Press: Amsterdam, The Netherlands; Washington, DC, USA, 2007; 455p.
19. Li, D.H.; Wu, Q.Y.; Ji, J.J.; Liu, S.H.; Zhang, M.K.; Zhang, Y. Hemolysis in a continuous-flow ventricular assist device with/without chamfer. *Adv. Mech. Eng.* **2017**, *9*. [CrossRef]
20. Shimono, T.; Makinouchi, K.; Nose, Y. Total Erythrocyte Destruction Time - the New Index for the Hemolytic Performance of Rotary Blood Pumps. *Artif. Organs* **1995**, *19*, 571–575. [CrossRef] [PubMed]
21. Byrnes, J.R.; Wolberg, A.S. Red blood cells in thrombosis. *Blood* **2017**. [CrossRef] [PubMed]
22. Walton, B.L.; Byrnes, J.R.; Wolberg, A.S. Fibrinogen, red blood cells, and factor XIII in venous thrombosis. *J. Thromb. Haemost.* **2015**, *13* (Suppl. 1), S208–S215. [CrossRef] [PubMed]
23. Lowe, G.D.; Lee, A.J.; Rumley, A.; Price, J.F.; Fowkes, F.G. Blood viscosity and risk of cardiovascular events: The Edinburgh Artery Study. *Br. J. Haematol.* **1997**, *96*, 168–173. [CrossRef] [PubMed]
24. Holley, L.; Woodland, N.; Hung, W.T.; Cordatos, K.; Reuben, A. Influence of fibrinogen and haematocrit on erythrocyte sedimentation kinetics. *Biorheology* **1999**, *36*, 287–297. [PubMed]
25. Rampling, M.W. The binding of fibrinogen and fibrinogen degradation products to the erythrocyte membrane and its relationship to haemorheology. *Acta Biol. Med. Ger.* **1981**, *40*, 373–378. [PubMed]
26. Neeves, K.B.; Onasoga, A.A.; Hansen, R.R.; Lilly, J.J.; Venckunaite, D.; Sumner, M.B.; Irish, A.T.; Brodsky, G.; Manco-Johnson, M.J.; Di Paola, J.A. Sources of variability in platelet accumulation on type 1 fibrillar collagen in microfluidic flow assays. *PLoS ONE* **2013**, *8*, e54680. [CrossRef] [PubMed]
27. Hosokawa, K.; Ohnishi, T.; Kondo, T.; Fukasawa, M.; Koide, T.; Maruyama, I.; Tanaka, K.A. A novel automated microchip flow-chamber system to quantitatively evaluate thrombus formation and antithrombotic agents under blood flow conditions. *J. Thromb. Haemost.* **2011**, *9*, 2029–2037. [CrossRef] [PubMed]
28. Westein, E.; de Witt, S.; Lamers, M.; Cosemans, J.M.; Heemskerk, J.W. Monitoring in vitro thrombus formation with novel microfluidic devices. *Platelets* **2012**, *23*, 501–509. [CrossRef] [PubMed]

Article

Numerical Study of Pressure Fluctuation and Unsteady Flow in a Centrifugal Pump

Ling Bai [1], Ling Zhou [1,*], Chen Han [1], Yong Zhu [1,*] and Weidong Shi [1,2]

1 National Research Center of Pumps, Jiangsu University, Zhenjiang 212013, China; lingbai@ujs.edu.cn (L.B.); hanchen0622@outlook.com (C.H.); wdshi@ujs.edu.cn (W.S.)
2 School of Mechanical Engineering, Nantong University, Nantong 226019, China
* Correspondence: lingzhou@hotmail.com or lingzhou@ujs.edu.cn (L.Z.); zhuyong@ujs.edu.cn (Y.Z.)

Received: 9 May 2019; Accepted: 4 June 2019; Published: 9 June 2019

Abstract: A pump is one of the most important machines in the processes and flow systems. The operation of multistage centrifugal pumps could generate pressure fluctuations and instabilities that may be detrimental to the performance and integrity of the pump. In this paper, a numerical study of the influence of pressure fluctuations and unsteady flow patterns was undertaken in the pump flow channel of three configurations with different diffuser vane numbers. It was found that the amplitude of pressure fluctuation in the diffuser was increased gradually with the increase in number of diffuser vanes. The lower number of diffuser vanes was beneficial to obtain a weaker pressure fluctuation intensity. With the static pressure gradually increasing, the effects of impeller blade passing frequency attenuated gradually, and the effect of diffuser vanes was increased gradually.

Keywords: centrifugal pump; pressure fluctuation; unsteady flow; numerical simulation

---

1. Introduction

Centrifugal pumps are of vital importance for a variety of commercial and industrial applications [1], which could handle a great quantity of liquids at relatively high pressures [2]. For those applications that require extremely high pressures, multistage centrifugal pumps, which usually have more than two or even up to dozens of stages, are used. Owing to the high reliability requirements for a long shaft, the hydraulic design of the multistage centrifugal pumps is more complex than that of the single-stage pumps [3]. Recently, many researchers have conducted both numerical simulations and experiments on the efficiency of multistage centrifugal pumps to enhance their hydraulic design [4–6].

The operation process of multi-stage centrifugal pumps is often accompanied by vibration and noise. When the pump produces static pressure, it also produces dynamic pressure, which is also named as pressure fluctuation along with the periodic changing features like AC signal. Pressure fluctuations are inevitable during the operation of a pump, having strong unsteady characteristics that may lead to strong vibration and noise under some special working conditions [7].

In previous studies, a series of numerical investigations of pressure fluctuations within a complete single-stage centrifugal pump were undertaken. Khalifa et al. [8] investigated the pressure fluctuations and vibration in a single-stage centrifugal pump experimentally under different flow conditions. They found that under different flow rates, the vibrations have a close relationship with the pressure fluctuations, and the vibration could be controlled by optimizing the gap between the volute tongues and the impeller. Guelich et al. [9] studied the physical mechanisms behind the pressure fluctuations under the design flow conditions. They found that the pressure fluctuations mainly depend on the geometry of the diffuser and the impeller by performing 36 tests with different pump designs. Spence et al. [10] used the multi-block, structured mesh CFD (Computational Fluid Dynamics) code TASC flow to study the pressure fluctuations in a centrifugal pump for three different flow rates.

They found that the volute gap and the arrangement of vanes have the greatest effect on the different monitored positions. In a previous work of ours [11], we simulated the transient flow field of a centrifugal pump by employing ANSYS-CFX software based on the frozen rotor technique and utilizing the Shear Stress Transport (SST) k-ω turbulence model. The results shown that the pressure fluctuation is mainly affected by the number of impeller blades, but the influence decreased gradually when the water flowed into the diffuser. Pei et al. [12] focused on the numerical simulation of unsteady flow under part-load conditions, the results shown that the intensity of pressure fluctuation is larger than that of design flow condition because of leakage flow. Jiang et al. [13] studied the clocking effect and unsteady radial force in a centrifugal pump by numerical and experimental methods, they found that the unsteady flow and pump performance mainly effected by the relative position between the diffuser the volute. Wang et al. [14] carried out an experimental measurement to study the influence of rotating speed and flow rate on the pressure fluctuation in a double-suction centrifugal pump. They found that the pressure fluctuation under zero flow rate is two times that of the pressure fluctuation under the design flow rate. Posa and Lippolis [15] used Large Eddy Simulations (LES) to investigate the influence of diffuser setting angle and operating conditions on the pressure fluctuation and unsteady flow in a centrifugal pump. They found that the larger radial gaps between diffuser and impeller could lead to lower pressure fluctuations, but the lower flow rate could lead to higher pressure fluctuations.

While numerous studies have investigated the pressure fluctuation and unsteady characteristics in different kinds of pumps, which includes the effects of working conditions [14,16], inlet guide vanes [17], tip clearance [18], and volute curvature [19–21], but the influence of the number of diffuser vanes on the pressure fluctuation in the centrifugal pump is unclear and further investigation is needed. So, in this paper a numerical approach was adopted in a centrifugal pump with a different number of diffuser vanes, and the pressure fluctuations are analyzed and compared in detail. The results could supply a basis for further design improvement of the pump body structure and working reliability.

2. Pump Geometry

In this study, we have chosen a standard Electrical Submersible Pump (ESP) as the research objective. The main design parameters of this ESP are as follows: design flow rate Q_{des} = 36 m^3/h; rotating speed n = 2850 r/min; total head H = 75 m; number of stages N = 7; specific speed n_s = 175.8; design efficiency η = 72%; and maximum power P = 11 kW. More details about the parameters of pump geometry could be found in one of our previous paper [22].

This kind of ESP is usually manufactured with plastic materials, in order to ensure there is enough space for the unloading of casting, so the impeller was designed with five blades. In most cases, the number of diffuser vanes is selected to be relative prime number with the number impeller blades. In order to study the effects of the diffuser vanes number on the unsteady flow and pressure fluctuation, the same profile was used for the diffuser vanes and their number was adjusted to 6, 7, and 8 for three different configurations. The solid model of the impeller is shown in Figure 1, and the three configurations with different diffuser vanes number are shown in Figure 2.

Figure 1. Solid model of impeller with five blades.

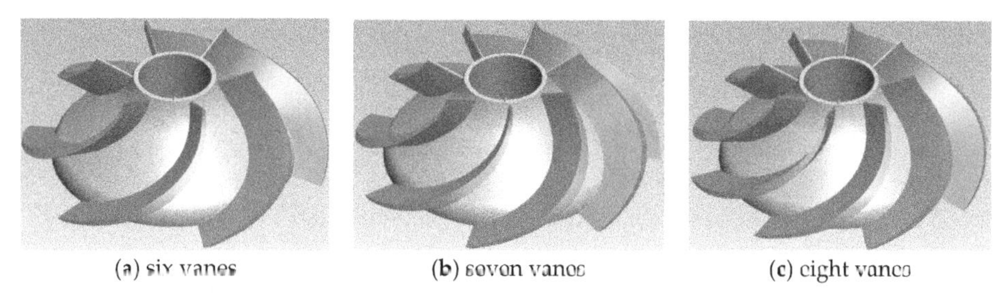

(**a**) six vanes (**b**) seven vanes (**c**) eight vanes

Figure 2. Solid model of three configurations with different diffuser vanes number.

3. Mathematical Model

The Navier–Stokes equation (N–S) and continuity equation of an incompressible fluid are [23]:

$$\frac{\partial u_i}{\partial t} + u_j \frac{\partial u_i}{\partial x_j} + \frac{1}{\rho}\frac{\partial p}{\partial x} - \nu \nabla^2 u_i = 0, \tag{1}$$

$$\frac{\partial u_i}{\partial x_i} = 0. \tag{2}$$

where ∇^2 is Laplace operator, u_i is the flow velocity component in i direction, $i = 1, 2, 3$ respectively represent the flow velocity component in the three directions of x, y, z.

Carrying out partial derivative operations on the Equation (1) we get:

$$\frac{\partial^2 u_i}{\partial t \partial x_i} + \frac{\partial u_j}{\partial x_i}\frac{\partial u_i}{\partial x_j} + u_j \frac{\partial^2 u_i}{\partial x_i \partial x_j} + \frac{1}{\rho}\frac{\partial^2 p}{\partial x_i^2} - \nu \nabla^2 \frac{\partial u_j}{\partial x_i} = 0. \tag{3}$$

Using formula (2) into Equation (3) we get:

$$\frac{\partial^2 p}{\partial x_i^2} = \nabla^2 p = -\rho\left(\frac{\partial u_i}{\partial x_j}\frac{\partial u_j}{\partial x_i}\right). \tag{4}$$

If $u_i = u_i + u_i'$, $u_j = u_j + u_j'$, $p = p + p'$, put them into Equation (4) and Equation (2) to get:

$$\nabla^2 \overline{p} + \nabla^2 p' = -\rho \frac{\partial^2}{\partial x_i \partial x_j}\left(\overline{u_i u_j} + \overline{u_i} u_j' + u_i' \overline{u_j} + u_i' u_j'\right), \tag{5}$$

$$\frac{\partial \overline{u_i}}{\partial x_i} + \frac{\partial u_i'}{\partial x_i} = 0, \tag{6}$$

where $\overline{u_i}$, $\overline{u_j}$, $\overline{p}$ represent time-averaged value, u_i', u_j', p' represent fluctuating value, $u_i = \overline{u_i}$, $u_j = \overline{u_j}$, $p = \overline{p}$, $\overline{u_i'} = 0$, $\overline{u_j'} = 0$, $\overline{p'} = 0$.

Taking the average of Equations (5) and (6) separately we get:

$$\nabla^2 \overline{p} = -\rho \frac{\partial^2}{\partial x_i \partial x_j}\left(\overline{u_i u_j} + \overline{u_i' u_j'}\right), \tag{7}$$

$$\frac{\partial \overline{u_i}}{\partial x_i} = 0. \tag{8}$$

Substituting Equation (7) from Equation (5) to obtain the expression of pressure fluctuation equation:

$$\nabla^2 p' = -\rho\left[\frac{\partial^2\left(\overline{u_i}u_j' + u_i\overline{u_j}\right)}{\partial x_i \partial x_j} + \frac{\partial^2}{\partial x_i \partial x_j}\left(u_i' u_j' - \overline{u_i' u_j'}\right)\right]. \tag{9}$$

Equation (6) minus Equation (8) we obtain the expression of pressure fluctuation continuous equation:

$$\frac{\partial u_i'}{\partial x_i} = 0. \tag{10}$$

Using Equation (10) into Equation (9) we finally get:

$$\nabla^2 p' = -\rho\left[2\frac{\partial \overline{u_i}}{\partial x_j}\frac{\partial u_j'}{\partial x_i} + \frac{\partial^2}{\partial x_i \partial x_j}\left(u_i' u_j' - \overline{u_i' u_j'}\right)\right], \tag{11}$$

where ρ is the density of water, p' is pressure fluctuation, u_i' is fluctuating velocity, $\overline{u_i}$ is time-averaged velocity.

Equation (11) indicates that the pressure fluctuation is mainly caused by the fluctuation of velocity, the first term on the right-hand side of the equation represents the interaction between mean velocity and fluctuating velocity, which is the mean velocity field distortion and generally called a "quick response term" [24,25]. The time-averaged velocity gradient expresses mean shear stress term, therefore, it is also known as the "turbulence-shear". The second term on the right side of the equation is the change of pressure fluctuation caused by the nonlinear effect in the fluctuating velocity field, commonly known as "turbulence-turbulence". Equation (11) can be understood as the basic equation of pressure fluctuation, in which we can see that pressure fluctuation is equal to the comprehensive effects of the two categories of the effects in brackets of the right side: the first category is the interaction of "shear-turbulence", and the second category is the interaction of "turbulence-turbulence". At the same time, it can be seen from the Equation (11) that there is no direct relation between the viscosity of the fluid and the pressure fluctuation. The viscosity of the fluid only plays an indirect role in pressure fluctuation [25].

4. Numerical Setup

4.1. Computational Domain

The solid models were created and assembled to form the computational domain in Pro/E Wildfire 5.0 software, which include the inlet section, impeller, diffuser, lateral cavity, and outlet section [26]. In order to balance the total grid number of points, calculation time, and the computer capability [27], only one stage model was chosen in this study, as shown in Figure 3.

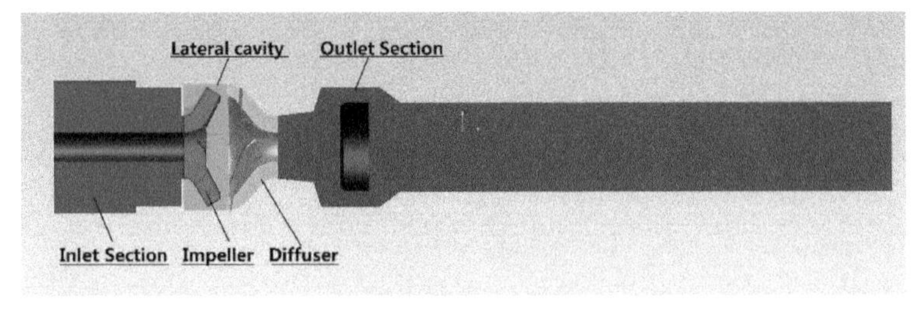

Figure 3. Computational domain.

4.2. Meshes

ANSYS-ICEM-CFD 14.5 software [28] was used to build the structured meshes associated with the block topology of Y-type and O-type. The mesh growth rate of boundary layer were considered to ensure the overall mesh thickness and density. The appropriate mesh density was decided according to the gird independence test [29,30]. Table 1 summarizes five schemes of different mesh numbers. By comparing the pump performance predicted by the same numerical methods we found that with the increase of mesh number, the numerical value of single-stage head, single-stage power, and efficiency show a stable trend. Between the schemes No. 4 and No. 5, increasing mesh has no influence on the simulation results, which means that the scheme No. 4 has enough accuracy with enough mesh density. Therefore, by considering the grid independence test and the computer capability, scheme No. 4 was selected as the optimal scheme of mesh for the numerical investigation. For the whole single-stage computational domain, the total of mesh elements is almost 4.9 million. Meanwhile, it was ensured that the entire computational domain $30 < y^{+} < 60$, which indicates that the near-wall mesh nodes are in a log-law layer. Figure 4 is the final mesh used for the impeller and the diffuser.

Table 1. Grid independence test.

Cases No.	Grid Number		Numerical Results		
	Impeller	Diffuser	Single-Stage Head (m)	Single-Stage Power (kW)	Efficiency (%)
1	43,695	62,564	10.41	1.5739	64.81
2	139,200	316,540	10.54	1.4697	70.29
3	313,750	509,208	10.53	1.4021	73.58
4	516,780	937,440	10.56	1.4033	73.77
5	950,895	1,297,380	10.57	1.4042	73.74

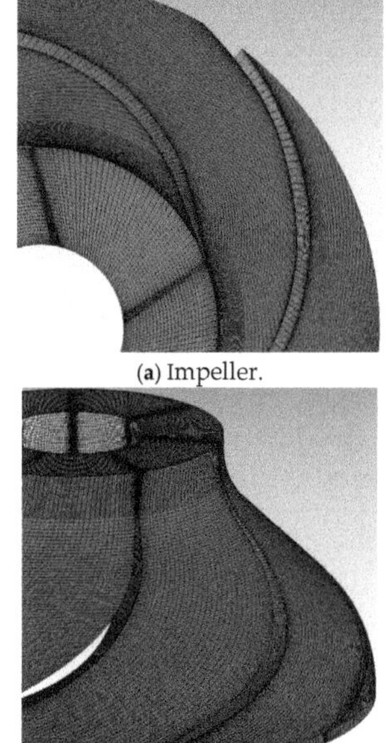

(**a**) Impeller.

(**b**) Diffuser.

Figure 4. Structured mesh of the impeller and diffuser.

4.3. Numerical Settings

ANSYS-Fluent 14.5 software [31] was used to run the numerical simulation for this study, in which the SST k-ω turbulence model, and the SIMPLEC algorithm was used to solve the discrete difference equations of the second-order upwind scheme. As shown in Figure 5, no-slip boundary conditions were used on the wall surface, inlet boundary conditions of mass flow were used on the pump inlet, and outlet boundary conditions of pressure were used on the outlet of the pump. For instance, the mass flow rate for inlet was set as 10 kg/s at the design flow rate, and the total pressure for the outlet was set as 10,135 Pa. Using sliding grid technology, the impeller region in the flow channel was set to be a rotating grid coordinate system provided by ANSYS-Fluent, while the diffuser and other hydraulic components were set to be a stationary coordinate system. After reaching convergence of steady calculation based on the fixed rotor coordinate system (Frozen Rotor), the flowfield was used as the initial value to start the unsteady calculation. The Principle of a Frozen Rotor is to set the impeller part to be a rotating reference coordinate system, and the static and dynamic interface is coupled by combing with sliding mesh technology [32,33].

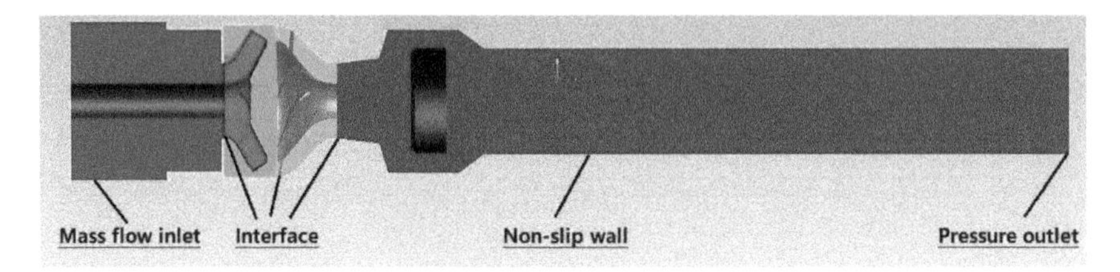

Figure 5. Boundary conditions.

4.4. Time Step

In general, in the periodic numerical simulation, the time step needs to satisfy the Courant number criterion [34], which is expressed as:

$$C_o = \frac{|\overline{v}|\Delta t}{l} \leq 100. \quad (12)$$

In the above formula, $|\overline{v}|$ is the absolute value of the estimated mean velocity, l is the smallest size of the grid, and C_o is required to be no more than 100. When the numerical convergence is not good, it is appropriate to take smaller values.

According to the requirement of sampling theorem, 2.56–4 times of the signal maximum frequency is selected generally for the time step. In this study, the impeller has five blades, the rotational speed is 2850 r/min, so the blade passing frequency (BPF) is 332.5 Hz. If there are at least seven points in a BPF cycle, and the desirable maximum value of the time step is 4.2966×10^{-4} s accordingly.

If the time step size is too large, some minor changes of the pressure will not be captured, but time steps that are too small will also lead to a significant increase in computing time [35,36]. Therefore, considering the computer configuration, the time step Δt was chosen as 5.8478×10^{-5} s, the impeller rotated one degree for one time step, 360 time steps were required for each rotation of the impeller. For this model, after 3600 time steps, that is, after 10 cycles of impeller rotation, the pressure fluctuation was periodic and the convergence condition was determined [37,38]. Defining this time as $t = 0$, and selecting a whole circle of the impeller rotation (360 time steps) as the analysis period for unsteady flow.

5. Results and Discussion

5.1. Pump Performance Validation

To validate the numerical method, the prototype pump with seven diffuser vanes was manufactured and measured in an open test bed. The detailed schematic diagram of this test bed can be found in our previous paper [22,38]. The comparison of pump performance in terms of predicted and experimental results are shown in Figure 6. Actually, the model ESP used in the performance test has seven stages, so the single-stage and power single-stage head were converted by the total power and total head, and then compared with the CFD predicted performance. As shown in Figure 6, the experimental results are slightly lower than the CFD results, but they have the same changing trend under different flow rates. The pump efficiency predicted by CFD has good agreement with the experimental efficiency. The comparison confirms the precision and accuracy of the numerical method used in this study. For the sources of error between the CFD and the test, one of the most important reasons is the one stage computational domain was used by considering the balance of computer capability and meshes number, while the tested prototype pump had seven stages. Generally speaking, the flow at the impeller inlet of the first stage is irrotational flow, but the flow at the impeller inlet of other stages is rotational flow, so in most cases the pump power and head of the first stage is higher than that of the other stages [22,38]. Another reason is that the leakage flow was not considered in this study, which means the volume losses caused by leakage flow was neglected. The real pump model with seven stages could be simulated in a powerful computer in a future study to obtain more arcuate results.

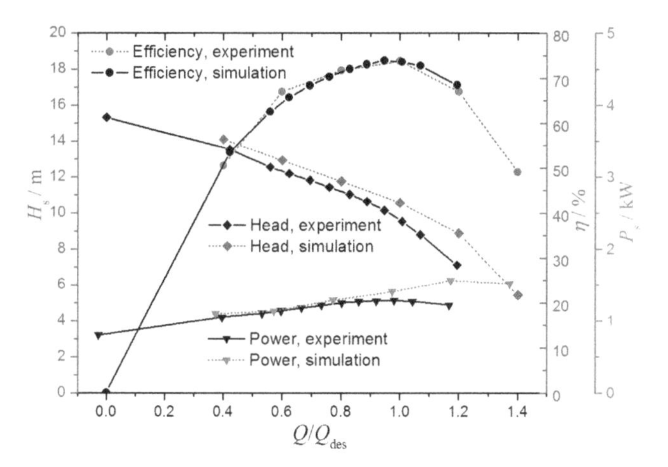

Figure 6. Comparison of pump performance between CFD and experimental results.

5.2. *Pressure Fluctuation Analysis*

In this study, the pressure coefficient is used to measure the pressure fluctuation. The pressure coefficient is defined as:

$$C_p = \frac{p - \bar{p}}{\frac{1}{2}\rho u^2}, \tag{13}$$

where p is the static pressure at the location of the monitoring points; u is the circumferential velocity of impeller outlet; and $\bar{p}$ is mean static pressure at the monitoring points during the rotation of the impeller. After 10 cycles of impeller rotation, the pressure fluctuations at the inlet and outlet of the impeller show the regular periodicity. In the process of the impeller rotating 360 degrees (360 time steps), the pressure fluctuation at each monitoring point was collected to analyze the unsteady flow in the flow channel.

To compare the transient pressure fluctuation characteristics in the flow channel of pump, seven monitoring points were arranged in the same position in three scenarios of space diffuser, 6P7–6P1 represents the pump flow channel of the six-vane configuration, 7P7–7P1 represents the pump flow channel of the seven-vane configuration, 8P7–8P1 represents the pump flow channel of the eight-vane configuration. The position of the seven monitoring points in the flow channel is shown in Figure 7.

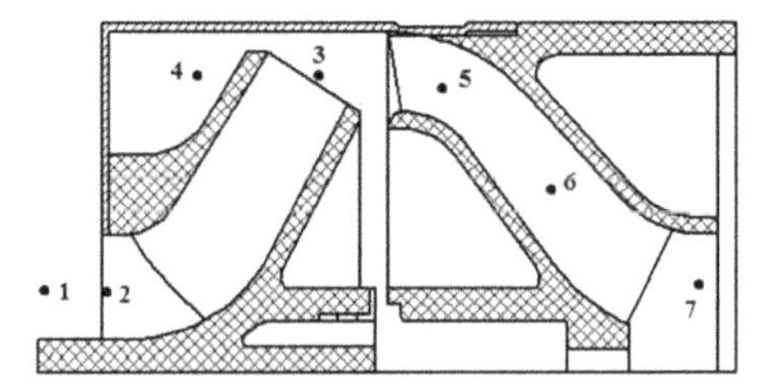

Figure 7. Location of monitoring points.

Figure 8 presents the pressure fluctuation for each monitoring point in the pump flow channel of the six-vane configuration. Generally speaking, the pressure fluctuation shows stable periodicity at the impeller outlet and inlet. During the process of impeller rotating 360 degrees, five peaks and five troughs are seen, which is equal to the number of impeller blades. The pressure fluctuation amplitude at P1 located at the impeller inlet is the lowest one. The pressure fluctuation amplitude at P3 is the most intense one, which is located at the intersection of the diffuser vanes and the impeller blades. The static pressure changes along with the rotation of the impeller. In particular, when the impeller blade rotates from the pressure surface to the suction surface, the static pressure changes rapidly from the smallest value to the largest value. By comparing the amplitude of each monitoring points, we could find that the pressure fluctuation signal is generated at the impeller outlet, and in the process of its transfer to the pump outlet, its intensity gradually decays.

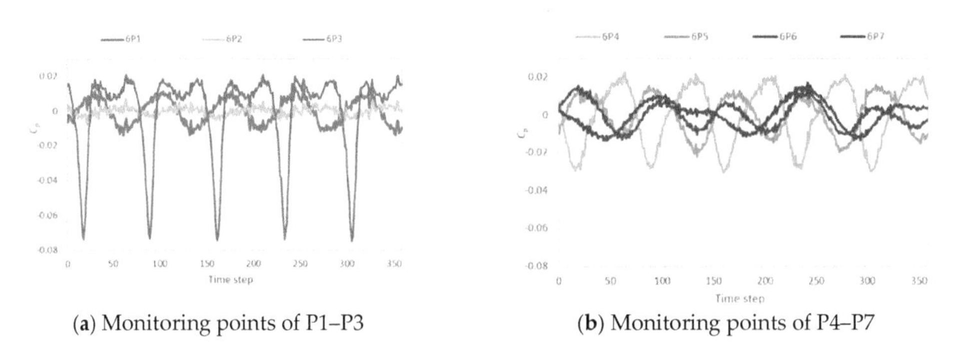

(**a**) Monitoring points of P1–P3 (**b**) Monitoring points of P4–P7

Figure 8. Pressure fluctuation of the six-vane configuration.

Figures 9 and 10 show time domain diagram of pressure fluctuation at each monitoring point for the configurations of 7 vanes and 8 vanes, respectively. It can be seen that pressure fluctuation at the monitoring point in the impeller inlet (P1, P2) and the impeller outlet (P3, P4) is similar to that of the six-vane configuration, exhibiting again five peaks and five troughs. It shows that at the impeller outlet and the impeller inlet, the pressure fluctuation is mainly affected by the impeller blade number, and the diffuser vane number has no effect on the pressure fluctuation at these locations. However, after entering the diffuser vane, the pressure fluctuations at P5, P6, and P7 change with the increase of the diffuser vane number and the pulse intensity increases with the increase of the vane number. Especially for the eight-vane configuration, the pressure fluctuation amplitude at the diffuser outlet is relatively large.

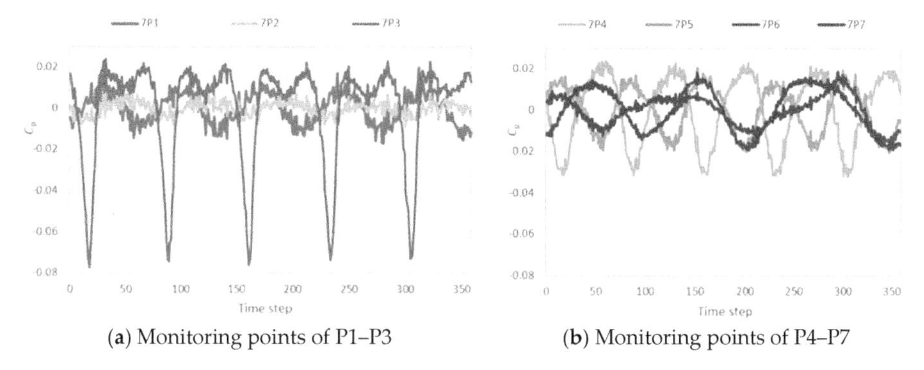

(**a**) Monitoring points of P1–P3 (**b**) Monitoring points of P4–P7

Figure 9. Pressure fluctuation of the seven-vane configuration.

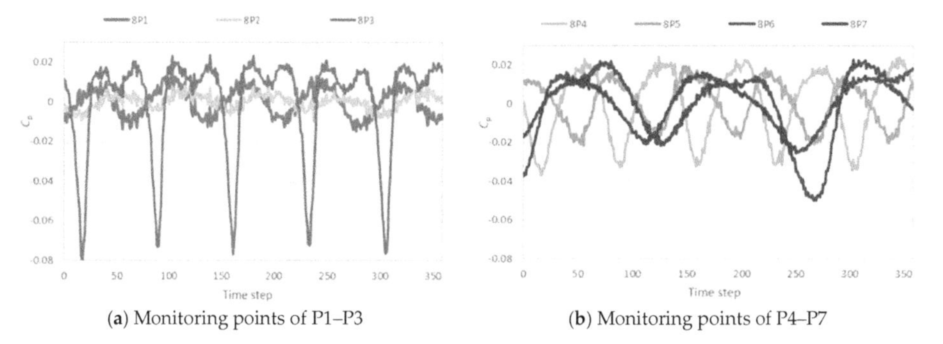

(**a**) Monitoring points of P1–P3 (**b**) Monitoring points of P4–P7

Figure 10. Pressure fluctuation of the eight-vane configuration.

5.3. *Frequency Domain Analysis*

In the rotating coordinate system, the static and dynamic interference between the diffuser vanes and the impeller blades is similar to the pressure fluctuation of sinusoid and cosine waves. This transient periodic fluctuation can be decomposed into Fourier series. In this study, we used Fast Fourier Transform (FFT) to analyze the frequency domain characteristics of the pressure fluctuation between the pump stages. The fundamental frequency of the internal pressure fluctuation is generally shaft frequency (f_s), so the frequency domain is analyzed by comparing with the shaft frequency f_S. The frequency is divided by the fundamental frequency to get a multiple of the fundamental frequency as a horizontal coordinate. The rotating speed of the investigated pump is 2850 r/min, so the shaft frequency is 47.5 Hz.

Figure 11 presents the spectrum of pressure fluctuation at each monitoring point in the pump flow channel of the six-blade configuration. The frequency distribution at each monitoring point is basically the same, and fluctuating frequency appears at the double leaf frequency of Blade Passing Frequency (BPF, f/f_s = 5) and its harmonic frequencies. It is further explained that the static and dynamic interference at the outlet of the impeller is the main cause of the pressure fluctuation, and the pressure fluctuation signal is gradually fading during the process of transferring at the inlet side of the impeller and the outlet side of the pump.

Figures 12 and 13 show the spectrum of pressure fluctuation at each monitoring point in the seven-vane and eight-vane configurations, respectively. Compared to the six-vane configuration, the frequency domain characteristics of pressure fluctuation are very similar. The closer to the impeller, the greater the pressure coefficient pulsation amplitude, and the fluctuating amplitude shows an obvious tendency of obviously increasing. As the number of the diffuser vanes increases, the amplitude of pressure fluctuation at each monitoring point in the diffuser vane increases gradually, which indicates that the velocity of the fluid is converted to pressure energy by the diffuser vanes. With the increase of static pressure, the influence of BPF decays gradually, and the effect of the diffuser becomes gradually obvious. The lower number of diffuser vanes is beneficial to obtain a weak pressure fluctuation intensity.

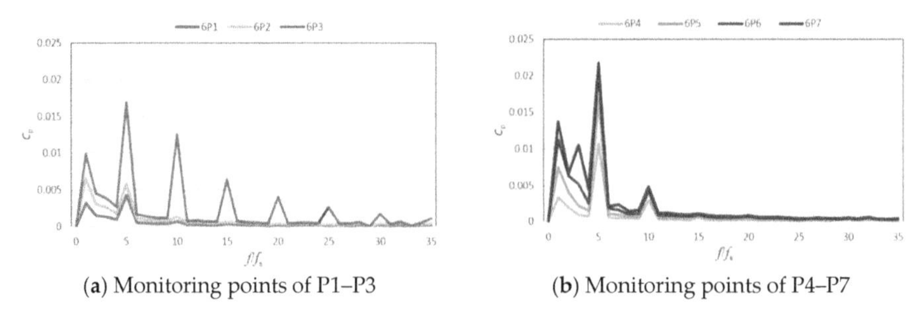

(**a**) Monitoring points of P1–P3 (**b**) Monitoring points of P4–P7

Figure 11. Frequency domain analysis of the six-vane configuration.

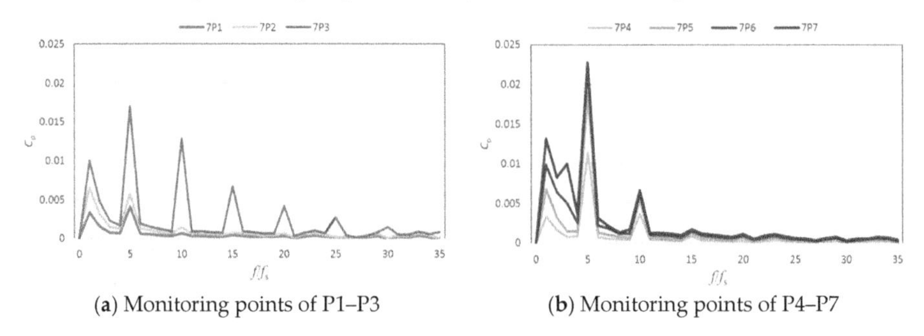

(**a**) Monitoring points of P1–P3 (**b**) Monitoring points of P4–P7

Figure 12. Frequency domain analysis of the seven-vane configuration.

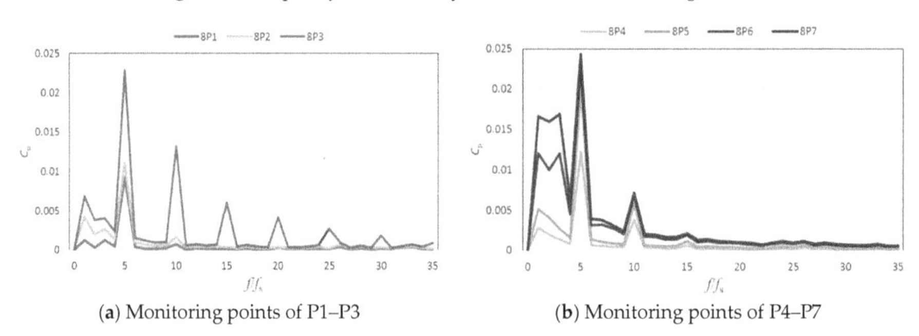

(**a**) Monitoring points of P1–P3 (**b**) Monitoring points of P4–P7

Figure 13. Frequency domain analysis of the eight-vane configuration.

5.4. Static Pressure Distributions

Figures 14–16 compare the static pressure distribution on the same diameter section in the pump flow channel of different blade configurations. It can be seen that the static and dynamic interference between the diffuser and the impeller makes the pressure field periodically change. At the impeller outlet, the blade is no longer working on the fluid, the flow rate begins to decline, flow separation appears gradually, and two different regions appear at the pressure fields of the impeller outlet. A low-pressure area is located at the outlet of the impeller blades, while the other one is in the mainstream area downstream from the blade pressure surface, which is also known as the wake-jet, as shown in Figure 14. Wake changes gradually along with the rotating impeller, while the location of the vortex core remains unchanged, and always appears at the exit of the blades. The generation of static and dynamic interference process can be attributed to the jet and wake flow that then flows into the diffuser vanes.

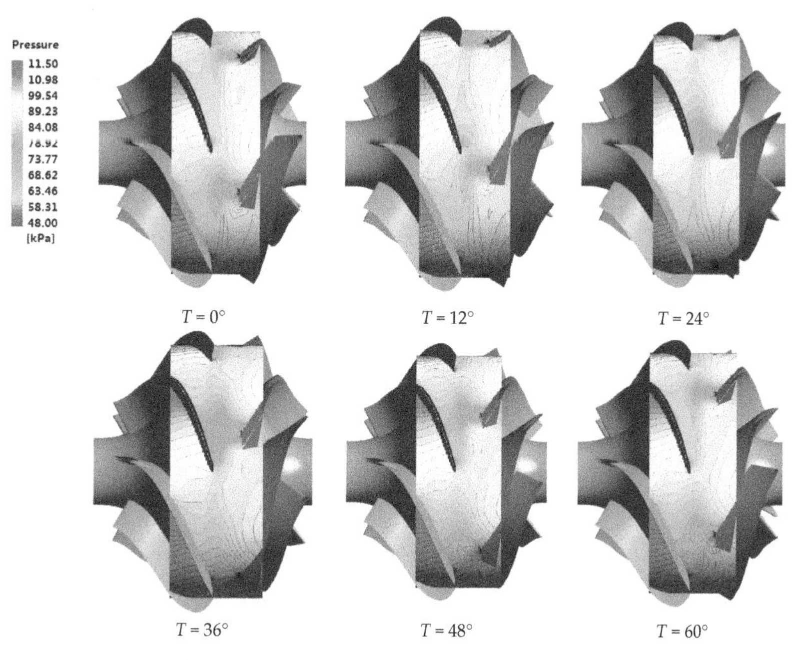

Figure 14. Static pressure distribution of the six-vane configuration.

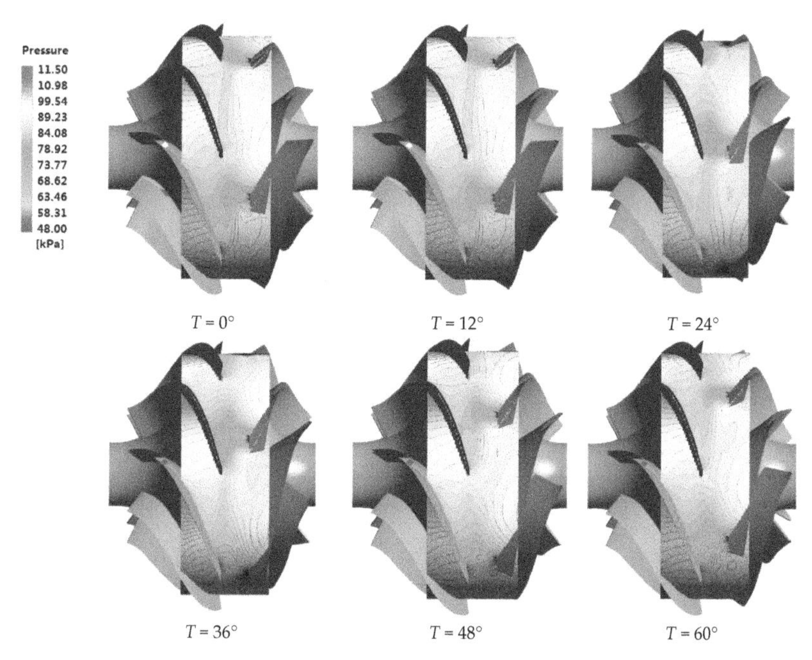

Figure 15. Static pressure distribution of the seven-vane configuration.

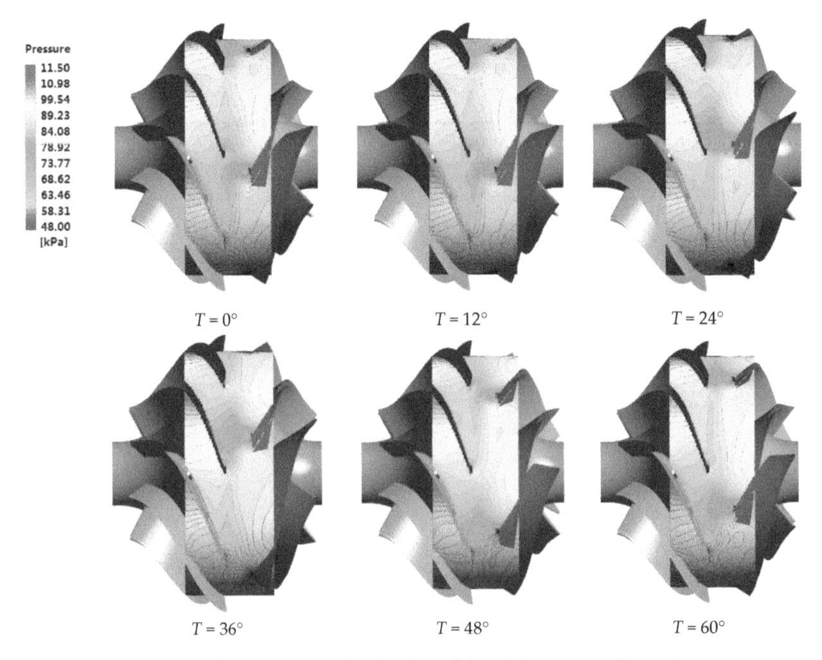

Figure 16. Static pressure distribution of the eight-vane configuration.

As shown in Figure 15, with the increase of diffuser vane number from six to seven, the flow channel area of diffuser decreases. Accordingly, the frequency of the dynamic and static interference between impeller blades and diffuser vanes increases, and the amplitude of pressure fluctuation increases obviously. Meanwhile, Figure 16 shows that the evolution period of static pressure in the flow channels of three configurations is about 72 degrees, which is the same as the rotation period of the impeller blades. When the outlet side of impeller blade converges with the inlet side of the diffuser vanes, the rotor-stator interaction is the most intense. Simultaneously, there is an obvious low-pressure zone at the outlet of impeller blade, which is close to the pressure surface. With the rotation of the impeller, when the outlet side of the impeller blade is gradually close to the inlet side of the diffuser vanes, the low pressure zone begins to diffuse from the pressure surface of the impeller blade to the suction surface of the diffuser. Then with the interleaving between the impeller channel and the diffuser channel, wake, and jet at the impeller outlet flows into the diffuser vanes successively, the static pressure distribution in the diffuser has a periodic evolution, the area of the low pressure region also changes with the rotation of the impeller blades. When the inlet of the diffuser vanes overlaps with the outlet of impeller blade, the area of low-pressure region generated by the wake is the smallest, and the pressure fluctuation amplitude inside the diffuser is the smallest.

6. Conclusions

This paper introduces the basic theory and research methods of the pressure fluctuation, and the unsteady numerical simulation under multi-working conditions of three different configurations of a centrifugal pump corresponding to different numbers of diffuser vanes. The unsteady pressure fluctuation in the transient flow is obtained, and the frequency domain and the time domain distribution are analyzed and compared by means of Fast Fourier Transform. With the increase of the number of diffuser vanes, the amplitude of each monitoring point in the diffuser increases gradually, which indicates that the velocity of the fluid is converted to pressure due to the operation of diffuser. The lower number of diffuser vanes is beneficial to obtain the weaker pressure fluctuation intensity. It can be clearly seen that with the static pressure gradually increasing, the effects of the impeller blade passing frequency decay gradually, and the effect of the diffuser vanes increases gradually. The results could supply a basis for further design improvement of the pump body structure and working reliability.

Author Contributions: Conceptualization, L.B., L.Z. and Y.Z.; Methodology, L.Z.; Resources: W.S.; Writing-Original Draft Preparation, L.B.; Writing-Review & Editing, C.H.; Supervision, L.Z.

Funding: This research was funded by the National Natural Science Foundation of China (Grant No. 51609106 and 51805214), China Postdoctoral Science Foundation (Grant No. 2015M581737 and 2017T100331), and Jiangsu Province 333 Projects (Grant No. BRA2017353).

Conflicts of Interest: The authors declare no conflict of interest.

References

1. Gülich, J.F. *Centrifugal Pumps*; Springer: Berlin/Heidelberg, Germany, 2008.
2. Wang, C.; Shi, W.; Wang, X.; Jiang, X.; Yang, Y.; Li, W.; Zhou, L. Optimal design of multistage centrifugal pump based on the combined energy loss model and computational fluid dynamics. *Appl. Energy* **2017**, *187*, 10–26. [CrossRef]
3. Yang, J.; Pavesi, G.; Liu, X.; Xie, T.; Liu, J. Unsteady flow characteristics regarding hump instability in the first stage of a multistage pump-turbine in pump mode. *Renew. Energy* **2018**, *127*, 377–385. [CrossRef]
4. Iino, T.; Sato, H.; Miyashiro, H. Hydraulic axial thrust in multistage centrifugal pumps. *J. Fluids Eng.* **1980**, *102*, 64–69. [CrossRef]
5. Xue, Z.; Cao, X.; Wang, T. Vibration test and analysis on the centrifugal pump. *J. Drain. Irrig. Mach. Eng.* **2018**, *36*, 472–477.
6. Chen, H.; Zhang, T.; Wang, C. Structural characteristics of ultrahigh-pressure multistage centrifugal pump typed SDZ310. *J. Drain. Irrig. Mach. Eng.* **2018**, *36*, 567–572.
7. Berten, S.; Dupont, P.; Farhat, M.; Avellan, F. Rotor-stator interaction induced pressure fluctuations: CFD and hydroacoustic simulations in the stationary components of a multistage centrifugal pump. In Proceedings of the ASME/JSME 2007 5th Joint Fluids Engineering Conference, San Diego, CA, USA, 30 July–2 August 2007; pp. 963–970.
8. Khalifa, A.E.; Al-Qutub, A.M.; Ben-Mansour, R. Study of pressure fluctuations and induced vibration at blade-passing frequencies of a double volute pump. *Arab. J. Sci. Eng.* **2011**, *36*, 1333–1345. [CrossRef]
9. Guelich, J.F.; Bolleter, U. Pressure pulsations in centrifugal pumps. *J. Vib. Acoust.* **1992**, *114*, 272–279. [CrossRef]
10. Spence, R.; Amaral-Teixeira, J. A CFD parametric study of geometrical variations on the pressure pulsations and performance characteristics of a centrifugal pump. *Comput. Fluids* **2009**, *38*, 1243–1257. [CrossRef]
11. Zhou, L.; Shi, W.; Bai, L.; Lu, W.; Li, W. Numerical investigation of pressure fluctuation and rotor-stator interaction in a multistage centrifugal pump. In Proceedings of the ASME 2013 Fluids Engineering Division Summer Meeting, Incline Village, NV, USA, 7–13 July 2013.
12. Ji, P.; Yuan, S.; Li, X.; Yuan, J. Numerical prediction of 3-D periodic flow unsteadiness in a centrifugal pump under part-load condition. *J. Hydrodyn. Ser. B* **2014**, *26*, 257–263.
13. Jiang, W.; Li, G.; Liu, P.; Fu, L. Numerical investigation of influence of the clocking effect on the unsteady pressure fluctuations and radial forces in the centrifugal pump with vaned diffuser. *Int. Commun. Heat Mass Transf.* **2016**, *71*, 164–171. [CrossRef]
14. Wang, Z.; Qian, Z.; Lu, J.; Wu, P. Effects of flow rate and rotational speed on pressure fluctuations in a double-suction centrifugal pump. *Energy* **2019**, *170*, 212–227. [CrossRef]
15. Posa, A.; Lippolis, A. Effect of working conditions and diffuser setting angle on pressure fluctuations within a centrifugal pump. *Int. J. Heat Fluid Flow* **2019**, *75*, 44–60. [CrossRef]
16. Tan, L.; Zhu, B.; Wang, Y.; Cao, S.; Gui, S. Numerical study on characteristics of unsteady flow in a centrifugal pump volute at partial load condition. *Eng. Comput.* **2015**, *32*, 1549–1566. [CrossRef]
17. Wang, L.; Tan, L.; Zhu, B.; Cao, S.; Wang, B. Numerical investigation of influence of inlet guide vanes on unsteady flow in a centrifugal pump. *Proc. Inst. Mech. Eng. Part C J. Mech. Eng. Sci.* **2015**, *229*, 3405–3416.
18. Feng, J.; Luo, X.; Guo, P.; Wu, G. Influence of tip clearance on pressure fluctuations in an axial flow pump. *J. Mech. Sci. Technol.* **2016**, *30*, 1603–1610. [CrossRef]

19. Alemi, H.; Nourbakhsh, S.; Raisee, M.; Raisee, M.; Najafi, A.F. Effects of volute curvature on performance of a low specific-speed centrifugal pump at design and off-design conditions. *ASME J. Turbomach.* **2015**, *137*, 041009. [CrossRef]
20. Melzer, S.; Müller, T.; Schepeler, S.; Kalkkuhl, T.; Skoda, R. Experimental and numerical investigation of the transient characteristics and volute casing wall pressure fluctuations of a single-blade pump. *Proc. Inst. Mech. Eng. Part E J. Process Mech. Eng.* **2019**, *233*, 280–291. [CrossRef]
21. Shibata, A.; Hiramatsu, H.; Komaki, S.; Miyagawa, K.; Maeda, M.; Kamei, S.; Hazama, R.; Sano, T.; Iino, M. Study of flow instability in off design operation of a multistage centrifugal pump. *J. Mech. Sci. Technol.* **2016**, *30*, 493–498. [CrossRef]
22. Zhou, L.; Bai, L.; Shi, W.; Li, W.; Wang, C.; Ye, D. Numerical analysis and performance experiment of electric submersible pump with different diffuser vanes number. *J. Braz. Soc. Mech. Sci. Eng.* **2018**, *40*, 1–11. [CrossRef]
23. Naumann, H.; Yeh, H. Lift and pressure fluctuations of a cambered airfoil under periodic gusts and applications in turbomachinery. *J. Eng. Power* **1973**, *95*, 1–10. [CrossRef]
24. Gonzalez, J.; Fernández, J.; Blanco, E.; Santolaria, C. Numerical simulation of the dynamic effects due to impeller-volute interaction in a centrifugal pump. *J. Fluids Eng.* **2002**, *124*, 348–355. [CrossRef]
25. Zhang, J.; Xia, S.; Ye, S.; Xu, B.; Song, W.; Zhu, S.; Xiang, J. Experimental investigation on the noise reduction of an axial piston pump using free-layer damping material treatment. *Appl. Acoust.* **2018**, *139*, 1–7. [CrossRef]
26. *Pro/ENGINEER Wildfire 5.0*; Parametric Technology Corporation: Boston, MA, USA, 2008.
27. Qian, J.Y.; Chen, M.R.; Liu, X.L.; Jin, Z.J. A numerical investigation of the flow of nanofluids through a micro Tesla valve. *J. Zhejiang Univ. Sci. A* **2019**, *20*, 50–60. [CrossRef]
28. *ANSYS ICEM CFD User Manual*; ANSYS ICEM CFD 14.5; ANSYS, Inc.: Canonsburg, PA, USA, 2012.
29. Li, X.; Gao, P.; Zhu, Z.; Li, Y. Effect of the blade loading distribution on hydrodynamic performance of a centrifugal pump with cylindrical blades. *J. Mech. Sci. Technol.* **2018**, *32*, 1161–1170. [CrossRef]
30. Moore, J.J.; Palazzolo, A.B. Rotordynamic force prediction of whirling centrifugal impeller shroud passages using computational fluid dynamic techniques. *J. Eng. Gas Turbines Power* **2001**, *123*, 910–918. [CrossRef]
31. *ANSYS Fluent Theory Guide*; Release 14.5; ANSYS, Inc.: Boston, MA, USA, 2012.
32. Gonzalez, J.; Santolaria, C. Unsteady flow structure and global variables in a centrifugal pump. *J. Fluids Eng.* **2006**, *128*, 937–946. [CrossRef]
33. Majidi, K. Numerical study of unsteady flow in a centrifugal pump. In Proceedings of the ASME Turbo Expo 2004: Power for Land, Sea, and Air, Vienna, Austria, 14–17 June 2004; pp. 805–814.
34. Desbrun, M.; Gascuel, M.-P. Smoothed particles: A new paradigm for animating highly deformable bodies. In *Computer Animation and Simulation'96*; Springer: Vienna, Austria, 1996; pp. 61–76.
35. Belytschko, T.; Neal, M.O. Contact-impact by the pinball algorithm with penalty and Lagrangian methods. *Int. J. Numer. Methods Eng.* **1991**, *31*, 547–572. [CrossRef]
36. Kim, J.; Moin, P. Application of a fractional-step method to incompressible Navier-Stokes equations. *J. Comput. Phys.* **1985**, *59*, 308–323. [CrossRef]
37. Zhang, S.; Li, X.; Zhu, Z. Numerical simulation of cryogenic cavitating flow by an extended transport-based cavitation model with thermal effects. *Cryogenics* **2018**, *92*, 98–104. [CrossRef]
38. Zhou, L.; Bai, L.; Li, W.; Shi, W.; Wang, C. PIV validation of different turbulence models used for numerical simulation of a centrifugal pump diffuser. *Eng. Comput.* **2018**, *35*, 2–17. [CrossRef]

Article

Numerical Study on the Gas-Water Two-Phase Flow in the Self-Priming Process of Self-Priming Centrifugal Pump

Chuan Wang [1,3], Bo Hu [2], Yong Zhu [1,3,*], Xiuli Wang [3], Can Luo [1] and Li Cheng [1,*]

[1] School of Hydraulic, Energy and Power Engineering, Yangzhou University, Yangzhou 225002, China; wangchuan198710@126.com (C.W.); luocan@yzu.edu.cn (C.L.)
[2] Department of Energy and Power Engineering, Tsinghua University, Beijing 100084, China; tigerbohu87@163.com
[3] Research Center of Fluid Machinery Engineering and Technology, Jiangsu University, Zhenjiang 212013, China; jsuwxl@163.com
* Correspondence: zhuyong@ujs.edu.cn (Y.Z.); chengli@yzu.edu.cn (L.C.); Tel.: +86-0511-88799918 (Y.Z.); +86-0514-87921191 (L.C.)

Received: 30 April 2019; Accepted: 29 May 2019; Published: 1 June 2019

Abstract: A self-priming centrifugal pump can be used in various areas such as agricultural irrigation, urban greening, and building water-supply. In order to simulate the gas-water two-phase flow in the self-priming process of a self-priming centrifugal pump, the unsteady numerical calculation of a typical self-priming centrifugal pump was performed using the ANSYS Computational Fluid X (ANSYS CFX) software. It was found that the whole self-priming process of a self-priming pump can be divided into three stages: the initial self-priming stage, the middle self-priming stage, and the final self-priming stage. Moreover, the self-priming time of the initial and final self-priming stages accounts for a small percentage of the whole self-priming process, while the middle self-priming stage is the main stage in the self-priming process and further determines the length of the self-priming time.

Keywords: self-priming pump; gas-water two-phase flow; computational fluid dynamics

1. Introduction

A pump is a kind of general machine with tremendous variety and extensive application fields, and it can be said that pumps serve in all places with liquid flow [1–4]. According to statistics, the power consumption of pumps account for 22% of the power generation and the oil consumption accounts for about 5% of the total oil consumption [5–10]. Self-priming centrifugal pumps, or self-priming pumps, have no bottom valve in the inlet pipe. The pump structure is changed such that it can store some water after stopping and exhaust the air in the inlet pipe at the start by the mixture and separation of air and water. The water returns to the appropriate position in the pump via the backflow channel, and the above process is repeated, thereby realizing the self-priming. The process requires a water supply only at the initial stage; no water supply is necessary in subsequent start-ups. After a short operating period, the pump itself can suck up the water and be put into normal operation. A self-priming pump is easy to operate and has stronger adaptability than an ordinary centrifugal pump. Such a pump is extremely suitable for situations with frequent start-ups or difficult liquid irrigations [11–16]. Self-priming is an important parameter in order to evaluate the performance of self-priming pumps, and it determines the normal operation of pumps. According to the Chinese standard JB/T6664-2007, the self-priming time of a 5 m vertical pipe should be controlled within 100 s. Given that the self-priming process of a self-priming pumps is a complicated, unsteady gas–liquid flow phenomenon, studying the self-priming performance of pumps is difficult. Moreover, many challenges emerge in the study of the influencing factors of the self-priming time of self-priming pumps.

Many studies on the self-priming process of self-priming centrifugal pumps have been conducted by theoretical calculation, numerical calculation, and test measurement; certain research results have also been obtained. Using theoretical calculations, Zhao et al. [17] deduced the formulas of self priming time and exhaustion rate according to fluid mechanics, thermodynamics, air dynamic equation, and energy invariant equation cited in their studies of vertical self-priming centrifugal pumps. Yi [18] summarized the data about self-priming time and the specific speed of 31 modes of external-mixture self-priming centrifugal pumps and explored the corresponding rules and trends. However, the self-priming time gained from the relevant calculation formula had a great range and only had statistical significance. Although the proposed theoretical calculation of self-priming time is not highly accurate and has narrow applicability, the theoretical calculation process indicates that moving bubbles are the relevant results of the main media accomplished by self-priming. Using numerical calculation, Wang et al. [19,20] adopted an inlet void fraction of 15% for the self-priming process of rotational flow of a self-priming pump using Fluent software. They found that the liquid phase drove the gas phase flow by phase interaction during the self-priming process and finally obtained the self-priming time. Using the Fluent software, Liu et al. [21,22] made a numerical calculation of the gas-water two-phase flow in the self-priming process of a single-stage self-priming centrifugal pump, and the pressure, velocity and gas distribution of the flow field under different assumed void fraction conditions were obtained. However, the exact void fraction in the pump inlet was unknown. Li et al. [23,24] simulated the gas–liquid states in a pump at different moments (initial, middle, and late stages) of the self-priming process using the quasi-steady method with a decreased void fraction, and then estimated the time required for the entire self-priming process. However, such a quasi-steady method is markedly different from the real self-priming process. For the numerical calculation of the self-priming time of self-priming pumps, scholars hypothesized that either the inlet void fraction of a pump comprised several fixed numerical values (5%, 10%, and 15%) and the velocity inlet was set or gas filled the entire pump inlet and the gas inlet velocity was a mean value. The former hypothesis was not based on the self-priming numerical calculation of self-priming pumps but on the numerical simulation of ordinary gas–liquid flow pumps. The latter hypothesis was close to the real self-priming situation, but it assumes that the velocity inlet was an average, which obviously contradicted the inversely V-shaped variation of self-priming speed in the self-priming process. The study of Huang et al. [25] was the closest to the real simulation of the self-priming time of a single-stage self-priming pump because they did not set the velocity inlet or mass outlet in the simulation process. However, the relevant calculation model only hypothesized that the self-priming height was 0.25 m. This result was inconsistent with the vertical self-priming height of 3 or 5 m in the real self-priming process, thereby failing to reflect the flow rule of the whole self-priming process. With regard to experimental measurements, existing studies mainly focused on the influences of structural improvement [26], volume of fluid reservoir [27], area and position of backflow hole [28], and tongue gap [29] of self-priming pumps on self-priming time.

In summary, the existing numerical calculations of the self-priming process in self-priming pumps are not precise, and relevant experimental studies on the self-priming process are scarce. Thus, it's very necessary to study the unsteady flow of the self-priming process through numerical methods. In the current study, a typical self-priming centrifugal pump was designed, and a numerical calculation of the gas-water two-phase flow in the self-priming process was performed using the ANSYS CFX.

2. Methodology

2.1. Three-Dimensional Model of the Impeller and Diffuser

In this study, a motor direct connection mode in the self-priming pump is used with a compact structure and easy installation and operation. The entire pump is composed of inlet and outlet pipes, gas–liquid mixture cavity, self-priming cover plate, impeller, diffuser, outer casing, shaft, gas–liquid separation cavity, and a motor, among others (Figure 1). Moreover, the core components of the self-priming pump include the impeller and diffuser, whose geometric parameters are calculated by using the velocity coefficient method, which is shown in Table 1. The three-dimensional models and practical pictures of the impeller and diffuser are shown in Figure 2, respectively.

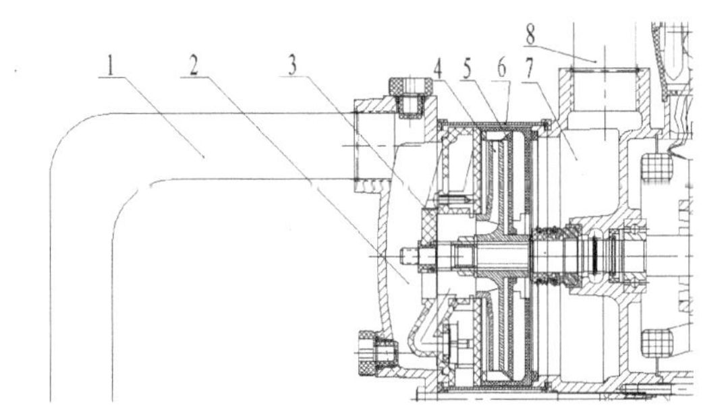

Figure 1. Assembly diagram of the self-priming centrifugal pump. 1. Inlet section; 2. Gas-water mixture cavity; 3. Self-priming cover plate; 4. Impeller; 5. Diffuser; 6. Outer casing; 7. Gas-water separation cavity; 8. Outlet section.

Table 1. Basic geometrical parameters of the pump by using the velocity coefficient method.

Geometric Parameter	Value	Geometric Parameter	Value
Inlet diameter of the impeller D_1 (mm)	20	Outlet width of the impeller b_2 (mm)	3.1
Hub diameter of the impeller D_{hb} (mm)	33.5	Number of the outward diffuser blades Z_p	9
Outlet diameter of the impeller D_2 (mm)	108	Number of the return diffuser blades Z_n	9
Inlet angle of the impeller blade β_1 (°)	40	Inlet diameter of the outward diffuser D_3 (mm)	109
Outlet angle of the impeller blade β_2 (°)	15	Inlet angle of the outward diffuser α_3 (°)	5
Wrap angle of the impeller blade θ_w (°)	150	Outlet angle of the return diffuser α_6 (°)	50
Number of the impeller blades Z	8	Rotational speed n (r/min)	2800

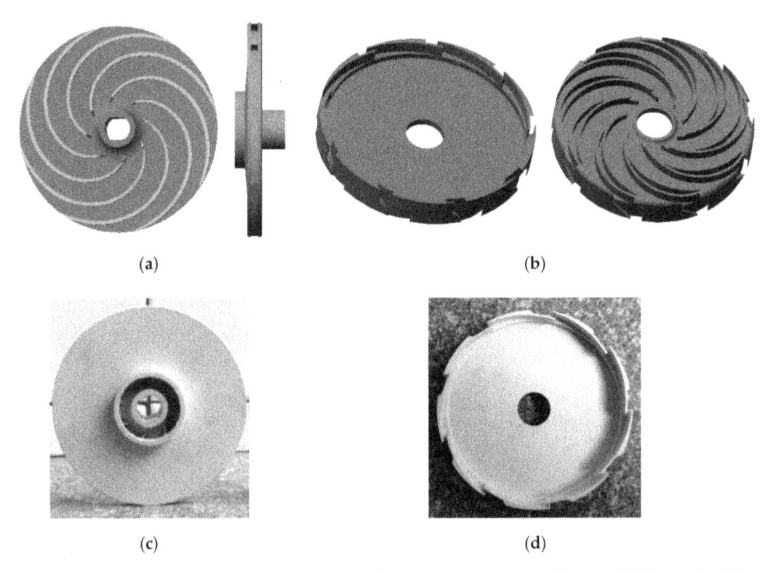

Figure 2. Three-dimensional models and practical pictures of the impeller and diffuser. (**a**) Plane and axial projection of the impeller; (**b**) Outward and return vane of the diffuser; (**c**) Practical impeller; (**d**) Practical diffuser.

2.2. Establishing the Calculation Domain

As shown in Figure 3, the calculation domain of the multistage self-priming pump includes inlet section, gas-water mixture cavity, self-priming cover plate, impeller, pump cavity, diffuser, gas-water separation cavity, backflow channel and outlet section. The inlet and outlet pipe are too long, so they are not completely shown.

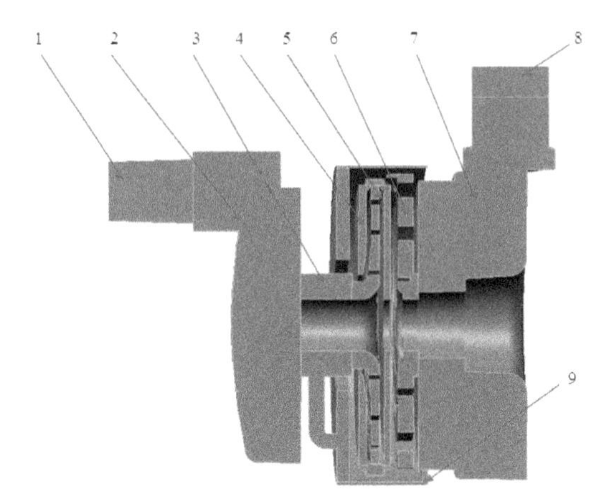

Figure 3. Calculation domain of the self-priming centrifugal pump by using Pro/Engineering software (Pro/E 5.0, Parametric Technology Corporation, Boston, MA, USA). 1. Inlet section; 2. Gas-water mixture cavity; 3. Self-priming cover plate; 4. Pump cavity; 5. Impeller; 6. Diffuser; 7. Gas-water separation cavity; 8. Outlet section; 9. Backflow channel.

2.3. Grid Information

The calculation domain should be discretized before grid-based simulation, and grid quality affects calculation accuracy and time. Generally, dividing a calculation model with a complex geometry and boundary by using hybrid grids is reasonable. The calculation domain was divided into hybrid grids by Gambit software in this study because the numerical model of the self-priming pump and the boundary conditions of the self-priming calculation were complex. Given that performing a grid-independence analysis for the unsteady self-priming calculation is inconvenient due to the large amount of calculation time, the number of grids was made as large as possible but still within the computing capability of the workstation. The main grid information is shown in Figure 4 and Table 2, and the total number of grids is more than four million.

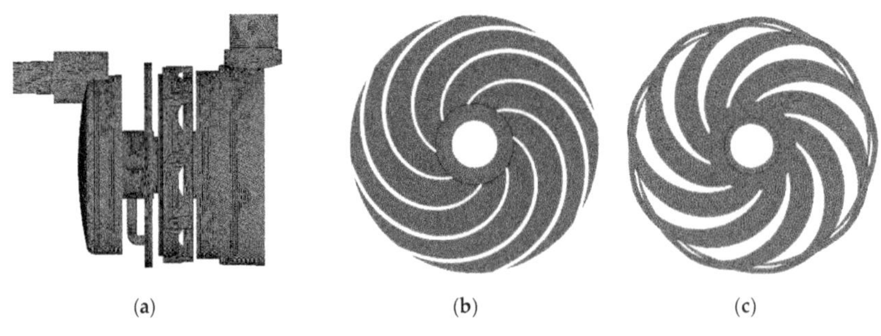

Figure 4. *Cont.*

(d)

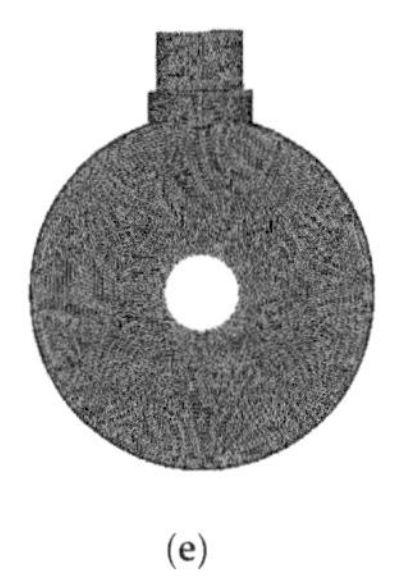

(e)

Figure 4. Grid of the self-priming centrifugal pump by using Gambit software (Gambit 2.4.6, ANSYS Corporation, Pittsburgh, PA, US.). (**a**) Total pump; (**b**) Impeller; (**c**) Diffuser; (**d**) Gas-water mixture cavity; (**e**) Gas-water separation cavity.

Table 2. Grid information of the self-priming centrifugal pump including the size, number, quality and type of the grid.

Name	Grid Size (mm)	Grid Number	Grid Quality	Grid Type
Inlet section	1.5	1,165,924	0.237	T-Grid
Gas–water mixture cavity	1.5	375,146	0.248	T-Grid
Self-priming cover plate	1	363,612	0.235	T-Grid
Impeller	1	169,777	0.143	T-Grid
Pump cavity	1	282,211	0.144	T-Grid
Diffuser	1.3	131,258	0.147	T-Grid
Gas-water separation cavity	1.5	513,275	0.214	T-Grid
Backflow cavity	0.5	510,742	0.158	T-Grid
Outlet section	2	623,942	0.231	T-Grid
Total	-	4,135,887	-	T-Grid

2.4. Time Step Independence

In general, in the periodic numerical simulation, the time step needs to satisfy the Courant number criterion [30], which is expressed as:

$$C_o = v\Delta t/l < 100 \tag{1}$$

In the formula, v is the absolute value of the estimated mean velocity, m/s; l is the smallest size of the grid, m; Δt is the time step, s; C_o is the Courant number criterion and required no more than 100. When the numerical convergence is not good, it is appropriate to take smaller values.

If the time step size is too large, the value of Courant number will also be large; however, too small a time step will also lead to a significant increase in computing time [31,32]. Therefore, considering the computer configuration, the time step Δt was chosen as 5×10^{-3} s. Moreover, the value of v is within 10 m/s, and the value of l is more than 5×10^{-3} m (according to Table 1); therefore the value of Courant number is within 10.

2.5. Setting of the Boundary Condition

ANSYS Computational Fluid X software (ANSYS CFX 14.5, ANSYS Corporation, Pittsburgh, PA, USA) was used to perform numerical calculation in the self-priming process of the self-priming centrifugal pump. The impeller and shroud in the pump cavity were based on the rotating reference frame, whereas the other sub-domains were based on the stationary reference frame. Moreover, the pressure inlet and the opening in the outlet were selected as inlet and outlet boundaries in order to approximate the actual self-priming condition as much as possible. The gas-water volume fraction contour of the multistage self-priming pump in the initial state is shown in Figure 5. The height of the inlet elbow was 1.5 m, and the height of the outlet pipe was 1 m. The inlet elbow was placed in the water. The elbow above the water surface was filled with gas (1 m in height), the elbow below the water surface was filled with water (0.5 m in height), and the outlet pipe was filled with gas (1 m in height). A check valve was installed in the pump inlet, and the entire pump was also filled with water. Given that the length of the inlet elbow below the water surface was 0.5 m, the initial pressure at the inlet was set to 5 kPa (gage pressure), and the pressure was reduced progressively to 0 kPa until at the water surface. The initial pressure at the inlet and outlet gas sections was 0 kPa. The initial pressure distribution of the entire self-priming pump is shown in Figure 6.

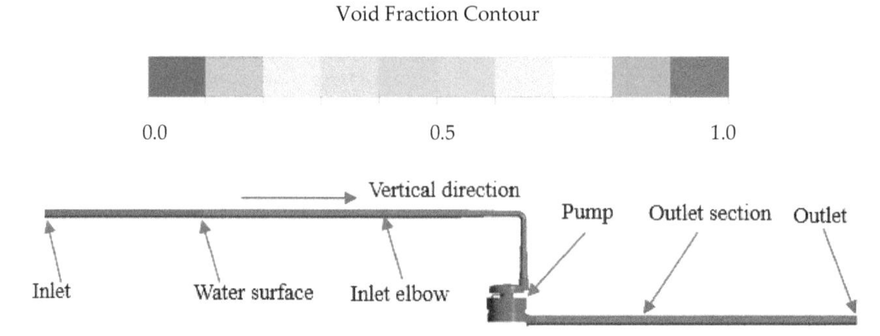

Figure 5. Void fraction contour of the self-priming pump in the initial state.

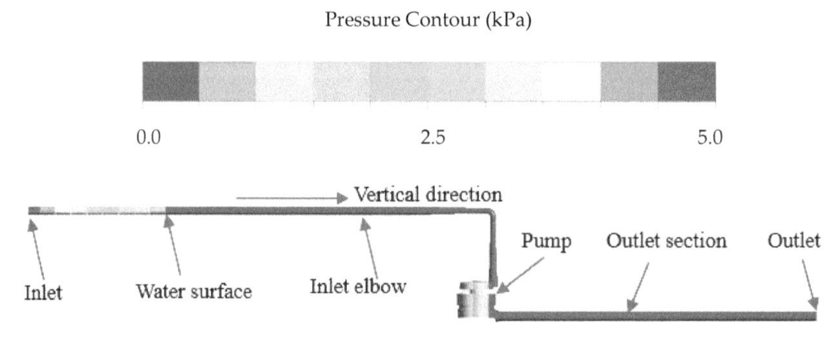

Figure 6. Static pressure contour of the self-priming pump in the initial state.

3. Results and Discussion

3.1. Gas-Water Two-Phase Distribution in the Inlet and Outlet Sections of the Pump

Figure 7 presents the gas-water two-phase distribution in the inlet and outlet sections of the pump at several moments of the entire self-priming process based on CFD. The red area represents gas, and the blue area represents water. a–f show the gas-water two-phase distribution at t = 0, 0.1, 0.2, 0.3, 0.4 and 0.5 s after the start of the pump. Due to the rotation of the impeller, the water in the pump flows rapidly to the outlet section, and the gas in the inlet section rapidly rushes to the pump, and the water column in the inlet and outlet sections rises continuously. Since the void fraction of the impeller is getting larger and larger, the work capacity of the impeller is gradually weakened, resulting in a slower rise in the water column in the inlet and outlet sections (see A_0 to A_5 and B_0 to B_5). At t = 0.5 s, the height of the water column in the inlet and outlet sections remains essentially constant (see A_4 to A_5 and B_4 to B_5). At this time, the self-priming process due to the impeller rotational role is substantially complete. Moreover, the gas in the inlet section begins to enter into the outlet section at t = 0.2 s.

Figure 7g–j show the gas-water two-phase distribution at t = 1, 2, 3, 4 s after the self-priming pump is started. It can be seen that the water column in the inlet section is rising and close to the gas-water mixture cavity (see A_7 to A_{10}). That's because some gas-water mixture in the gas-water mixture cavity enters into the gas-water separation cavity successively through the impeller and diffuser. In the gas-water separation cavity, the gas-water mixture is in a free-projecting state. Under the action of buoyancy, the lighter gas flows upwardly into the outlet section, and the heavier water flows downward, and flows back to the impeller inlet through the backflow channel for the next cycle. Repeatedly, the total amount of gas in the pump body steadily decreases, and the gas in the inlet section is continuously replenished to the inside of the pump, thereby causing the water column in the inlet section to continuously rise. Compared to the previous self-priming process, the rising rate of the water column in the current self-priming process is significantly smaller, mainly because the maximum self-priming rate in the previous self-priming process is approximately equal to the pump's internal maximum flow rate, while the gas and the water are repeatedly mixed and separated, and the gas is exhausted a little bit in the current self-priming process. When $0.5 \text{ s} < t \leq 2 \text{ s}$, the gas in the pump has not completely escaped from the water column in the outlet section. As the gas in the water column increases, the water column rises continuously (see B_5 to B_7). When $2 \text{ s} < t \leq 4$, the gas escapes from the water column of the outlet section. The current self-priming process is a gas-suction stage due to the role of gas-water mixture and gas-water separation.

Figure 7k–m shows the gas-water two-phase distribution at $t = 5, 6, 7$ s. It can be seen that the water column in the inlet section rises continuously and enters the pump at $t > 4$ s, so the impeller void fraction decreases continuously and the work capacity of the impeller is enhanced, leading to a continuous rise in the water column in the outlet sections (see B_9 to B_{11}). Compared with the previous self-priming process, the exhaustion rate of the gas in the current self-priming process is significantly larger and most of the gas in the pump is exhausted in a short time. Finally, the entire self-priming process ends and the self-priming centrifugal pump enters the normal working condition.

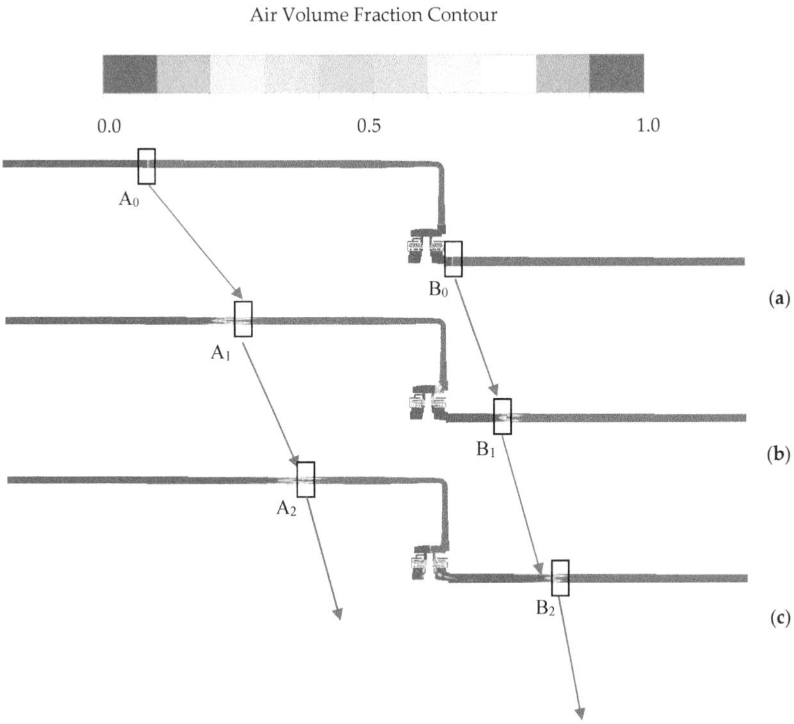

Figure 7. *Cont.*

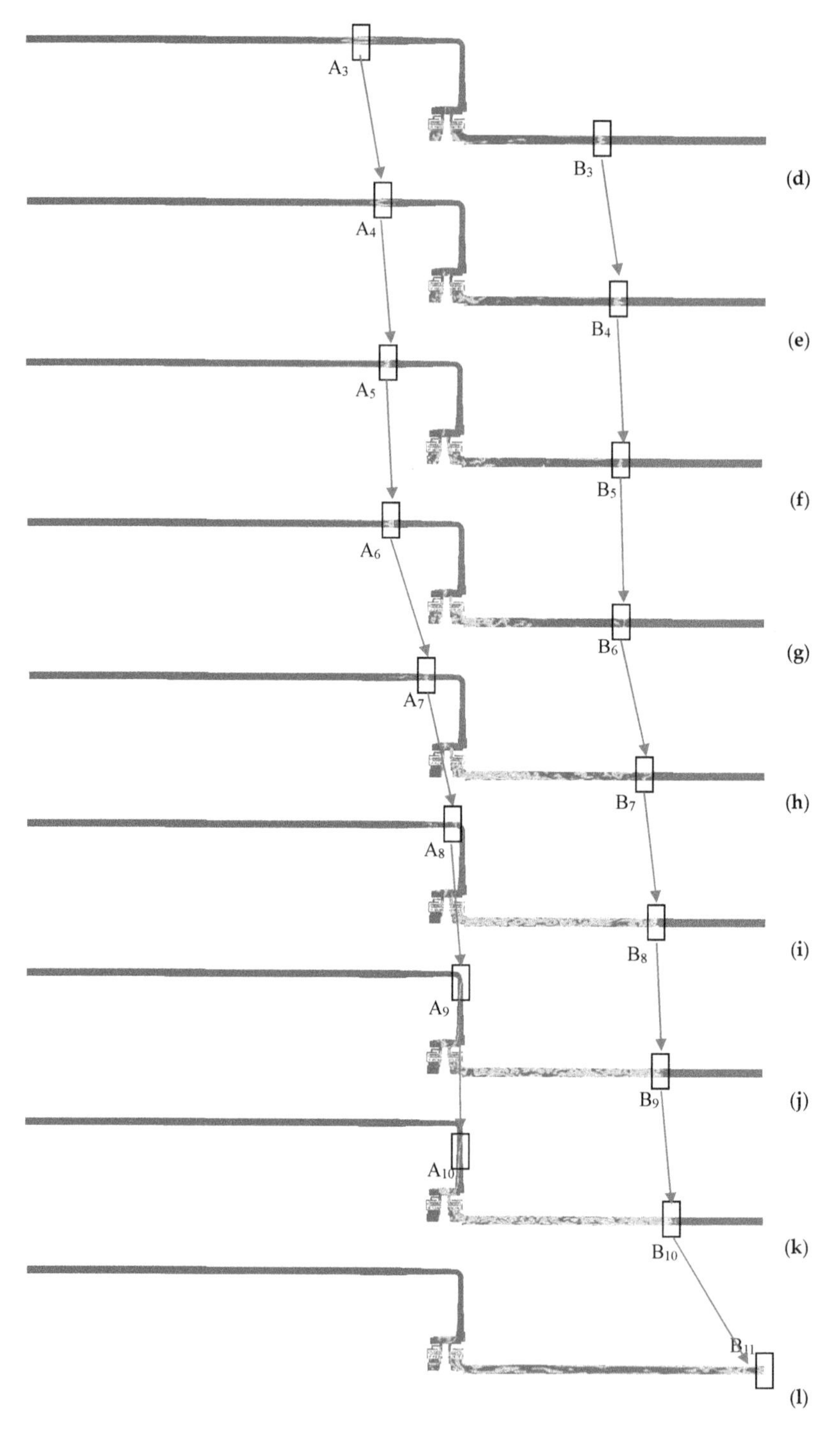

Figure 7. *Cont.*

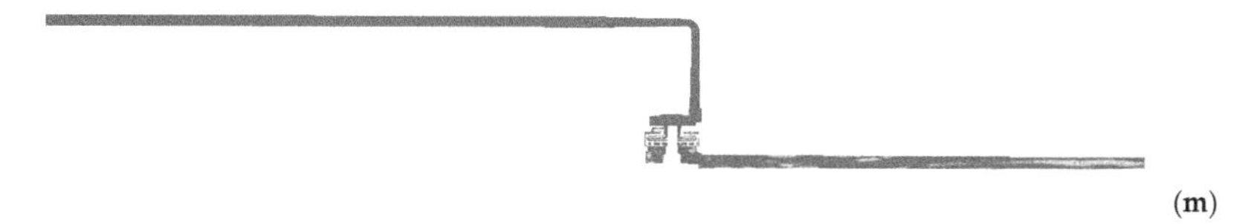

Figure 7. Gas–water two-phase distribution in the inlet and outlet sections of the pump at several moments of the entire self-priming process. (**a**) $t = 0$ s; (**b**) $t = 0.1$ s; (**c**) $t = 0.2$ s; (**d**) $t = 0.3$ s; (**e**) $t = 0.4$ s; (**e**) $t = 0.5$ s; (**f**) $t = 1$ s; (**g**) $t = 2$ s; (**h**) $t = 3$ s; (**i**) $t = 3$ s; (**j**) $t = 4$ s; (**k**) $t = 5$ s; (**l**) $t = 6$ s; (**m**) $t = 7$ s.

3.2. *Streamline and Gas–Water Two-Phase Distribution in the Middle Section of the Pump*

From the previous analysis, it can be found that the whole self-priming process of the self-priming pump is divided into three stages: the gas-suction stage due to the impeller's rotating role in the initial self-priming stage, gas-suction stage due to gas-water mixing and the role of gas–water separation in the middle self-priming stage, and gas-suction stage due to the water flowing from the inlet elbow into the pump in the final self-priming stage.

In order to comprehensively study the gas-liquid flow of the pump in the self-priming process, the middle section of the entire pump that best reflects the flow characteristics was selected. Figure 8 shows the streamline and gas-water two-phase distribution in the middle section of the pump at several moments of the self-priming process. The following three self-priming stages of the multistage self-priming pump were obtained through numerical calculation: initial ($t \leq 1$ s), middle ($1 \text{ s} < t < 5$ s), and final ($t \geq 5$ s) self-priming stages. Three moments were selected for analysis for each self-priming process. As shown in Figure 8a,b, the time that the impeller rotates for a cycle is about 0.02 s, and the rotating impeller worked on the water to make it flow quickly to the outlet pipe in the initial self-priming process. Hence, the gas in the inlet section entered the gas-water mixture cavity, mixed with the water, then continuously flowed through the impeller and diffuser (A_1, A_2, A_3). The upper part of the gas-water mixture from the diffuser flowed to the outlet pipe (B_1, B_2, B_3), and the lower part flowed to the lower side of the gas-water separation cavity (C_1, C_2, C_3).

Figure 8d–f shows the streamline and gas-water two-phase distribution in the middle section of the pump at $t = 2$, 2.02 and 2.04 s. It can be seen that compared with the initial self-priming stage, the change of gas-water two-phase distribution in the middle self-priming is slow, and the gas-water two-phase boundary line in the pump is based on the uppermost end of the lower half of the impeller (D_1–D_2). The upper half part is mainly the gas phase, and the lower part is mainly the water phase, mainly because the amount of water flowing out from the impeller is basically the same as that from the backflow channel into the impeller, and the amount of exhausting gas is basically the same as the amount of inhaling gas, resulting in that the gas-water two-phase boundary is basically stable. Figure 8g,i indicate that in the final self-priming stage, the entire pump was mainly full of water. A large amount of gas was exhausted from the pump within a short period. In addition, since the backflow channel is only in communication with the lower half of the self-priming cover, a small portion of the gas is sealed in the upper half of the backflow channel and a large resting bubble region is formed (E_1, E_2, E_3). The gas is finally exhausted and even a small amount of gas is not exhausted.

Void Fraction Contour

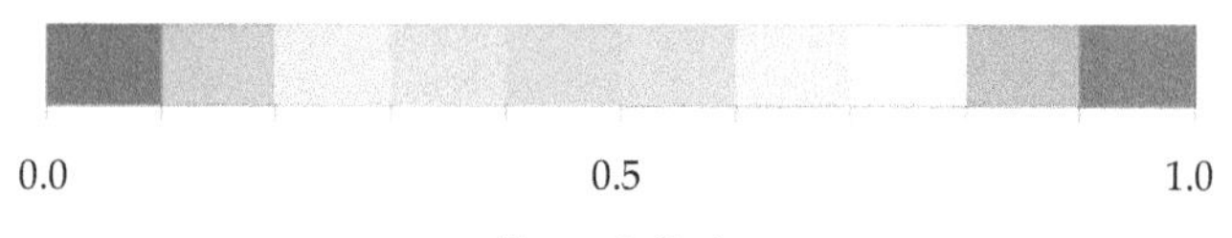

Figure 8. *Cont.*

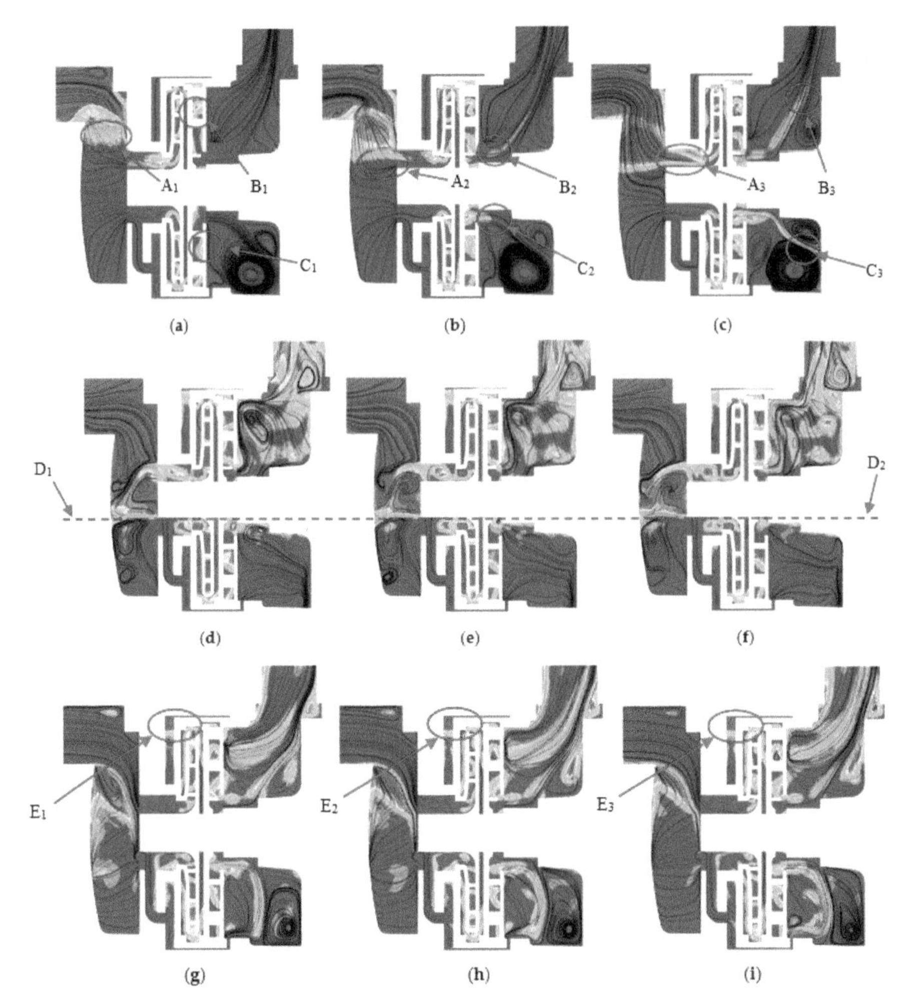

Figure 8. Streamline and gas–water two-phase distribution in the middle section of the pump at several moments of the self-priming process. (**a**) t = 0.1 s; (**b**) t = 0.12 s; (**c**) t = 0.14 s; (**d**) t = 2 s; (**e**) t = 2.02 s; (**f**) t = 2.04 s; (**g**) t = 6 s; (**h**) t = 6.02 s; (**i**) t = 6.04 s.

3.3. Vector and Pressure Distribution in the Middle Section of the Backflow Channel

The backflow channel is a key component for water circulation in the self-priming pump. Figure 9 presents the vector and pressure distribution in the middle section of the pump in the three self-priming stages at t = 0.2, 2, and 6 s. It can be seen that due to the work of the rotating impeller, the pressure on the right side of the backflow channel is higher than the left side (B_1 and F_1), so that the water in the gas-water separation cavity can flow back to the impeller inlet through the backflow channel. In the initial self-priming stage (t = 0.2 s), the gas-water mixture flowing out of the diffuser enters the lower part of the gas-liquid separation cavity (A_1). The gas-water mixture presents three flow states. Firstly, the water in the gas-water mixture flows to the backflow hole by gravity (B_1). Secondly, the gas in the water in the gas-water mixture flow freely (C_1). Thirdly, because the backflow hole is located on the left side of the gas-liquid separation cavity and the backflow capacity of the backflow channel is limited, resulting in a large vortex region on the right side (D_1). In the middle self-priming stage (t = 2 s), the flow region of the whole gas-liquid separation cavity is divided into two parts (E_1–E_2).

In the lower region, the water flows down to the backflow hole, while in the upper region, the gas flows upward to the pump outlet. In the final self-priming stage (t = 6 s), the pressure difference between the two sides of the backflow channel increases.

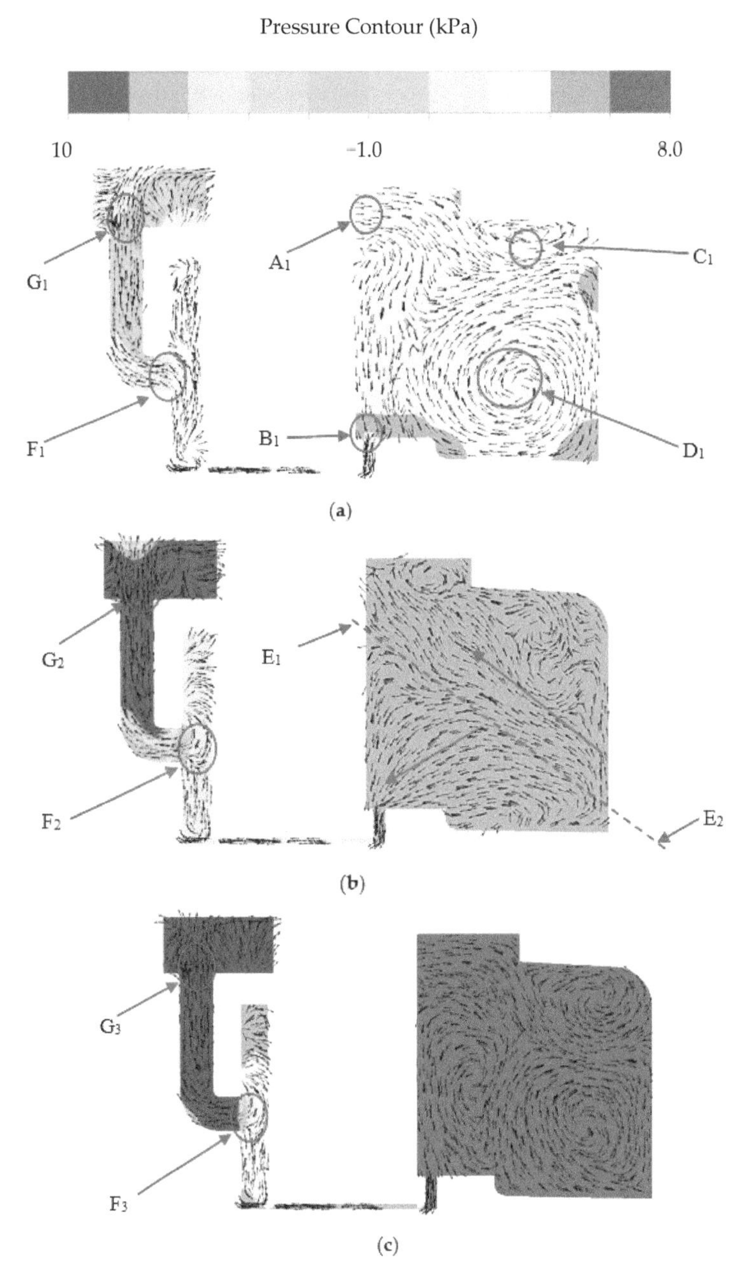

Figure 9. Vector and pressure distribution of the backflow channel in the self-priming process. (**a**) t = 0.2 s; (**b**) t = 2 s; (**c**) t =6 s.

In the initial self-priming stage, the water flowing back from the backflow channel is diverted at the intersection of the self-priming cover (F_1). Some water flows to the impeller inlet (G_1) and the remainder flows to the upper side of the self-priming cover. In the middle and final self-priming stages, as the water column in the inlet section rises and the pressure at the impeller inlet decreases (G_1, G_2, G_3), the pressure difference between point F and point G rises continuously. The water flowing back from the backflow channel merges with the water on the upper side of the self-priming cover at point F, and flows to the inlet of the impeller. The above results show fully that the backflow capacity of the self-priming pump is gradually enhanced in the three self-priming stages.

3.4. Streamline and Gas-Water Two-Phase Distribution in the Middle Section of the Impeller and Diffuser

The impeller and diffuser are the core components of the pump. A radial diffuser with outward and return vanes was used, and the middle section of the impeller coincided with the middle section of the outward diffuser. Figure 10 presents the streamline and gas–water two-phase distribution in the middle section of the impeller and outward diffuser in the three self-priming stages. As shown in Figure 10a,b, the void fraction of the impeller and outward diffuser increased rapidly in the initial self-priming stage. When t = 0.2 s, most of the space of the first-stage impeller was filled with gas. That is, only a small amount of gas–water mixture was present at the inlet and pressure surface of the impeller blade (A_1 and B_1), whereas plenty of gas existed near the suction surface of the impeller that had a large number of vortices, similar to a type of "dead water zone" (D_1). The gas-water mixture from the impeller inlet flowed along the pressure surface to the impeller outlet. Upon entering the outward diffuser, the velocity of the gas–water mixture decreased, so more water existed at the impeller outlet and outward diffuser (C_1). When t = 0.4 s, the void fraction of the impeller and outward diffuser is further reduced, and the region containing the gas-water mixture at the inlet and pressure surface of the impeller blade is significantly decreased (A_1 and A_2). As shown in Figure 10c,d, in the middle self-priming stage, most of the space in the impeller was filled with gas, and a small amount of water was present at the inlet and pressure side of the impeller blade. Eventually, the exhausting and inhaling rates gradually reached a state of dynamic equilibrium, and the void fraction of the impeller and outward diffuser is basically stable. Compared with the initial self-priming stage, and the vortex region in the impeller is slightly reduced in the middle self-priming stage (D_1 and D_2). As shown in Figure 10f,g, the water in the inlet section has begun to enter the impeller in the final self-priming stage, and the void fraction of the impeller is drastically reduced. When t = 6 s, the impeller is basically filled with water; however, there is still some gas in some flow channels of the impeller, indicating that the exhausting process of the impeller has unsteady characteristics. In summary, the key to the self-priming of the pump is that the rotating impeller forces a small amount of gas-water mixture at the impeller inlet to flow along the blade pressure surface to the impeller outlet and into the outward diffuser.

The return diffuser is an important component of the diffuser. It not only introduces the liquid into the outlet section, but also eliminates the rotational component of the liquid. Figure 11 presents the flow line and gas-water two-phase distribution in the middle section of the return diffuser in the three self-priming stages. As shown in Figure 11a,b the void fraction of the return diffuser increases sharply (A_1 and A_2) in the initial self-priming stage, and the return diffuser is substantially filled with gas at t = 0.4 s, and there are many vortices in the gas region. As can be seen from Figure 11c,d, compared with the initial self-priming stage, the void fraction of the return diffuser is decreased (A_3). As shown in Figure 11e,f, a large amount of water enters the return diffuser in the final self-priming stage, and the void fraction of the return diffuser is further decreased (A_4). In general, the streamline and gas-water mixture in the return diffuser do not evenly distributed along the circumference. The gas first gathers in the region close to the suction surface of the return diffuser (B_1), while the water first gathers near the pressure surface (see C_1), and the gas-liquid mixture from the outward diffuser flows along the pressure surface of the return diffuser, which is consistent with the gas-water two-phase distribution in the impeller.

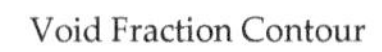

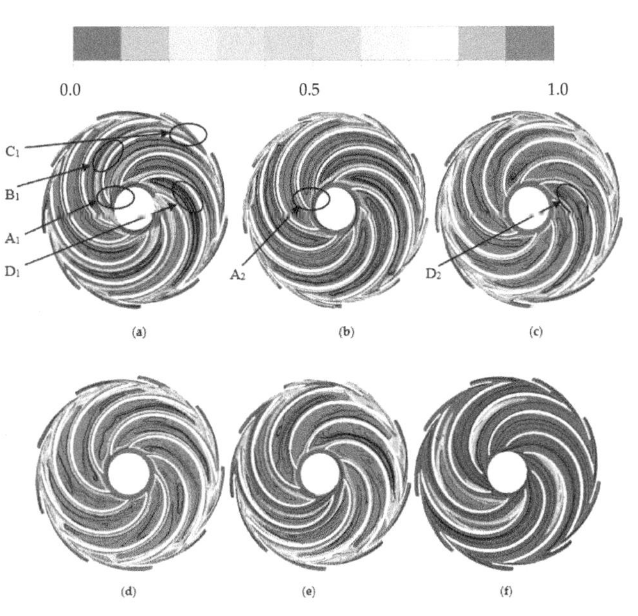

Figure 10. Streamline and gas–water two-phase distribution of the impeller and outward diffuser at several moments of the self-priming process. (**a**) $t = 0.2$ s; (**b**) $t = 0.4$ s; (**c**) $t = 1$ s; (**d**) $t = 2$ s; (**e**) $t = 5$ s; (**f**) $t = 6$ s.

Void Fraction Contour

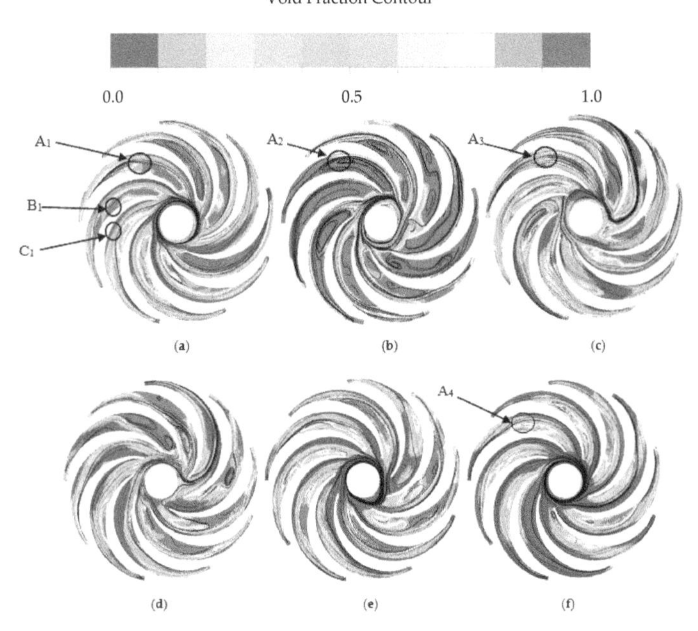

Figure 11. Streamline and gas-water two-phase distribution of the return diffuser at several moments of the self-priming process. (**a**) $t = 0.2$ s; (**b**) $t = 0.4$ s; (**c**) $t = 1$ s; (**d**) $t = 2$ s; (**e**) $t = 5$ s; (**f**) $t = 6$ s.

3.5. Vector and Gas-Water Two-Phase Distribution in the Middle Section of the Gas-Water Mixture and Separation Cavities

As an important part of the self-priming pump, the gas-water mixture cavity mainly provides space to promote the mixing of gas and water. Figure 12 shows the vector and gas-water two-phase distribution of the gas-water mixture cavity in the self-priming process. As shown in Figure 12a, there is a clear boundary layer in the gas-water mixture cavity in the initial self-priming stage, and the gas is in the upper layer, while the water is in the lower layer. Moreover, when t = 0.4 s, a large amount of gas from the inlet section enter the gas-water mixture cavity and flows downward in three directions (A_1, B_1 and C_1). The gas in the middle direction flows directly downward, the gas in the left and right directions flows down the wall of the gas-liquid mixture cavity, and some of the gas flows back along the boundary layer to the middle portion of the gas-water mixture cavity (B_2 and C_2), and finally merges with the inflowing gas in the middle direction (A_2), and the other gas is mixed with water at the boundary layer to enter the water region (B_3 and C_3). As can be seen from Figure 12b, the gas-water boundary layer becomes unclear in the middle self-priming stage. The gas from the inlet section enter the gas-water mixture cavity in the middle direction (D_1), one part is mixed with water at the boundary layer to enter the water region (D_2), and the other part is refluxed in the left and right direction (D_3 and D_4), forming a large number of vortices. Since the inhaling rate of the gas in the middle self-priming stage is much smaller than that in the initial stage, three parts of gas are combined in the middle of the gas-water mixture cavity in the initial self-priming stage, while a strand of gas is divided into three parts in the middle self-priming stage. As shown in Figure 12c, the gas-water boundary layer has disappeared in the final self-priming stage. Compared with the water region, the flow field in the gas region is more disordered, indicating that the vortex is more likely to be generated in the gas region.

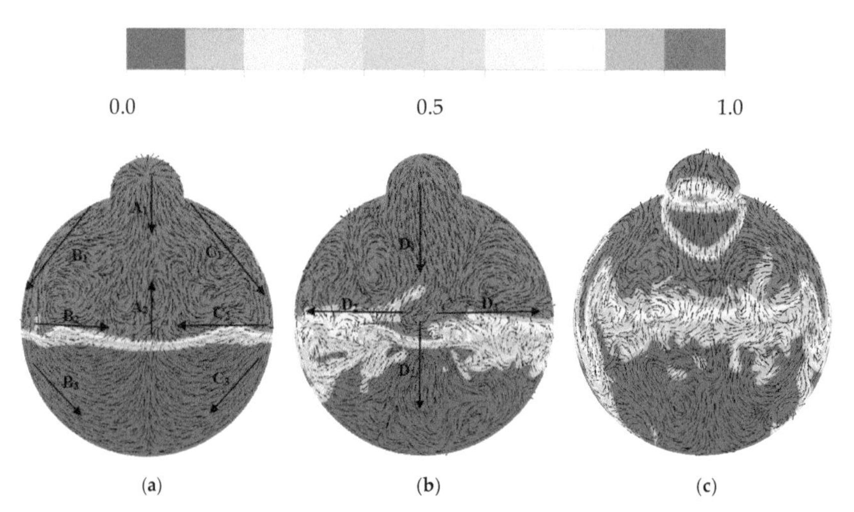

Figure 12. Vector and gas-water two-phase distribution of the gas-water mixture cavity in the self-priming process. (**a**) t = 0.2 s; (**b**) t = 2 s; (**c**) t = 5 s.

The gas-water separation cavity mainly provides space to promote the separation of gas-water mixture. Figure 13 illustrates the vector and gas-water two-phase distribution of the gas-water separation cavity in the self-priming process. As shown in Figure 13a, the void fraction of the gas-water separation cavity is not high overall and exhibits a non-uniform distribution in the initial self-priming stage. Under the influence of the rotation of the impeller, the gas-water mixture rotates counterclockwise in the annular cavity and escapes from the separation cavity, which is consistent with the rotating direction of the impeller (A_1, B_1, C_1 and D_1). As can be seen from Figure 13b, the gas-liquid separation cavity plays the role of the separation of gas and water in the middle self-priming stage. The gas is basically distributed in the upper half of the gas-water separation cavity due to the influence of gravity. Although the internal flow field of the entire gas-water separation cavity is disordered, the flow direction of the annular cavity is counterclockwise. As shown in Figure 13c, since the water from the inlet section starts to enter the pump, a large amount of gas-water mixture flows in the gas-water separation cavity, which not only causes a decrease in the void fraction, but also causes the gas to spread throughout the gas-water separation cavity.

Void Fraction Contour

0.0 0.5 1.0

(a) (b) (c)

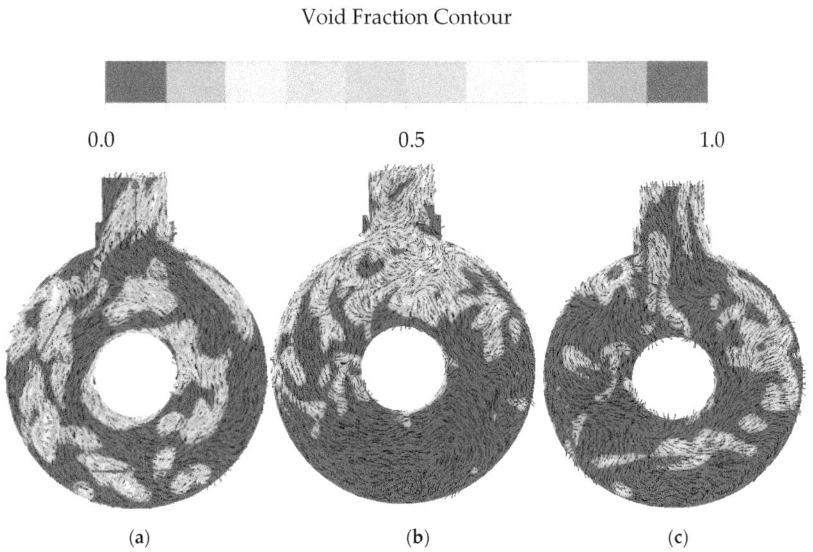

Figure 13. Vector and gas-water two-phase distribution of the gas-water separation cavity in the self-priming process. (**a**) $t = 0.2$ s; (**b**) $t = 2$ s; (**c**) $t = 5$ s.

3.6. Data Analysis of the Self-Priming Pump Based on CFD

In the self-priming process of a self-priming centrifugal pump, the gas in the section is exhausted to the outlet pipe through the operation of the pump. In Figure 14, the inlet surface of the inlet section is defined as "In"; the inlet surface of the gas-liquid mixing cavity is defined as "In2" and the outlet surface of the gas-liquid separation cavity is defined as "Out2". The outlet surface of the outlet pipe is defined as "Out". In order to illustrate the self-priming capability of the pump, the gas volume coefficient, gas flow coefficient, water flow coefficient, impeller void fraction, and diffuser void fraction are defined in Formulae (2) to (4), respectively.

$$C_V = V_g/V_{gt}, \tag{2}$$

$$C = Q_g/Q_d \tag{3}$$

$$C' = Q_w/Q_d \tag{4}$$

where C_V is the gas volume coefficient in the pump; V_g is the sum of the internal gas volume of the pump's components, except for the outlet pipe, in m^3; V_g is the gas volume in the pump in the initial state, which is equal to the gas volume in the inlet section, in m^3; C is the gas flow coefficient in the pump; C' is the water flow coefficient in the pump; Q_g is the gas flow in the self-priming process in m^3/h; Q_w is the water flow in the self-priming process in m^3/h; Q_d is the water flow at the rated condition after the pump operates normally in m^3/h.

Figure 14. Definition of several typical surfaces of the self-priming pump.

Figure 15 shows the gas volume coefficient C_V and the gas flow coefficient C_{out2} on the "Out2" surface during the self-priming process. A negative gas flow coefficient means that the gas flows out of the pump, whereas a positive one indicates that the gas flows into the pump. In accordance with the previous analysis, the self-priming process of the pump was divided into three stages as follows: initial ($t \leq 0.5$ s), middle (0.5 s $< t <$ 4 s), and final (4 s $\leq t$). These self-priming stages are shown as I, II, and III in Figure 14. When 0 s $< t \leq$ 0.15 s, the amplitude of C_{out2} was close to 0, and the value of C_V remained 1, indicating that the gas in the inlet section began to reach the "Out2" surface at t = 0.15 s. When $t >$ 0.15 s, the distribution of C_{out2} showed a serious fluctuation, which shows the unsteady characteristics of the self-priming process of the pump. When 0.15 s $< t \leq$ 0.5 s, the amplitude of C_{out2} increased initially then decreased; afterward, it reached the maximum of 0.15. When 0.5 s $< t \leq$ 4 s, it decreased and reached the minimum of 0.02. When 4 s $< t \leq$ 5.8 s, it increases again, reaching the maximum of 0.56. When 5.8 s $< t \leq$ 8 s, it decreases again and finally approached 0, indicating that the gas in the pump was basically exhausted. In the entire self-priming process, the value of C_V continuously decreased, and the decrement rate was affected by C_{out2}.

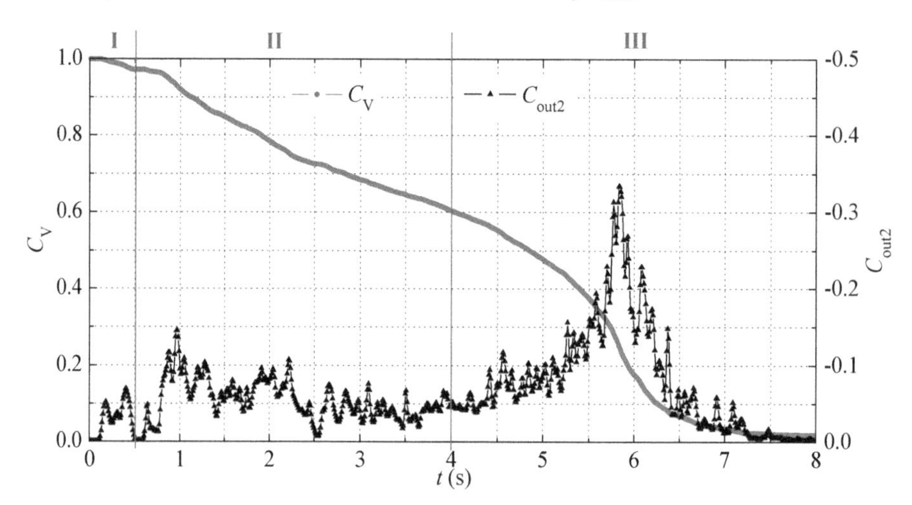

Figure 15. Gas volume coefficient C_V and gas flow coefficient C_{out2} on the "Out2" surface during the self-priming process.

The impeller and diffuser are the core components of the self-priming pump. Figure 16 displays the gas flow coefficient C_{ip_in} and C_{ip_out} of impeller inlet and outlet in the self-priming process. When 0 s $< t \leq$ 0.05 s, the value of C_{ip_in} increased from 0 to 0.8, while that of C_{ip_out} increased from 0 to 0.65, indicating that part of the gas remained in the impeller. When 0.08 s $< t \leq$ 0.5 s, the working capacity of the impeller was weakened due to the increase of the impeller void fraction in the initial self-priming stage, and the amplitudes of C_{ip_in} and C_{ip_out} decreased and approached 0, showing a certain positive correlation between them. When 0.5 s $< t \leq$ 4 s, the amplitudes of C_{ip_in} and C_{ip_out} fluctuated continuously but increased first and then decreased on the whole. The reason for the increment was that the amplitudes of C_{ip_in} and C_{ip_out} were close to 0 at t = 0.5 s, while the reason for the decrement was that the lower the pressure at the impeller inlet, the slower the rising rate of the water column in the inlet section, and the slower the suction rate of the impeller. When $t > 4$ s, the water in the inlet section flowed into the impeller in the final self-priming stage, causing the drainage capacity of the impeller to increase sharply. The amplitudes of C_{ip_in} and C_{ip_out} continued to increase, and the maximum value was close to 0.2. Then, as the total amount of gas in the pump decreased, the amplitudes of C_{ip_in} and C_{ip_out} continued to decrease, and finally close to 0.

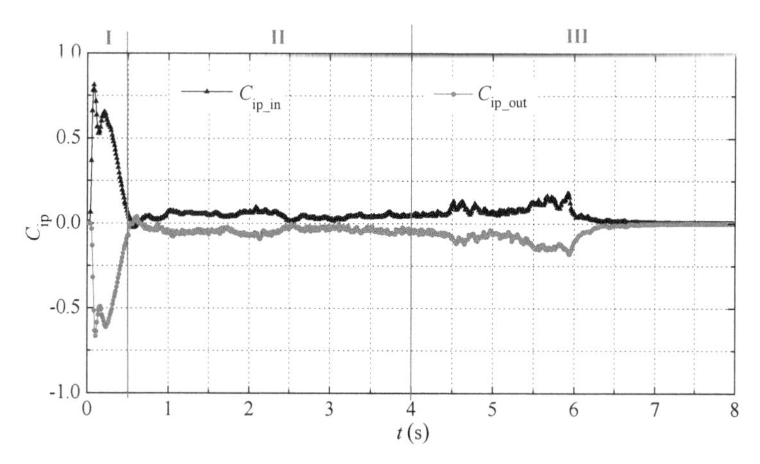

Figure 16. Gas flow coefficient C_{ip} of impeller inlet and outlet in the self-priming process.

Figure 17 shows the gas flow coefficient C_{df_in} and C_{df_out} of the diffuser inlet and outlet in the self-priming process. It can be seen that the changing law of C_{df_in} and C_{df_out} was basically the same as that of C_{ip_in} and C_{ip_out}. In the initial, middle, and final self-priming stages, the value of C_{df_in} and C_{df_out} first increased and then decreased.

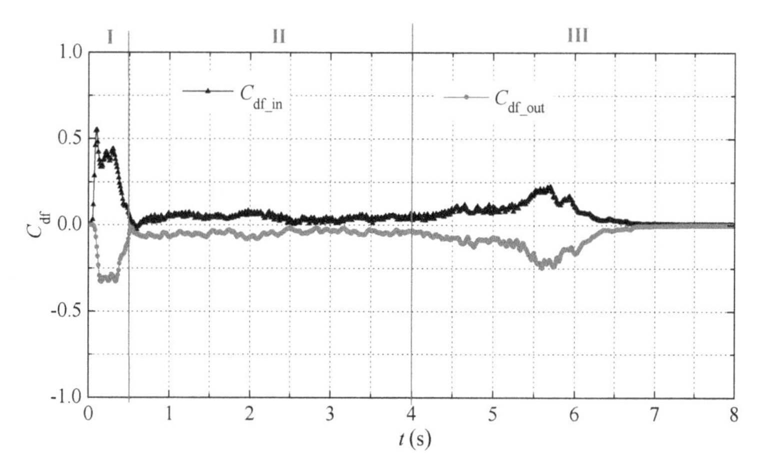

Figure 17. Gas flow coefficient C_{df} of diffuser inlet and outlet in the self-priming process.

As a standard characteristic parameter of the pump, pressure is one of the important factors affecting the self-priming capability of the self-priming centrifugal pump. Figure 18 presents the pressure P_{in} on the "In" surface and pressure P_{out2} on the "Out2" surface in the self-priming process. It can be seen that the value of P_{in} was basically kept at 5 kPa, due to that the "In" surface was 0.5 m under water and the pressure value was only affected by atmospheric pressure. In addition, the value of P_{out2} was rapidly increased from 0 kPa to 5 kPa in the initial self-priming stage, it remained stable in the middle self-priming stage, and rapidly increased to 10 kPa in the final self-priming stage.

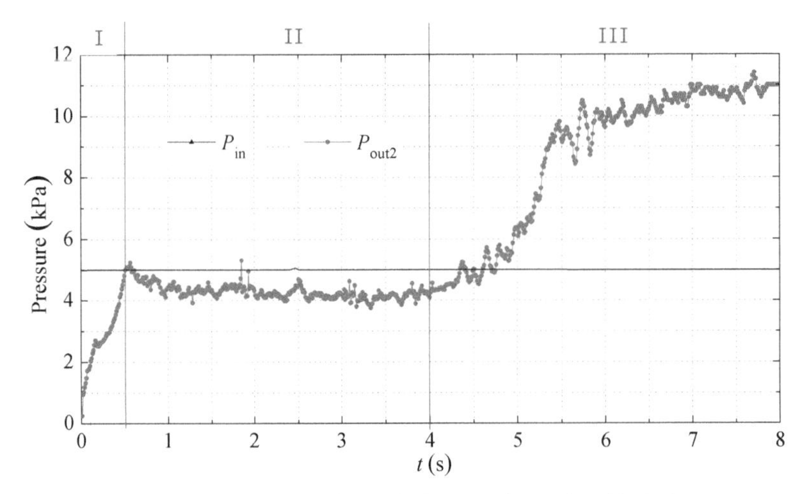

Figure 18. Pressure on the "In" surface and "Out2" surface in the self-priming process.

Figure 19 illustrates the pressure P_{ip_in} and P_{ip_out} of the impeller inlet and outlet in the self-priming process. It can be seen that when 0 s < $t \leq$ 0.05 s, after the impeller filled with water started to rotate, the value of P_{ip_in} rapidly decreased from 0 kPa to −27 kPa and then increased to 7 kPa, while the value of P_{ip_out} increased to 45 kPa. When 0.05 s < $t \leq$ 0.5 s, with the rapid increase of the impeller void fraction, the working capability of the impeller was sharply weakened, the value of P_{ip_in} and P_{ip_out} decreased to −2.5 kPa and 3.5 kPa, respectively. When 0.5 s < $t \leq$ 4 s, since the gas in the inlet section was slowly exhausted out of the pump, the water column in the inlet section raised continuously, and the value of P_{ip_in} showed a downward trend, while the value of P_{ip_out} was basically stable. When 4 s < $t \leq$ 8 s, the value of P_{ip_in} dropped sharply and the value of P_{ip_out} raised significantly due to the sharp decrease of the impeller void fraction. Finally, when the impeller void fraction was close to 0, the value of P_{ip_in} and P_{ip_out} remained at −25 kPa and 42 kPa.

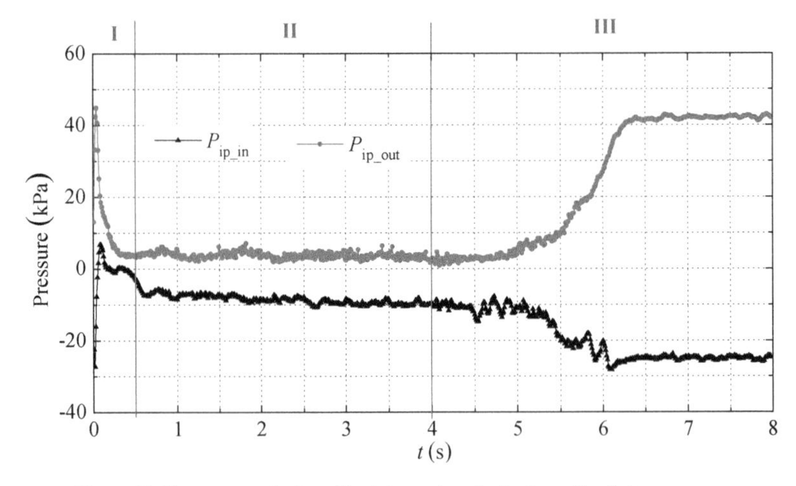

Figure 19. Pressure on the impeller inlet and outlet in the self-priming process.

Figures 20 and 21 present the pressure difference between the two ends of backflow channel P_{bc} and water flow coefficient C'_{bc} in the whole self-priming process and middle self-priming stage. It can be seen that in the initial self-priming stage, the value of P_{bc} and C'_{bc} reached the maximum rapidly, then decreases rapidly. In the middle self-priming stage, the value of P_{bc} and C'_{bc} constantly fluctuated. In the final self-priming stage, the values of P_{bc} and C'_{bc} increased sharply and remained essentially stable. Moreover, the value of C'_{bc} had a positive correlation with P_{bc}, which was especially evident in the middle self-priming stage, as shown in Figure 20. In the middle self-priming stage of t = 2.65 s, the maximum value of P_{bc} reached 2.8 kPa, while the maximum value of C'_{bc} was 0.095.

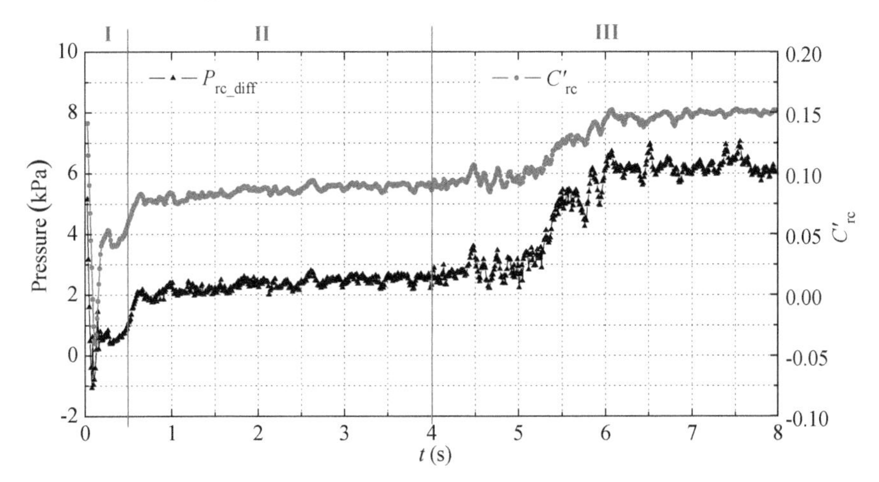

Figure 20. Pressure difference between the two ends of backflow channel P_{bc} and water flow coefficient C'_{rc} in the whole self-priming process.

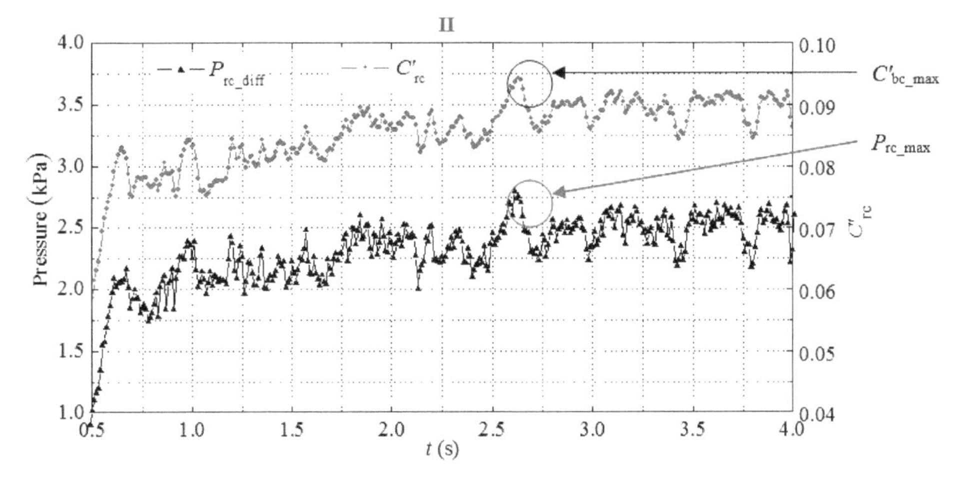

Figure 21. Pressure difference between the two ends of backflow channel P_{bc} and water flow coefficient C'_{bc} in the middle self-priming stage.

4. Conclusions

(1) The entire self-priming process of the self-priming centrifugal pump can be divided into three stages: gas-suction stage due to the impeller's rotating role in the initial self-priming stage, gas-suction stage due to the role of gas-water mixture and gas-water separation in the middle self-priming stage, and gas-suction stage due to the water flowing from the inlet section into the pump in the final

self-priming stage. Moreover, the self-priming time of the initial and final self-priming stages accounts for a small percentage of the entire self-priming process. The middle self-priming stage is the main stage in the self-priming process and determines the length of self-priming time.

(2) In the initial self-priming stage, the self-priming centrifugal pump is mainly based on drainage, and the gas is mixed with water before going outside the pump; it takes a certain amount of time for the gas to move from the inlet section to the outlet section. With the increase in the void fraction of the impeller, the drainage capacity of the self-priming pump is weakened, and the gas-exhausting rate is fast then slow. After entering the middle self-priming stage, due to the formation of a large negative pressure at the impeller inlet, the decreasing rate of the pressure continues to slow down on this basis, and the gas-suction rate of the self-priming centrifugal pump also slows down. In the final self-priming stage, the water in the inlet section goes into the impeller, whose power capability is enhanced obviously. The water mixed with the gas goes to the outlet section, so the gas-suction rate of the self-priming pump increased and reached the maximum value at a certain time. Afterward, the gas-suction rate decreased and finally approaches 0, due to the fact that the amount of gas in the pump decreases continuously.

(3) In the initial self-priming stage, the region near the suction surface of the impeller blade (low-pressure region) is easily occupied by gas. In the final self-priming stage, the region near the pressure surface of the impeller blade (high-pressure region) is easily occupied by water. In the middle self-priming stage, the impeller void fraction is high and a small amount of gas-water mixture exists in a small region close to the pressure surface of the impeller blade and impeller outlet. The key to successful self-priming is that the impeller's rotation forces a small amount of water mixed with some gas to flow along the pressure side of the impeller blade from the impeller inlet to the impeller outlet. Then, the mixture goes into the outward diffuser and flows along the pressure surface of the return diffuser, which shows that the gas-water mixture always flows along the high-pressure side of the pump in the self-priming process. The diffuser void fraction varies like the impeller void fraction, but its overall amplitude is lower.

(4) In the whole self-priming process, the backflow function of the backflow channel is the key to self-priming, and the pressure difference at both ends of the backflow channel is the main reason for the backflow function.

Author Contributions: B.H. and Y.Z. conceived and designed the experiments; X.W. performed the experiments; C.L. analyzed the data; C.W. and L.C. wrote the paper.

Acknowledgments: This research was funded by National Natural Science Foundation of China (Grant No.51609105), Jiangsu Province Science Foundation for Youth (Grant No. BK20170507), China Postdoctoral Science Foundation (Grant No.2016M601738 and 2018T110458). And the APC was funded by Priority Academic Program Development of Jiangsu Higher Education Institutions(PAPD).

Conflicts of Interest: The authors declare no conflict of interest.

References

1. Li, X.; Jiang, Z.; Zhu, Z.; Si, Q.; Li, Y. Entropy generation analysis for the cavitating head-drop characteristic of a centrifugal pump. *Proc. Inst. Mech. Eng. Part C: J. Mech. Eng. Sci.* **2018**, *232*, 4637–4646. [CrossRef]
2. Wang, C.; He, X.; Shi, W.; Wang, X.; Qiu, N. Numerical study on pressure fluctuation of a multistage centrifugal pump based on whole flow field. *Am. Inst. Phys. Adv.* **2019**, *9*, 035118. [CrossRef]
3. Zhang, S.; Li, X.; Hu, B.; Liu, Y.; Zhu, Z. Numerical investigation of attached cavitating flow in thermo-sensitive fluid with special emphasis on thermal effect and shedding dynamics. *Int. J. Hydrogen Energy* **2019**, *44*, 3170–3184. [CrossRef]
4. Wang, C.; Chen, X.; Qiu, N.; Zhu, Y.; Shi, W. Numerical and experimental study on the pressure fluctuation, vibration, and noise of multistage pump with radial diffuser. *J. Braz. Soc. Mech. Sci. Eng.* **2018**, *40*, 481. [CrossRef]
5. Li, X.; Gao, P.; Zhu, Z.; Li, Y. Effect of the blade loading distribution on hydrodynamic performance of a centrifugal pump with cylindrical blades. *J. Mech. Sci. Technol.* **2018**, *32*, 31161–31170. [CrossRef]

6. Wang, C.; Shi, W.; Wang, X.; Jiang, X.; Yang, Y.; Li, W.; Zhou, L. Optimal design of multistage centrifugal pump based on the combined energy loss model and computational fluid dynamics. *Appl. Energy* **2017**, *187*, 10–26. [CrossRef]
7. Gao, R.; Li, Y.; Yu, M.; Liu, W. Effects of heat pump drying parameters on the volatile flavour compounds in silver carp. *J. Aquat. Food Prod. Technol.* **2016**, 735–744. [CrossRef]
8. Xie, F.; Xuan, R.; Sheng, G.; Wang, C. Flow characteristics of accelerating pump in hydraulic-type wind power generation system under different wind speeds. *In. J. Adv. Manuf. Technol.* **2017**, *92*, 189–196.
9. Xia, C.; Cheng, L.; Luo, C.; Jiao, W.; Zhang, D. Hydraulic Characteristics and Measurement of Rotating Stall Suppression in a Waterjet Propulsion System. *Trans. FAMENA* **2018**, *4*, 85–100. [CrossRef]
10. Qian, J.; Chen, M.; Liu, X.; Jin, Z. A numerical investigation of the flow of nanofluids through a micro Tesla valve. *J. Zhejiang Univ. SCIENCE A* **2019**, *20*, 50–60. [CrossRef]
11. Lu, Z.; He, X.; Wang, C. Influencing factors of self-priming time of multistage self-priming centrifugal pump. *DYNA* **2018**, *93*, 630–635. [CrossRef]
12. Chang, H.; Shi, W.; Li, W.; Liu, J. Energy loss analysis of novel self-priming pump based on the entropy production theory. *J. Therm. Sci.* **2018**, *28*, 1–13. [CrossRef]
13. Wang, C.; He, X.; Zhang, D.; Hu, B. Numerical and experimental study of the self-priming process of a multistage self-priming centrifugal pump. *Int. J. Energy Res.* **2019**, 1–19. [CrossRef]
14. Chang, H.; Li, W.; Shi, W.; Liu, J. Effect of blade profile with different thickness distribution on the pressure characteristics of novel self-priming pump. *J. Braz. Soc. Mech. Sci. Eng.* **2018**, *40*, 518. [CrossRef]
15. Qian, J.; Gao, Z.; Liu, B.; Jin, Z. Parametric study on fluid dynamics of pilot-control angle globe valve. *ASME J. Fluids Eng.* **2018**, *140*, 111103. [CrossRef]
16. Zhang, J.; Xia, S.; Ye, S.; Xu, B.; Song, W.; Zhu, S.; Xiang, J. Experimental investigation on the noise reduction of an axial piston pump using free-layer damping material treatment. *Appl. Acoust.* **2018**, *139*, 1–7. [CrossRef]
17. Zhao, X.; Xu, Y.; Lei, Q. Theoretical and experimental research on characteristics of vertical self-priming centrifugal pump. *J. East China Univ. Sci. Technol.* **1996**, *22*, 455–461.
18. Yi, Q. Statistics and exploration on the self-priming time and specific speed of the self-priming pump. *Drain. Irrig. Mach.* **1992**, *4*, 8–11.
19. Wang, C.; Wu, Z.; Si, Y.; Yi, T. Gas-liquid two-phase flow numerical simulation of a vortex flow self-priming pump. *Drain. Irrig. Mach.* **2009**, *27*, 163–167.
20. Wang, C.; Si, Y.; Zheng, H.; Peng, N.; Zhao, B.; Zhang, H. Numerical simulation of rotational interior flow in self-priming pump. *Drain. Irrig. Mach.* **2008**, *26*, 31–35.
21. Liu, J.; Su, Q.; Xu, Y.; Wang, D. Numerical simulation of internal flow field of self-priming irrigation pump. *Drain. Irrig. Mach.* **2009**, *27*, 347–351.
22. Liu, J.; Su, Q. Simulation on gas-liquid two-phase flow in self-priming pump. *Trans. Chin. Soc. Agric. Mach.* **2009**, *40*, 73–76.
23. Li, H.; Xu, D.; Tu, Q.; Cheng, J. Numerical simulation on gas-liquid two-phase flow of self-priming pump during starting period. *Trans. Chin. Soc. Agric. Eng.* **2013**, *29*, 77–83.
24. Lu, T.; Li, H.; Zhan, L. Transient numerical simulation and visualization of self-priming process in self-priming centrifugal pump. *J. Drain. Irri. Mach. Eng.* **2016**, *34*, 927–933.
25. Huang, S.; Su, X.; Guo, J.; Yue, Y. Unsteady numerical simulation for gas–liquid two-phase flow in self-priming process of centrifugal pump. *Energy Convers. Manag.* **2014**, *85*, 694–700. [CrossRef]
26. Hubbard, B. Self-priming characteristics of flexible impeller pumps. *World Pumps* **2000**, *405*, 19–21.
27. Yan, H. Test on the effects of gas-water separation chamber volume of self-priming pump on self-priming performance. *Fluid Mach.* **1996**, *24*, 39–40.
28. Zhang, X. Effect of backflow hole on the self-priming pump. *Gen. Mach.* **2004**, *3*, 41–46.
29. Chen, M. Design and Test of High-head self-priming centrifugal pump. *Fluid Mach.* **1998**, *26*, 7–11.
30. Desbrun, M.; Gascuel, M.P. Smoothed particles: A new paradigm for animating highly deformable bodies. In *Computer Animation and Simulation'*; Springer: Vienna, Austria, 1996; pp. 61–76.

31. Belytschko, T.; Neal, M.O. Contact-impact by the pinball algorithm with penalty and Lagrangian methods. *Int. J. Numer. Methods Eng.* **1991**, *31*, 547–572. [CrossRef]
32. Kim, J.; Moin, P. Application of a fractional-step method to incompressible Navier-Stokes equations. *J. Comput. Phys.* **1985**, *59*, 308–323. [CrossRef]

Article

A Simulation-Based Multi-Objective Optimization Design Method for Pump-Driven Electro-Hydrostatic Actuators

Longxian Xue [1,2], Shuai Wu [3,*], Yuanzhi Xu [3] and Dongli Ma [1]

1 School of Aeronautic Science and Engineering, Beihang University, Beijing 100091, China; desertseas@163.com (L.X.); madongli@buaa.edu.cn (D.M.)
2 Chengdu Aircraft Design and Research Institute, Chengdu 610000, China
3 School of Automation Science and Electrical Engineering, Beihang University, Beijing 100091, China; xuyuanzhi@foxmail.com
* Correspondence: wushuai.vip@gmail.com

Received: 26 March 2019; Accepted: 3 May 2019; Published: 9 May 2019

Abstract: A pump-driven actuator, which usually called an electro-hydrostatic actuator (EHA), is widely used in aerospace and industrial applications. It is interesting to optimize both its static and dynamic performances, such as weight, energy consumption, rise time, and dynamic stiffness, in the design phase. It is difficult to decide the parameters, due to the high number of objectives to be taken into consideration simultaneously. This paper proposes a simulation-based multi-objective optimization (MOO) design method for EHA with AMESim and a python script The model of an EHA driving a flight control surface is carried out by AMESim. The python script generates design parameters by using an intelligent search method and transfers them to the AMESim model. Then, the script can run a simulation of the AMESim model with a pre-set motion and load scenario of the control surface. The python script can also obtain the results when the simulation is finished, which can then be used to evaluate performance as the objective of optimization. There are four objectives considered in the present study, which are weight, energy consumption, rise time, and dynamic stiffness. The weight is predicted by the scaling law, based on the design parameters. The performances of dynamic response energy efficiency and dynamic stiffness are obtained by the simulation model. A multi-objective particle swarm optimization (MOPSO) algorithm is applied to search for the parameter solutions at the Pareto-front of the desired objectives. The optimization results of an EHA, based on the proposed methodology, are demonstrated. The results are very useful for engineers, to help determine the design parameters of the actuator in the design phase. The proposed method and platform are valuable in system design and optimization.

Keywords: electro-hydrostatic actuator; multi-objective optimization; weight; energy consumption; rise time; dynamic stiffness

---

1. Introduction

Using a pump as the driver in an electro-hydrostatic actuator (EHA) has the advantages of compact integration, high output force, and ease of maintenance [1]. Therefore, it has become part of a developing trend in fluid power transmission. They have been used in aerospace applications, to replace traditional hydraulic actuators in more-electric aircraft (MEA) Rongjie et al. [2], such as the A380 Van Den Bossche [3]. It is also emerging as a preferable solution for industrial applications, as their design combines the best of both electro-mechanical and electro-hydraulic technologies.

The schematic diagram of a typical EHA is illustrated in Figure 1. The basic function of an EHA is as a servo motor to drive a bi-directional hydraulic pump to generate a pressured cyclic flow rate, to control the cylinder extend/retract. The EHA has the advantage of being electric-powered, which can eliminate heavy, messy, and fault-liable hydraulic pipes; which makes them preferable in more-electric aircraft (MEA). Therefore, they can can make the system much more easy to maintain and reduce the system weight. In Kulshreshtha and Charrier [4], it was demonstrated, in the A380, that using more electric actuators saved over 450 kg. EHA have become a hot topic, and have been developed significantly in many aspects. However, they still have some issues which to be solved before they can be applied with high performance and reliability, such as high speed hydraulic pumping [5], and over-heating. In order to overcome the over-heating problem, some interesting works have been reported recently, such as using a load-sense pump in Chao et al. [6], and a novel control method with energy feedback in Shang et al. [7].

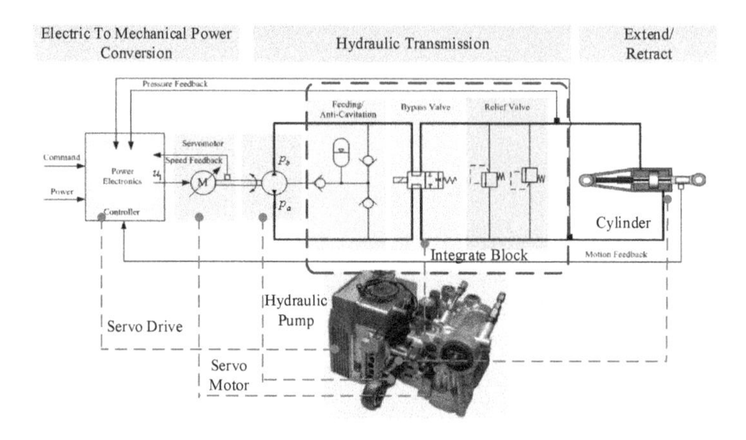

Figure 1. Schematic diagram of an electro-hydrostatic actuator (EHA).

In the early phases of an EHA design project, only a few design parameters are available, but a lot design parameters have to be decided and several properties should be considered simultaneously. It is well-known that early verification and virtual validation of the system design in the preliminary design phase, based on advanced simulations and computational tools, can significantly reduce the cost and enhance the quality of the design process. Some researchers have studied simulation-based preliminary design methods for aircraft actuator design, sizing, analysis, and optimization in recent years. A simulation-based preliminary design and optimization method of an electro-mechanical and hydraulic actuation system for an aircraft flight control surface was proposed by Fraj et al. [8], where the weight was the optimized objective and the weight estimation models of major components of actuator were presented. An improved integrated methodology for the preliminary design of electromechanical actuators in a redundant electro-mechanical nose-gear steering system was published by Liscouët et al. [9]. In order to obtain the properties of an electro-mechanical actuator (EMA) for multi-objective optimization in preliminary design, the estimation models for power size, thermal balance, dynamics, and reliability were studied by Budinger et al. [10]. After that, Budinger et al. [11] presented a methodology for the optimal preliminary design of EMAs. A MATLAB/Simulink-based methodology for the sizing, simulation, analysis, and optimization of both EHAs and EMAs, for the primary and secondary control surfaces of a more-electric aircraft (MEA), was proposed by Chakraborty et al. [12]. After that, in Chakraborty et al. [13], an electric control surface actuator design optimization and allocation for MEA was studied. Recently, a multi-level virtual prototyping of EMAs, using bond-graph modeling method, was proposed by Fu et al. [14], FU et al. [15]. However, most of these studies did not take advantage of intelligent optimization methods to find optimal design parameters; therefore, they cannot provide strong support for designers.

The most important properties of EHA includes light-weight, less energy consumption, quick response, and high stiffness to disturbance. These objectives usually conflict with each other and are hard to balance. Therefore, using a multi-objective optimization (MOO) method to design an EHA leads to more preferable solutions. MOO methods usually obtain a set of Pareto-optimal solutions, instead of a single optimal solution Marler and Arora [16]. The Pareto front is able to indicate the relations between the design parameters and the desired performance, which is very useful for the engineer to achieve an optimal design. In order to offer support for preliminary design of MEA, Wu et al. [17] proposed estimation models for the weight and efficiency of EHA, and a multi-objective optimization algorithm to get the Pareto front of considered performances. In the following work by Yu et al. [18], an estimation model of stiffness was considered and a synthesis decision-making method, based on an analytic hierarchy process (AHP), was used to choose the best solution in the Pareto front. This method can offer significant support to engineers in system design. However, it didn't integrate with a simulation tool and, thus, could not evaluate the dynamic performance.

The present work aims to offer an efficient and powerful simulation-based multi-objective optimization design method for fast and easy preliminary design of EHA. This methodology can search the design parameters automatically, which will make the EHA have Pareto-front performances. This study considered the four important indexes of weight, energy consumption, dynamic response, and stiffness. The weight of the actuator is predicted, based on a scale law according to the design parameters. The performances of energy consumption, dynamic response, and stiffness are obtained by dynamic simulation with AMESim. The intelligent optimization program sent the parameters to the AMESim model, and then ran the simulation and analyzed the dynamic performances (according to the simulation results) automatically. The MOO method of the program in present study is the multi-objective particle swarm optimization (MOPSO) method. The results present the mapping between parameters to the performance and the relations between these objectives. It also illustrated that the proposed method has the significant benefits in the design of EHA and other similar systems.

2. MOO Design Method of EHA

For the design of systems with high integration and complex structure, such as EHA actuation systems, choosing the design parameters to get a satisfactory performances is a difficult task, as there ae typically many objectives to be considered, such as light weight, high energy efficiency, good dynamic performance, and stiffness. Furthermore, these objectives usually conflict with each other. This is a MOO problem which usually needs a method to find the Pareto front of all the objectives intelligently.

Modeling and simulation are important tools in modern design, which can obtain the performances quickly by using a computer. There are some powerful modeling and simulation tools for electrical and hydraulic system design, such as AMESim. In AMESim, an EHA system can be modeled and simulated, with high precision and without too much effort. However, setting and optimizing the parameters is a daunting task, since there are a number of parameters which should be decided and the engineer has to go through a lot of simulation curves to judge the performances, which usually needs expert experience and takes a lot of labor time. If the model calculation does not obtain an ideal control state, the designer needs to find the improper parameters and simulate again and again, until the results are satisfactory. However, a "satisfying" result is usually not the optimal solution, since the designer can not continue to optimize the parameters for a long time. In summary, the manual simulation-based design methods rely on the experience of the designer and increases the time cost of system development. Obviously, the traditional approach to designing EHA is outdated, in the current era of "automation and intelligence". Therefore, improving design efficiency is a problem for these highly complex system design methods.

2.1. Description of the Proposed Method

Fortunately, the interface of python is provided by AMESim. Python is a powerful, well-scalable program language in artificial intelligence. It has a lot of libraries for scientific analysis and optimization, which are easy to use. Therefore, a model-based intelligent optimization method with python is proposed in the present study. The model of an EHA driving a flight control surface is carried out by AMESim. The developed python script can generate design parameters by using an intelligent search method, and transfers them to the AMESim model. Then, the python script can run a simulation of the AMESim model with a pre-defined motion and load scenario of the control surface. The python script also can get the results when the simulation is finished, which can be used to evaluate the performance as the objective of optimization. Therefore, using the proposed intelligent design method cannot only greatly save labor costs, but also shorten the system development time.

The flow chart of the developed intelligent optimization method is shown in Figure 2. The intelligence MOO method will update the parameters, based on the search algorithm, to get the optimal solutions by iteration until the Pareto front of the design is obtained. The entire intelligent MOO design process can be described as:

1. Build the AMESim model of EHA and initialize the model parameters randomly in the defined range.
2. Simulate and analyze the static and dynamic performances of the EHA.
3. Determine whether the design requirements are met. If not, the multi-objective optimization (MOO) algorithm will search the design parameters by the intelligent search algorithm.
4. The MOO method generates the design parameters, based on the search algorithm, and sends these parameters to the AMESim model of EHA.
5. Jump to the step 3 and return the performance metrics to the MOO algorithm for comparison.
6. Cycle through steps 3 to 6, until the Pareto front of the designs is obtained.
7. Validate whether the design requirements are all satisfied; if not, jump to step 2 to re-define the optimization parameters range and conditions.

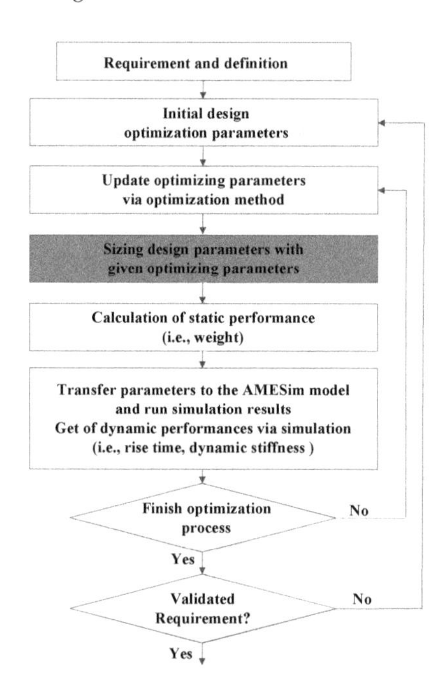

Figure 2. Flowchart of the intelligent optimization design process, based on python.

2.2. Modeling the EHA System with AMESim

A diagram of the location of an EHA in the aircraft wing profile is shown in Figure 3, and the AMESim model of a flight control surface actuation system by EHA is shown in Figure 4. It consists of an EHA, a flight control surface, and a controller. The model in the red-dashed box is the EHA, which contains all the major parts of an EHA (motor, pump, anti-cavitation device, safety valves, and selector valve). The model in the green-line box is the controller. Only a simple three-loop (displacement, velocity, and current) controller is used in the present work, since the controller is not the purpose. The controller generates the command (as voltage) to drive a direct current motor, where the shaft of the motor is connected to a hydraulic fixed-displacement pump. The pump drives the actuator through a hydraulic circuit, which consists of re-feeding valves with a compensating hydraulic accumulator for anti-cavitation, a mode selector valve to select the mode (damping or active), and two pressure relieve valves for safety. The piston of the actuator moves the flight control surface of the aircraft, and the control surface is modeled with a planar mechanical 2D library, which can simulate the external load more realistically. This model has all the major parameters of the EHA system in the application scenario and the performance under different parameters can be obtained and evaluated.

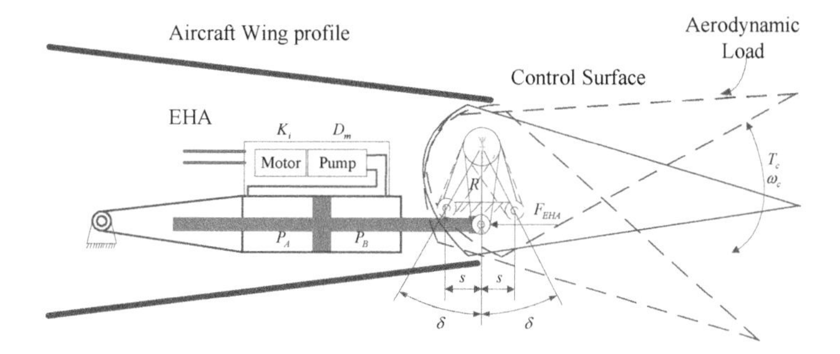

Figure 3. Diagram of the location of the EHA in the aircraft wing profile.

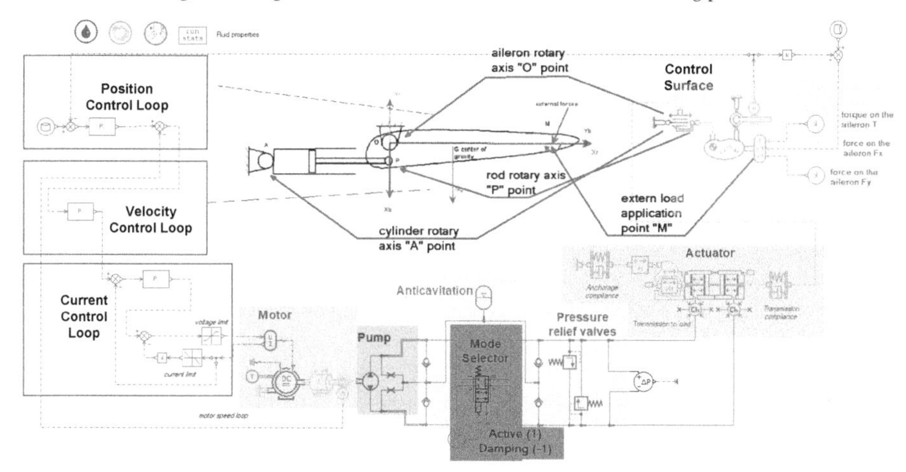

Figure 4. Model of an EHA actuation system in AMESim.

In order to evaluate the performances of the EHA system, a typical angle command and load scenario of flight control surface is constructed, which is shown in Figure 5. The aerodynamic load include two parts; elastic load (proportional to the angle) and disturbance load. The angle and load commands include different conditions to evaluate different control performances, which are:

1. A low-frequency sinusoid angle command is set in the first 5 s to evaluate control precision;
2. a step angle change is set from 5–11 s to obtain the rise time, in order to evaluate the dynamic performance;
3. a wind gust is simulated during 11–15 s, while the control surface is held at a constant-angle position, to evaluate the dynamic stiffness; and
4. some higher frequency sinusoid angle commands are also included, and the total input energy of electrical motor during the 20 s is used to evaluate the energy consumption performance.

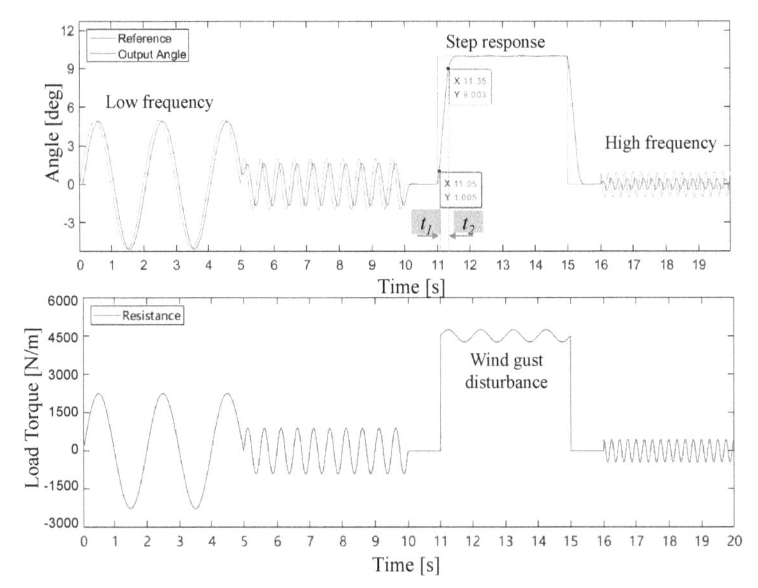

Figure 5. Dynamic response of a defined angle command and load scenario.

2.3. Sizing Design Parameters

The fourth step of the flow chart in Figure 2 is sizing the design parameters with given optimization parameters. This is an important process, and its diagram is illustrated in Figure 6. The inputs of the sizing process are the requirements of a control surface, which includes maximum hinge moment (T_c), maximum rotation velocity (ω_c), and deflection angle (δ_c). For a flight control application, the EHA drives the control surface deflection by a level. Therefore, the level length (R) is a key parameter that affects the parameters of the EHA actuation system significantly. The maximum force (F), maximum speed (v), and linear stroke (s) can be calculated by (R) and (T_c), (ω_c), and (δ_c). When the working pressure (P_s) of the EHA system is determined, the relevant parameters of the hydraulic cylinder can also be determined, such as the piston area (A_p) and the flow rate. The second key parameter that needs to be determined is the displacement of the hydraulic pump (D_m). As shown in Wu et al. [17], the maximum speed of the pump (ω_p) is limited by the displacement of the hydraulic pump. The torque of the motor (T_m) can be obtained by the product $D_m P_s$. Usually, the maximum current is limited by the servo motor driver, and then the torque constant of the motor (K_i) is known. The weight of the EHA can be estimated, based on the scaling law in Wu et al. [17] with the sizing parameters. The parameters also will be transferred to the AMESim model and the dynamic performances can be obtained after running a simulation. In summary, the selected optimization design variables in this paper are (R) and (D_m), and the optimization targets are weight, energy consumption, rise time, and dynamic stiffness.

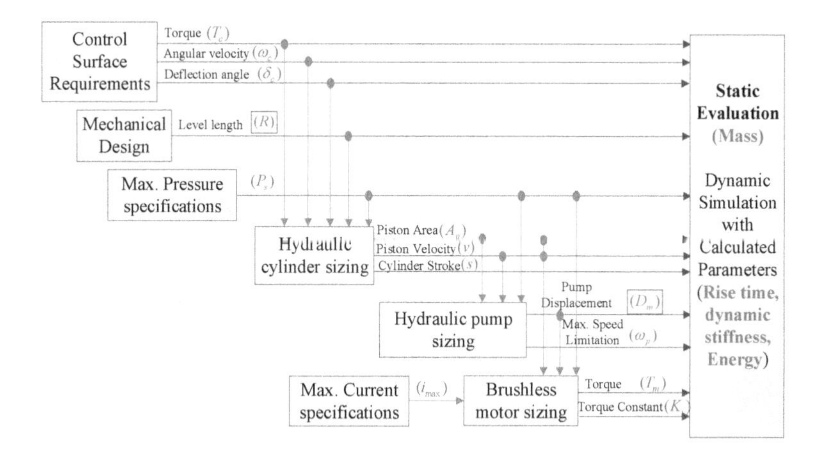

Figure 6. Diagram of the design process in the EHA parameter optimization.

3. Multi-Objective Optimization of EHA

In the present study, four objectives are considered for optimization. Therefore, the evaluation model of these four objectives should be integrated into the optimization program. These four objectives include both static and dynamic performances, requiring different evaluation methods which will discussed in the present section.

3.1. Weight Prediction Model of EHA

Weight is an important property which should be minimized for aerospace applications. As shown in Figure 1, a EHA mainly has five basic parts: Pump, motor, cylinder, manifold, and power electronics. The total weight of the EHA can be defined as the sum of the gross weights of the five parts. In order to evaluate the weight of the EHA with the major design parameters in the preliminary design phase, two weight estimate methods are employed in the present study. The weight prediction, with the design parameters, usually uses the scaling law. It has the advantage of requiring only one reference component for a complete estimation of a product range [8,10]. More details of mass prediction of other components can be found in [8,17].

3.2. Energy Consumption Index of EHA

Energy efficiency is another key characteristic for EHA. Some previous works used the design parameters to calculate the efficiency of EHA, but could not evaluate energy consumption well since it is related to the working condition. In the present study, the energy consumption is indexed by the total energy input of EHA to finish the pre-defined representative simulation task. The integration of input power of the electrical motor is the energy consumption of the EHA, which can be expressed as:

$$E_m = \int_0^T U(t)I(t)dt, \quad (1)$$

where $U(t)$ and $I(t)$ are the input voltage and current of the motor, which can be obtained from the dynamic simulation results; and T is the simulation time of the typical scenario.

3.3. Dynamic Response Performance Index of EHA

The dynamic performance is very important for EHA and it usually can be evaluated by rise time (t_r). The smaller t_r, the faster the system responds. The rise time is evaluated by the dynamic simulation. The reference angle has a step variation, and the EHA need to control the control surface to follow the reference signal. The time between 10% to 90% of step usually defined as the rise time. As can be seen from the Figure 5, the t_r can be expressed as:

$$t_r = t_2 - t_1 \tag{2}$$

where t_1 and t_2 are the time corresponding to the response curve being 10% and 90% of the steady state value, respectively.

3.4. Dynamic Stiffness of EHA

Dynamic stiffness is also an important indicator for evaluating the dynamic performance of EHA. Low dynamic stiffness can severely degrade the control accuracy of the system. Hydraulic spring effects are a major part of the EHA stiffness. Therefore, the stiffness of the hydraulic spring is considered as an objective in this study. The dynamic stiffness is evaluated by the root-mean square (RMS) of the angle of control surface offset from the given value, during 12–15 s in the simulation. When the EHA holds the control surface stalled at a constant position under wind gust disturbance, more stiffness will result in a smaller RMS value.

4. Multi-Objective Optimization Method Description

The design parameters are searched by the MOO method, instead of manually adjusting, to improve design efficiency. Various objectives usually conflict with each other, which means improving one objective usually comes at the expense of reducing the character of another. The MOO method can search for the solution set that can not improve any one objectives without reducing other objectives, which usually is called the Pareto front.

Various MOO methods have been proposed, including the multi-objective particle swarm optimization (MOPSO) algorithm Coello et al. [19], the fast non-dominated sorting genetic algorithm (NSGA-2) method Deb et al. [20], DEB [21], andthe Pareto-frontier differential evolution (PDE) approach Abbass et al. [22], among others. In the present study, the (MOPSO) method is employed as it has the advantage of very good global searching capability.

The pseudo-code of the MOPSO method in the present study can be described as:

1. Initialize each particle of the population.
2. Update (initialize in the first loop) the position of each particle with parameter values of ($R(i)$, $D_m(i)$), according to the multi-objective particle swarm optimization algorithm.
3. Evaluate each of the particles and get the weight and according dynamic parameters by parameters. Send the dynamic parameters to the simulation model and run a simulation with the updated parameters. Calculate the dynamic objectives and assign the fitness value to the particle.
4. Store the positions (parameters) of the particles that represent non-dominant vectors in the repository.
5. Select the non-dominated particles (according to the Pareto dominance relation) and put them into the non-dominated set.
6. Compute the speed of each particle using Equation (3) and compute the new position of the particles adding current parameter with the speed.

$$\begin{aligned} v_i(t+1) = wv_i(t) &+ c_1r_1(p_i^{\text{BEST}}(t) - x_i(t)) \\ &+ c_2r_2(R_h(t) - x_i(t)), \end{aligned} \tag{3}$$

where p_i^{BEST} is the best position in the whole search history; R_h is the selected leader from the repository; w represent the inertia coefficient of velocity; c_1 and c_2 are local and social coefficients, respectively; and r_1 and r_2 are two random values in the range $[0, 1]$. Then, the position is updated at each iteration based on the velocity.

$$S_i^{POP}(t) = S_i^{POP}(t) + v_i(t). \tag{4}$$

7. Determine whether the maximum count of iterations has been reached or not. If the maximum number of iterations has not been reached, go back to the step 2. Otherwise, stop the optimization process and return the Pareto front stored in the repository.

5. Optimization Results and Discussion

A case of EHA system optimization is presented, in order to validate the feasibility of the proposed method. The design requirements of a typical control surface are listed in the Table 1. The maximum voltage is $U_{\max} = 270$ VDC and maximum current is set to 50 A. The maximum pressure of EHA is set to 35 MPa. In the optimization process, the optimization method generates the radius of the lever (R) and the displacement of the pump (D_m). Then the parameters of the EHA (i.e., the area of the piston, the stroke of the cylinder, the rotational inertia of the motor and pump, and the torque constant of the motor) can be calculated. These parameters are transferred to the AMESim model and the simulation is run. Then, the simulation results are obtained and the dynamic objectives can be analyzed automatically by the program.

Table 1. Control surface requirements.

Specifications	Value [Unit]
Hinge moment	7200 [Nm]
Velocity rate	$60°/s$
Maximum deflection angle	$\pm 25°$

Consider the application condition, where the range of the two optimization parameters are set as shown in Table 2. These two parameters decide the major characteristics of the EHA. A longer lever means the force requirement of the EHA is small but the stroke should be longer, in a slender form. On the contrary, a shorter lever means a large force requirement and short stroke. A larger displacement of the pump requires low rotary speeds for a desired flow rate, which will make the motor and pump more heavy, but can improve the efficiency and may make the EHA response quicker. These conflicting objectives make it hard to find a solution to trade off all performances. It usually required of MOO design methods to find the Pareto Front to help the engineer in the design process.

Table 2. Range of optimization parameters.

Parameter	Range [Unit]
Level length	$R \in [50, 250]$ mm
Pump displacement	$D_m \in [0.6, 4]$ mL/rev

The parameters of MOPSO, in the present study, include the number of particle population $N_p = 100$, maximum repository capacity of the Pareto front is 100, $w = 0.9$, $c_1 = c_2 = 1.1$, and the maximum iteration count is 10. The Pareto fronts of each of the two objectives are shown in Figure 7. These figures show that the weight and energy consumption trade off against each other. The results also indicate that the rise time and weight also trade off against each other. The rise time and energy consumption are positively correlated. The relationships between dynamic stiffness and the other performances are more complicated, as they are not monotonous functions. These relationships between objectives are useful for designers to balance different performances. The results also indicate that the proposed method can search the Pareto front solutions intelligently, which can save huge effort in adjustment by the engineer.

The relationships between the parameters and the objectives are shown in Figure 8. The weights of the solutions in the Pareto front, with design parameters, are shown in the upper right of Figure 8.

The lightest solution is with the lever length about 125 mm, and a displacement of about 0.6 mL/rev. Compared to the energy consumption and rise time, the lightest solution had the biggest energy consumption and slowest response. The most efficient solution is locating at $D_m = 4$ mL/rev and lever length of 103 mm. However, this solution also had a high weight and low dynamic stiffness. The optimization results indicate that using a smaller displacement pump is beneficial for reducing weight and increasing dynamic stiffness. Using a bigger displacement pump will improve the efficiency and make the EHA response quicker, but will make the EHA more heavy. The designer can get all design parameters and performance indices, as in Figure 8, of the solutions in the Pareto front, then they can choose one solution which is most satisfactory for the application. Thus, the proposed method can offer significant support for the engineer.

The time domain performances of four typical solutions in the Pareto front (the lightest solution, the quickest solution, the most efficient solution, and the best stiffness solution) are simulated. The simulation results are illustrated in the Figure 9. All of these four solutions obtain an acceptable control performance, only having a little difference in dynamic response; but the weight and energy consumption had a large disparity. This means the weight and energy consumption can be reduced, if appropriate parameters are given. The Pareto front obtained by the proposed method is very helpful for the engineer to determine the design parameters.

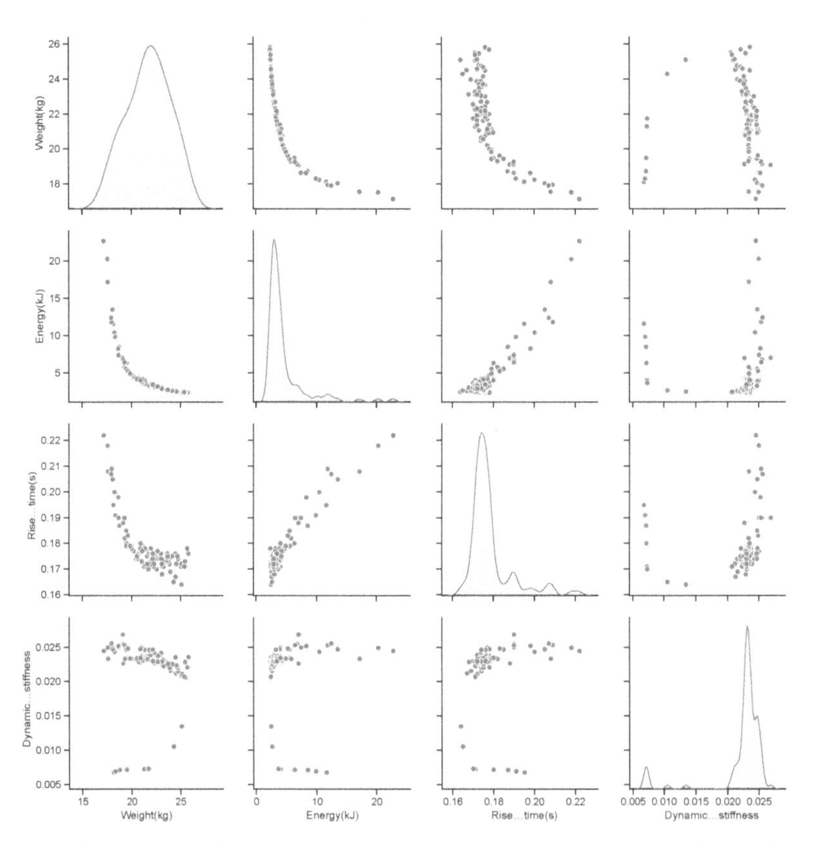

Figure 7. Scatter graph of the optimization results, including the Pareto front of each two objectives and the distribution of the each objective.

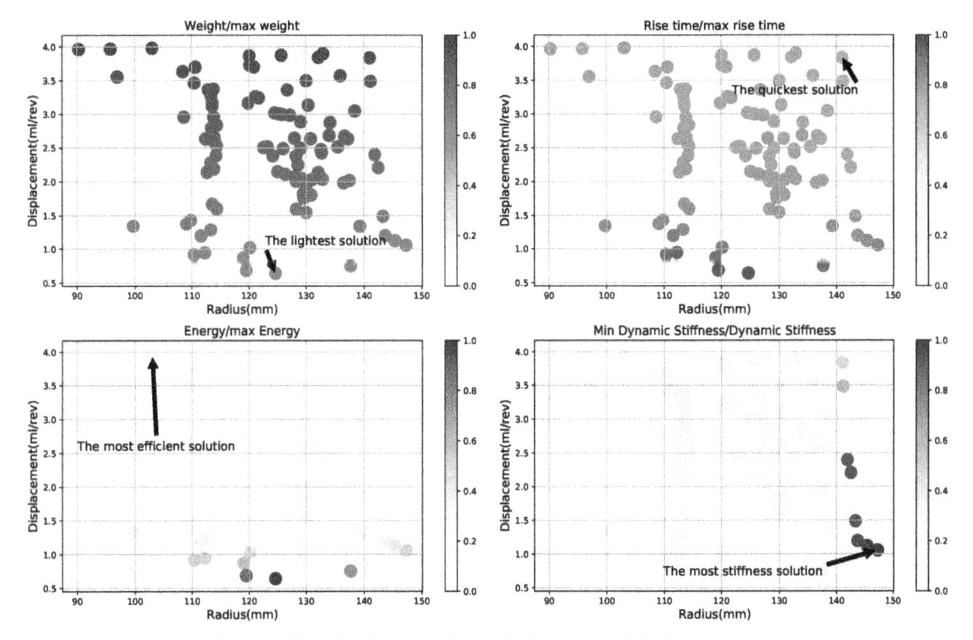

Figure 8. Objective locations in the design space of the Pareto front.

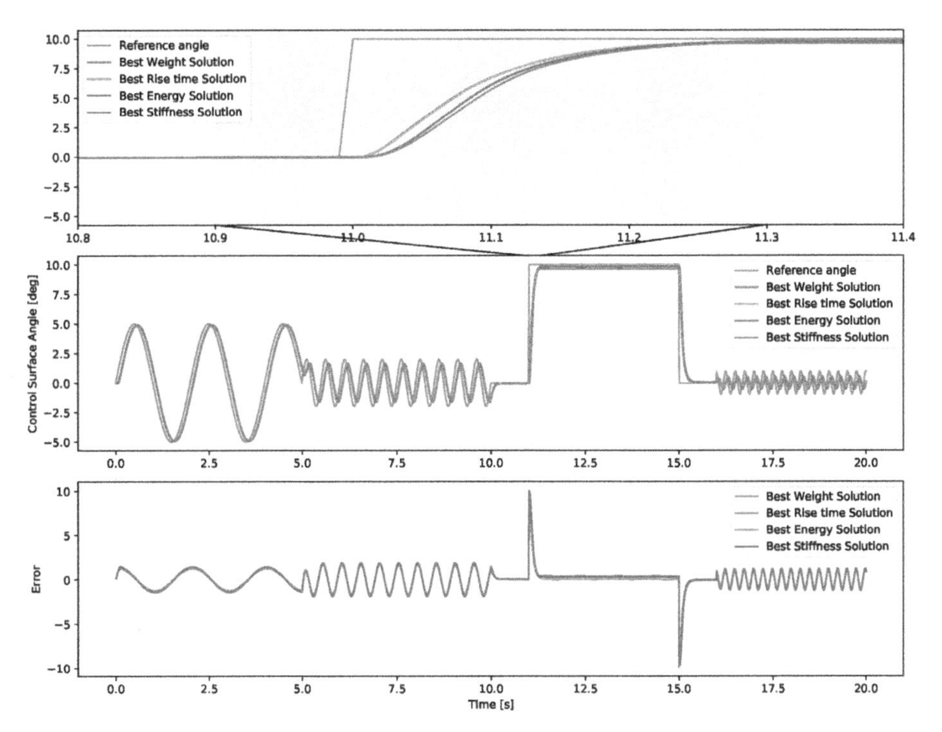

Figure 9. Dynamic simulation results of four types of typical design parameters.

6. Conclusions

This paper presents a simulation-based intelligent multi-objective optimization method of a pump-driven electro-hydrostatic actuator with AMESim and a python script. The model of an EHA driving a flight control surface is carried out by AMESim. The python script generates design parameters by using an intelligent search method and transfers them to the AMESim model. Then, the script can run a simulation of the AMESim model with a pre-defined motion and load scenario of a control surface. The python script also can obtain the results when the simulation is finished, which can then be used to evaluate the performance as the objective of optimization. The multi-objective particle swarm optimization (MOPSO) method is applied to obtain the Pareto front of solutions. In the present study, the design parameters of level length and pump displacement of the pump are optimized. An application case of optimizing an EHA driving a flight control surface is studied to validate the proposed method. Both the static objectives of weight and dynamic performances of energy consumption, rise time, and dynamic stiffness are considered. These four performances are very important for an EHA and should be optimized simultaneously. The Pareto front of these four objectives is obtained with the relevant design parameters. The results present the mapping between the design parameters to the performance and the relations between these objectives. This work indicated the proposed MOO method and this platform can be used in the design phase to help engineers to determine the design parameters according to the required performance. It is also envisaged that the proposed method can be used in similar design problems.

Author Contributions: Investigation, L.X.; Simulation and Analysis, L.X. and S.W.; Methodology, S.W.; Software, S.W.; Writing and Editing, L.X.; Validation, Y.X.; Project Administration D.M.

Funding: This research was funded by National Aviation Science Foundation (Grant No. 20160751003) and National Science Fundation of China (Grant No. 51890885, 51775014).

Conflicts of Interest: The authors declare that there is no conflict of interest regarding the publication of this paper.

Abbreviations

The following abbreviations are used in this manuscript:

EHA	Electro-hydrostatic Actuator
MEA	More Electric Aircraft
MOO	Multi-Objectives Optimization
MOPSO	Multi-Objectives Particle Swarm Optimization

References

1. Rosero, J.; Ortega, J.; Aldabas, E.; Romeral, L. Moving towards a more electric aircraft. *IEEE Aerosp. Electron. Syst. Mag.* **2007**, *22*, 3–9. [CrossRef]
2. Rongjie, K.; Zongxia, J.; Shaoping, W.; Lisha, C. Design and simulation of electro-hydrostatic actuator with a built-in power regulator. *Chin. J. Aeronaut.* **2009**, *22*, 700–706. [CrossRef]
3. Van Den Bossche, D. The A380 flight control electrohydrostatic actuators, achievements and lessons learnt. In Proceedings of the 25th Congress of the International Council of the Aeronautical Sciences, Hamburg, Germany, 3–8 September 2006.
4. Charrier, J.; Kulshreshtha, A. Electric actuation for flight and engine control; evolution and current trend & Future Challenges. In Proceedings of the 45th AIAA Aerospace Sciences Meeting and Exhibit, Reno, Nevada, 8–11 January 2007; Volume 1391.
5. Chao, Q.; Zhang, J.; Xu, B.; Huang, H.; Pan, M. A Review of High-Speed Electro-Hydrostatic Actuator Pumps in Aerospace Applications: Challenges and Solutions. *J. Mech. Des.* **2018**, *141*, 050801. [CrossRef]
6. Chao, Q.; Zhang, J.; Xu, B.; Shang, Y.; Jiao, Z.; Li, Z. Load-Sensing Pump Design to Reduce Heat Generation of Electro-Hydrostatic Actuator Systems. *Energies* **2018**, *11*, 2266. [CrossRef]

7. Shang, Y.; Li, X.; Wu, S.; Pan, Q. A Novel Electro Hydrostatic Actuator System with Energy Recovery Module for More Electric Aircraft. *IEEE Trans. Ind. Electron.* **2019**. [CrossRef]
8. Fraj, A.; Budinger, M.; El Halabi, T.; Maré, J.C.; Negoita, G.C. Modelling approachs for the simulation-based preliminary design and optimization of electromechanical and hydraulic actuation systems. In Proceedings of the 53rd AIAA/ASME/ASCE/AHS/ASC Structures, Structural Dynamics and Materials Conference, Orlando, FL, USA, 12–15 April 2012; In Proceedings of the 20th AIAA/ASME/AHS Adaptive Structures Conference, Orlando, FL, USA, 12–15 April 2012; In Proceedings of the 14th AIAA, Orlando, FL, USA, 12–15 April 2012; p. 1523.
9. Liscouët, J.; Maré, J.C.; Budinger, M. An integrated methodology for the preliminary design of highly reliable electromechanical actuators: Search for architecture solutions. *Aerosp. Sci. Technol.* **2012**, *22*, 9–18. [CrossRef]
10. Budinger, M.; Liscouet, J.; Hospital, F.; Mar, J.C. Estimation models for the preliminary design of electromechanical actuators. *Proc. Inst. Mech. Eng. Part G J. Aerosp. Eng.* **2012**, *226*, 243–259. [CrossRef]
11. Budinger, M.; Reysset, A.; El Halabi, T.; Vasiliu, C.; Maré, J.C. Optimal preliminary design of electromechanical actuators. *Proc. Inst. Mech. Eng. Part G J. Aerosp. Eng.* **2014**, *228*, 1598–1616. [CrossRef]
12. Chakraborty, I.; Jackson, D.; Trawick, D.R.; Mavris, D. Development of a Sizing and Analysis Tool for Electrohydrostatic and Electromechanical Actuators for the More Electric Aircraft. In Proceedings of the 2013 Aviation Technology, Integration, and Operations Conference, Los Angeles, CA, USA, 12–14 August 2013; pp. 1–17. [CrossRef]
13. Chakraborty, I.; Trawick, D.; Jackson, D.; Mavris, D. Electric Control Surface Actuator Design Optimization and Allocation for the More Electric Aircraft. In Proceedings of the 2013 Aviation Technology, Integration, and Operations Conference (AIAA Aciation), Los Angeles, CA, USA, 12–14 August 2013; p. 4283.
14. Fu, J.; Maré, J.C.; Fu, Y. Modelling and simulation of flight control electromechanical actuators with special focus on model architecting, multidisciplinary effects and power flows. *Chin. J. Aeronaut.* **2017**, *30*, 47–65. [CrossRef]
15. Fu, J.; Mare, J.C.; Yu, L.; Fu, Y. Multi-level virtual prototyping of electromechanical actuation system for more electric aircraft. *Chin. J. Aeronaut.* **2018**, *31*, 892–913. [CrossRef]
16. Marler, R.T.; Arora, J.S. Survey of multi-objective optimization methods for engineering. *Struct. Multidiscip. Optim.* **2004**, *26*, 369–395. [CrossRef]
17. Wu, S.; Yu, B.; Jiao, Z.; Shang, Y.; Luk, P.C.K. Preliminary design and multi-objective optimization of electro-hydrostatic actuator. *Proc. Inst. Mech. Eng. Part G J. Aerosp. Eng.* **2017**, *231*, 1258–1268. [CrossRef]
18. Yu, B.; Wu, S.; Jiao, Z.; Shang, Y. Multi-Objective Optimization Design of an Electrohydrostatic Actuator Based on a Particle Swarm Optimization Algorithm and an Analytic Hierarchy Process. *Energies* **2018**, *11*, 2426. [CrossRef]
19. Coello, C.; Pulido, G.; Lechuga, M. Handling multiple objectives with particle swarm optimization. *IEEE Trans. Evol. Comput.* **2004**, *8*, 256–279. [CrossRef]
20. Deb, K.; Agrawal, S.; Pratap, A.; Meyarivan, T. A Fast Elitist Non-dominated Sorting Genetic Algorithm for Multi-objective Optimization: NSGA-II. *Lect. Notes Comput. Sci.* **2000**, *1917*, 849–858.
21. Deb, K. A fast elitist non-dominated sorting genetic algorithm for multi-objective optimization: NSGA-2. *IEEE Trans. Evol. Comput.* **2002**, *6*, 182–197. [CrossRef]
22. Abbass, H.; Sarker, R.; Newton, C. PDE: A Pareto-frontier differential evolution approach for multi-objective optimization problems. In Proceedings of the 2001 Congress on Evolutionary Computation, Seoul, Korea, 27–30 May 2001; Volume 2, pp. 971–978. [CrossRef]

Article

A Numerical Analysis of Pressure Pulsation Characteristics Induced by Unsteady Blood Flow in a Bileaflet Mechanical Heart Valve

Xiao-gang Xu [1,2], Tai-yu Liu [1], Cheng Li [1], Lu Zhu [1] and Shu-xun Li [1,*]

[1] College of Petrochemical Technology, Lanzhou University of Technology, Lanzhou 730050, China; xuxiaogang@lut.edu.cn (X.-g.-X.); lty0108@126.com (T.-y.-L.); lutlicheng@163.com (C.L.); zlyx9323@126.com (L.Z.)
[2] Machinery Industry Pump and Special Valve Engineering Research Center, Lanzhou 730050, China
* Correspondence: lishuxun@lut.cn; Tel.: +86-931-782-3181

Received: 20 March 2019; Accepted: 18 April 2019; Published: 24 April 2019

Abstract: The leaflet vibration phenomenon in bileaflet mechanical heart valves (BMHVs) can cause complications such as hemolysis, leaflet damage, and valve fracture. One of the main reasons for leaflet vibration is the unsteady blood flow pressure pulsation induced by turbulent flow instabilities. In this study, we performed numerical simulations of unsteady flow through a BMHV and observed pressure pulsation characteristics under different flow rates and leaflet fully opening angle conditions. The pressure pulsation coefficient and the low-Reynolds k-ω model in CFD (Computational Fluid Dynamics) software were employed to solve these problems. Results showed that the level of pressure pulsation was highly influenced by velocity distribution, and that the higher coefficient of pressure pulsation was associated with the lower flow velocity along the main flow direction. The influence of pressure pulsation near the trailing edges was much larger than the data obtained near the leading edges of the leaflets. In addition, considering the level of pressure pulsation and the flow uniformity, the recommended setting of leaflet fully opening angle was about 80°.

Keywords: bileaflet mechanical heart valve; computational fluid dynamics; pressure pulsation; unsteady flow; hemodynamics

1. Introduction

Bileaflet mechanical heart valves (BMHVs) are designed and used to replicate the function of natural human heart valves to maintain a unidirectional blood flow, depending on the pressure difference in the upstream and downstream sides of the leaflets. More than 170,000 of the valve replacement operations that occur annually around the world use mechanical heart valves [1,2]. Among various mechanical heart valve types, BMHVs are the most popular and are often implanted to replace diseased heart valves because of their longer lifespan and reliable performance.

Despite the widespread clinical use of mechanical valve replacements, the BMHVs are far from perfect. The major potential complications that remain as drawbacks to mechanical heart valves include hemolysis, damage of blood elements, and thrombosis, as a result of the high-velocity jet flow through the narrow passage between the leaflets [3,4]. Another noticeable phenomenon associated with mechanical valves is unsteady, blood flow-induced leaflet vibration, which leads to the complex interaction of flow dynamics and leaflet kinematics. One of the main reasons for leaflet vibration is the unsteady blood flow pressure pulsation, which is induced by the turbulent flow instabilities [5,6]. Although the occurrence of such a phenomenon is quite low, it may lead to the fracture of BMHVs and life-threatening conditions to patients. In order to deal with the potential risks caused by the leaflet vibration phenomenon, it is of critical importance to understand the mechanism and characteristics of pressure pulsation induced by unsteady blood flow [7,8].

Previous studies mainly focus on flow characteristics and the flow structure interactions of BMHVs. For example, Matteo Nobili et al. [9] investigated the dynamics of a BMHV by means of a fluid–structure interaction method and an ultrafast cinematographic technique. The computational model captured the main features of the leaflet motion during the systole. Iman Borazjani et al. [10] performed high-resolution fluid–structure interaction simulations of physiologic pulsatile flow through a BMHV in an anatomically realistic aorta. Meanwhile, numerous in vitro and in silico studies on the characteristics of leaflet motion and the hemodynamic performances of BMHVs have been conducted. For example, L. Ge et al. [11] analyzed the results of 2D high-resolution velocity measurements, and a full 3D numerical simulation for pulsatile flow through a BMHV mounted in a model axisymmetric aorta, to investigate the mechanical environment experienced by blood elements under physiologic conditions. Redaelli et al. [12] analyzed the opening phase of a bileaflet heart valve under low flow rates and validated the leaflet motion experimentally. Cheng et al. [13] presented a three-dimensional unsteady flow analysis past a bileaflet valve prosthesis in the mitral position incorporating an FSI algorithm for leaflet motion during the valve closing phase.

As mentioned above, few studies have been performed regarding the mechanism and characteristics of pressure pulsation induced by unsteady blood flow. The mechanisms and influencing factors of pressure pulsation in BMHVs are not well understood at the present time. Hence, it is imperative to study the mechanism and characteristics of pressure pulsation induced by unsteady blood flow.

The objective of this study is, therefore, to numerically investigate the characteristics and influence factors of pressure pulsation in BMHVs under different conditions of flow rate and leaflet fully opening angle. Additionally, the non-dimensional coefficient of pressure pulsation was proposed to evaluate the impact on pressure pulsation induced by unsteady blood flow in a BMHV.

2. Numerical Methods and Modeling

2.1. Numerical Method

Ansys CFX commercial software based on finite volume method was used to run pulsatile flow simulations in this study. According to previous experimental results, especially in the studies of velocity and pressure distribution, blood flow was assumed incompressible, turbulent, and Newtonian [2,4].

The continuity equation and the Reynolds-averaged Navier–Stokes (RANS) equation can be written as follows:

$$\frac{\partial u_i}{\partial x_i} = 0 \tag{1}$$

$$\frac{\partial u_i}{\partial t} + \frac{\partial u_i u_j}{\partial x_j} = -\frac{1}{\rho}\frac{\partial P}{\partial x_j} + \frac{1}{\rho}\frac{\partial}{\partial x_j}\left(\mu\left(\frac{\partial u_i}{\partial x_j} + \frac{\partial u_j}{\partial x_i}\right) - \rho\overline{u_i' u_j'}\right) \tag{2}$$

where ρ is the blood density, u is the velocity, P is the pressure, and μ is the dynamic viscosity. These governing equations are numerically solved for the pressure and velocity profiles in the flow field.

$$p(x, y, z, t) = p_m(x, y, z, t) + p'(x, y, z, t) \tag{3}$$

$$v(x, y, z, t) = v_m(x, y, z, t) + v'(x, y, z, t) \tag{4}$$

As shown in Equations (3) and (4), all of the spontaneous variables are decomposed into average values and fluctuating components [14]. For example, where p is the transient pressure at some point in the flow field, p_{m} and p' represent the time-average pressure and time-average pulsating pressure, respectively.

In this study, the pressure pulsation of the BMHV internal flow field under different flow rate and opening angle conditions was analyzed by the non-dimensional coefficient of pressure pulsation.

$$\tau = p'(x, y, z, t)/(\rho u^2/2) \quad (5)$$

where τ is the coefficient of pressure pulsation (it was proposed to evaluate the impact on pressure pulsation induced by unsteady blood flow).

2.2. Geometry of BMHV

Figure 1 shows a three-dimensional simplified geometry of the St. Jude bileaflet mechanical heart valve investigated in this research, which was chosen to be similar to previous studies [14,15]. The model geometry mainly consists of four parts: valve leaflets, aortic sinus, inlet, and outlet sections [16]. In the previous experimental study, the phenomenon of leaflet fluttering was only observed when the leaflets were in the fully open position, so the leaflets were assumed to be fixed in the fully open position to simplify the calculation. The geometry of the aortic root is modelled as an axial-symmetric expansion. The aortic root diameter was 27 mm, and the leaflet opening angle was 80°. The hinge mechanism was simplified as solid cylinder. To obtain the fully developed flow and minimize the influence of outlet boundary conditions, the lengths of the upstream and downstream sections were added as 2D and 6D (where D is the inlet diameter), respectively. Three models with different fully opening angles (75°, 80°, and 85°) were created for the purpose of this study.

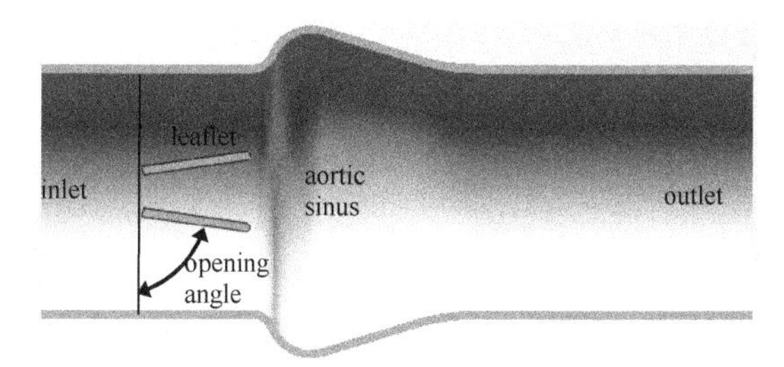

Figure 1. The perpendicular cut-plane view of the three-dimensional bileaflet mechanical heart valve (BMHV) geometry.

2.3. Meshing Configurations and Mesh Independency Test

ICEM-CFD software was employed to generate the mesh of fluid domain. The regions of valve and aortic sinus were discretized with locally refined tetrahedral elements. In the remaining domain (upstream and downstream sections), the hexahedral mesh was used in order to limit the numerical diffusion. One of the mesh independency tests is shown in Table 1. The maximum velocity value at central and lateral orifices were used to check the mesh independency in all cases [17,18]. The total number of cells chosen was equal to 850,131.

Table 1. Mesh independency test.

Number of Cells	Max. Velocity at Central Orifice (m/s)	Max. Velocity at Lateral Orifices (m/s)
368,935	1.16	0.91
655,422	1.19	0.92
850,131	1.21	0.95
1,020,373	1.21	0.96

2.4. Boundary Conditions

For the boundary conditions, the different flow rates (from 5 L/min to 25 L/min) measured in the vitro study [19,20] (Figure 2) were considered for the inlet condition (ventricular side), and the zero static pressure was set as outlet condition (aorta side) [21]. The density and dynamic viscosity of blood were set to 1055 kg/m^3 and 0.0035 Pa·s, respectively. Because the influence of the change of pressure and velocity boundary conditions on the phenomenon of leaflet fluttering is not clear, only part of the cardiac cycle was selected for simulation. The total transient simulation time was controlled, ranging from the 200th ms to the 500th ms in one cardiac cycle (Figure 2), and the 300 ms time was discretized with 1500 steps corresponding to a time step of 0.2 ms.

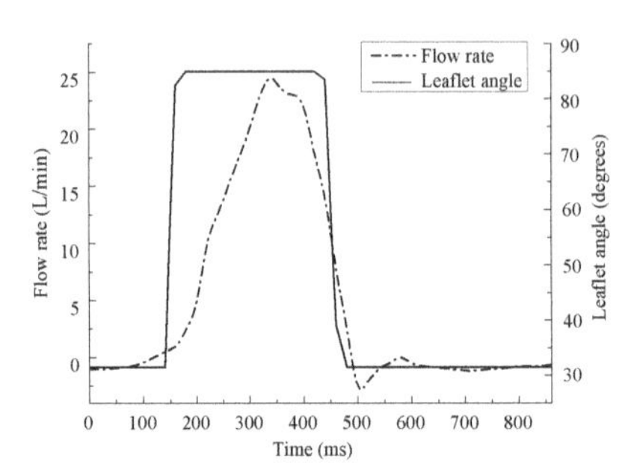

Figure 2. Flow rate and leaflet motion throughout one cardiac cycle.

2.5. Validation of the Computational Methods

For validation of the current calculation, the monitoring line was set at 2.54 mm cross-section after the leaflets, as shown in Figure 3a. The opening angle of 85° was used. Comparisons of the velocity profiles along the monitoring line calculated by three different turbulent models (standard k-ε, RNG k-ε, and k-ω) under a 25 L/min flow rate are shown in Figure 3b. The velocity profile observed in previous experimental studies that considered a BMHV with similar geometry and the same flow condition [15,22] is also shown in the same figure. The comparison of the results illustrates that the results calculated by the k-ω model, in particular the velocity distribution near the leaflets, were in good agreement with the experimental data. The main differences between all of the computational results and the previous experimental data are located in the vicinity of the trailing edges. Considering the fact that this study mainly focused on the flow field near the leaflets, the k-ω model was chosen for simulation.

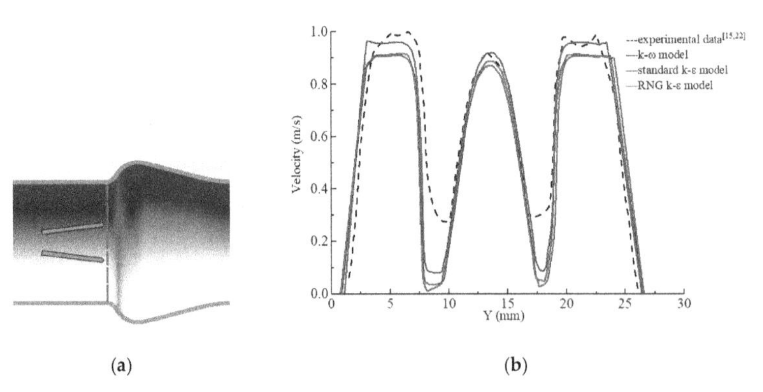

Figure 3. Comparisons of the velocity profiles between the simulation results by different turbulent models and the experiment results [15,22]. (**a**) Monitoring line at 2.54 mm cross-section after the leaflets. (**b**) Current simulation results and previous experimental data for velocity profiles.

3. Results and Discussion

3.1. Effect of Different Flow Rate Conditions

Figure 4a shows the blood flow patterns by velocity vector distribution of a BMHV at different flow rate conditions (5 L/min, 10 L/min, 15 L/min, 20 L/min, and 25 L/min), where the color represents the velocity magnitude [23,24]. The flow patterns developed in all cases were similar to each other because of the same opening degree of 80°. Because of the nonphysiologic geometries of mechanical valves, a three-jet configuration through the central and lateral orifices can show a relatively higher velocity and a low-velocity region located at the whole surfaces and trailing edges of both leaflets. The spread region of the trailing edges of the leaflets increases with the flow rate. As the flow rate changed from 5 L/min to 25 L/min, the maximum velocity increased from 0.25 m/s to 1.21 m/s in the central orifices and increased from 0.22 m/s to 0.95 m/s in the lateral orifices, respectively. Higher velocities at the central and lateral orifices were accompanied by higher level of velocity gradient in the vicinity of leaflet surfaces [25,26].

Figure 4b shows the vortex structure of the flow field at different flow rate conditions by means of the vortex core region, where the color represents the value of the Q criterion. In all cases, the vortex core region was mainly formed in the low-velocity region near both leaflet edges. As the flow rate changed from 5 L/min to 25 L/min, the value of the Q criterion in the vortex core region increased rapidly, and the maximum Q value increased from 1.65×10^4 s^{-2} to 3.24×10^5 s^{-2}. As the flow rate increased to more than 15 L/min, the velocity of the mainstream region increased and the vortices shed downstream from the trailing edges of the leaflets.

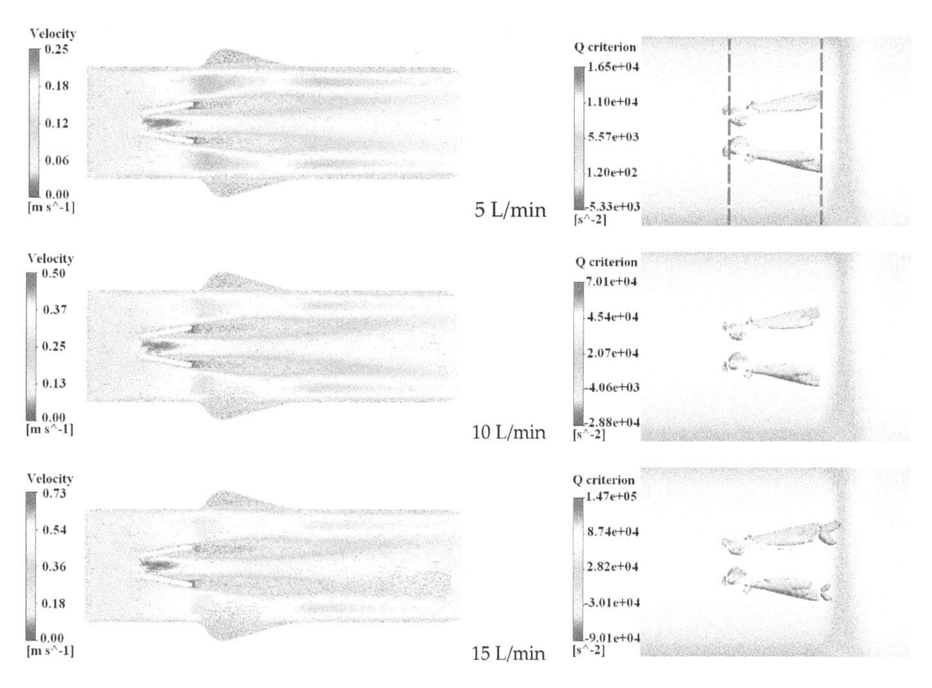

Figure 4. *Cont.*

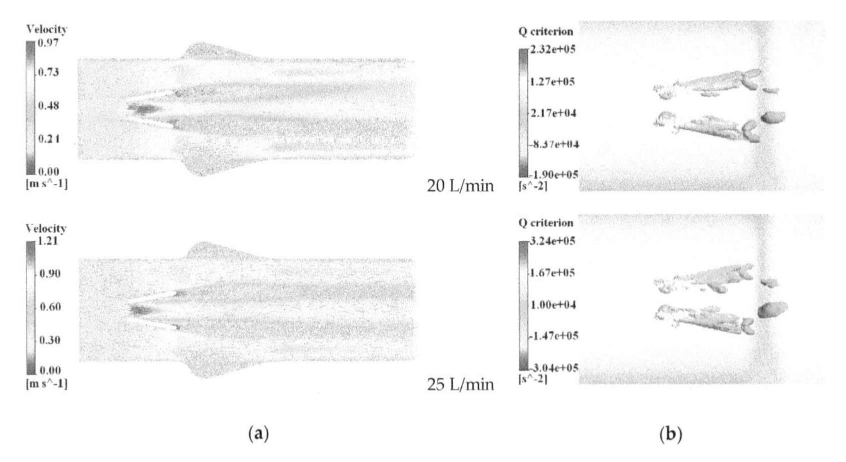

Figure 4. Velocity distribution and Q criterion of vortex core region of the 80° opening degree bileaflet mechanical heart valve (BMHV). (**a**) Velocity vector distribution and (**b**) Q criterion of vortex core region.

In order to study the pressure pulsation characteristics in the vicinity of the leaflets under different flow conditions, two monitoring lines perpendicular to the flow direction (Y direction) were set at the leading edge and trailing edge of the leaflets of the BMHV, respectively, as shown in Figure 4b. One hundred monitoring points were set on every monitoring line, which were evenly distributed in the central and lateral orifices. The transient flow calculation results showed that the pressure value became stable after the first calculation cycle. The pressure value data in the second calculation cycle at each monitoring point were recorded, and the time-average pressure pulsation at each point were calculated by Equation (5).

$$p' = (\int_{T}^{2T} |p - p_m| dt) / \Delta T \tag{6}$$

where ΔT is the time of one calculation cycle of 300 ms, from the 200th ms to the 500th ms in one cardiac cycle. p' is the time-average pulsating pressure, which represents the average pressure pulsation range at a certain point.

The average pressure pulsation at all monitoring points under five different flow conditions were measured and calculated, and the coefficients of pressure pulsation were calculated by equation (4). The coefficient distribution of pressure pulsation at the leading edge and trailing edge monitoring lines under different flow conditions are shown in Figures 5 and 6, respectively.

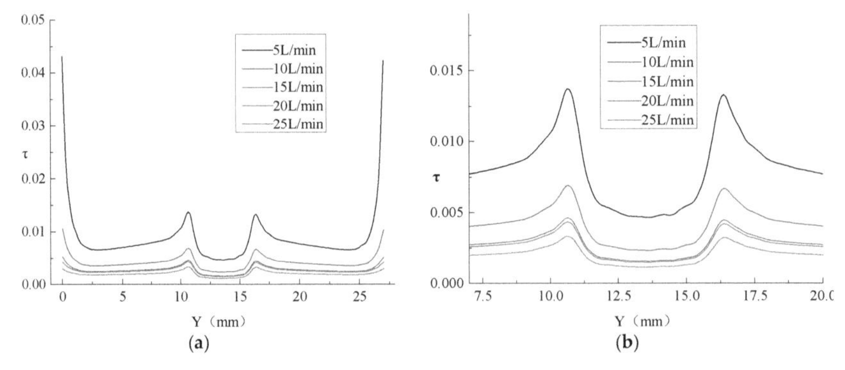

Figure 5. Coefficient distribution of pressure pulsation at the leading edge monitoring lines under different flow conditions at 80° opening degree. (**a**) The complete view. (**b**) The local enlarged view.

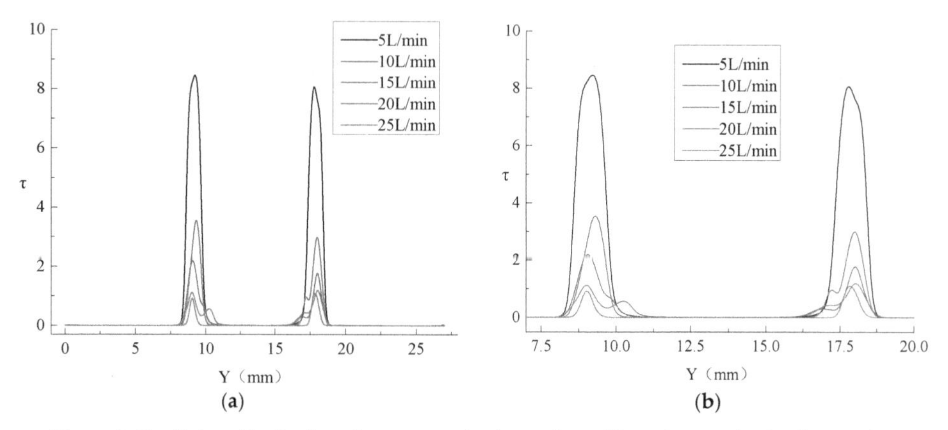

Figure 6. Coefficient distribution of pressure pulsation at the trailing edge monitoring lines under different flow conditions at 80° opening degree. (**a**) The complete view. (**b**) The local enlarged view.

Figure 5a is the complete view of the coefficient distribution of pressure pulsation at the leading edge monitoring lines at flow rates from 5 L/min to 25 L/min. Figure 5b is the local enlarged view. Under different flow conditions, the level of pressure pulsation coefficient near the wall and the leading edge of the leaflets was higher. The level of the pressure pulsation coefficient in central and lateral orifices was relatively lower. As the flow rate increased, the distribution of the pressure pulsation coefficient on the leading edge monitoring line showed a trend of gradual decline, and the maximum value of pressure pulsation coefficient near the leading edge decreased from 0.014 to 0.003. The results show that the lower the flow velocity on the leading edges of the leaflets, the larger the coefficient of pressure pulsation. The flow velocity near the leading edges of the leaflets increased with the increased flow rate. The vortices fell off downstream, and the influence of the pressure pulsation decreased.

Figure 6a is the complete view of the coefficient distribution of pressure pulsation at the trailing edge monitoring lines at flow rates from 5 L/min to 25 L/min. Figure 6b is the local enlarged view. Under different flow conditions, the coefficient of pressure pulsation near the trailing edge of the leaflets was relatively higher (the maximum value, equal to 8.4), which was reached at the trailing edges of the leaflets. At the main flow region, the coefficients of pressure pulsation in central and lateral orifices were close to zero. As the flow rate increased from 5 L/min to 15 L/min, the distribution of the pressure pulsation coefficient near the trailing edge decreased rapidly, with the maximum value decreasing from 8.4 to 2.1. As the flow rate increased from 20 L/min to 25 L/min, the maximum value of pressure pulsation coefficient gradually decreased from 1.4 to 1.2. The results show that the high-pressure pulsation region was mainly formed in the vicinity of the trailing edges of leaflets. As the flow rate increased, the coefficient of pressure pulsation showed a decreasing trend. Additionally, the level of the pressure pulsation coefficient near the trailing edges was much larger than the data obtained near the leading edges of leaflets. It could be considered that one of the main reasons for the result in leaflet vibration was the pressure pulsation located at the trailing edges of leaflets under low-velocity conditions.

3.2. *Effect of Different Fully Opening Angles of Leaflets*

In the same way, the pressure pulsation characteristics in the vicinity of the leaflets under different fully opening angle conditions, from 75° to 85°, were studied. Figure 7a shows the blood flow patterns by the velocity vector distribution of a BMHV at different opening degree conditions, where the color represents the velocity magnitude. All cases were simulated under the same flow rate condition of 5 L/min. The leaflets act as an obstruction to the blood flow through the BMHV, and coupled with the relatively high-velocity three-jet through the central and lateral orifices. As the leaflet opening degree increased from 75° to 85°, the maximum velocity decreased from 0.28 m/s to 0.22 m/s and from 0.24 m/s to 0.19 m/s in the central and lateral orifices, respectively. The results showed that, with the increase of the opening angle, the velocity distribution was more uniform and the average level of flow velocity decreased. Figure 7b shows the vortex structure of the flow field under different leaflet opening angle conditions by means of vortex core region, where the color represents the value of the Q criterion. In all cases, the vortex core region mainly formed in the vicinity of the surfaces of both leaflets. As the leaflet opening degree increased from 75° to 85°, the range of the vortex core region near the leaflets was gradually enlarged, and the maximum Q value decreased from 2.02×10^4 s^{-2} to 1.13×10^4 s^{-2}. The results showed that, because of the larger opening angle and the lower average velocity, the vortices were more likely to stay at the surfaces of the leaflets than to shed downstream from the trailing edges.

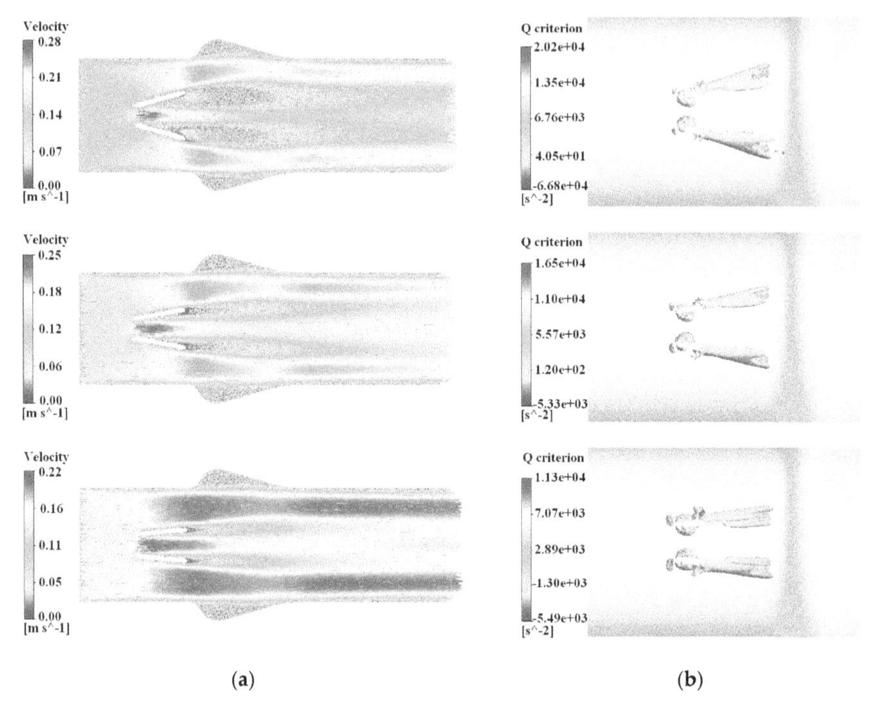

Figure 7. Velocity vector distribution and Q criterion of the vortex core region of the different opening angle bileaflet mechanical heart valves (BMHVs) at a flow rate of 5 L/min. (**a**) Velocity vector distribution. (**b**) Q criterion of vortex core region.

Figure 8 shows the coefficient distribution of pressure pulsation at the leading edge monitoring lines at different fully opening leaflet angles from 75° to 85°. Under different fully opening angle conditions, the coefficient distribution of pressure pulsation at the leading edge monitoring lines were similar. The level of the pressure pulsation coefficient near the wall and the leading edge was relatively higher, and relatively lower at the central and lateral orifices. As the fully opening angle increased from 75° to 85°, the level of the pressure pulsation coefficient in the region of the leading edges and the lateral orifices slightly decreased, and the maximum value of the pressure pulsation coefficient near the leading edge decreased from 0.02 to 0.017. Under the same conditions, the pressure pulsation coefficient in the central orifice gradually increased and the minimum value of the pressure pulsation coefficient at the central orifice increased from 0.007 to 0.011. The results showed that, at the same flow rate condition, the level of the pressure pulsation coefficient near the leading edges slightly decreased with the increased leaflet fully opening angle. It could be inferred that the main influencing factor of the pressure pulsation coefficient level near the leading edges was the different velocity distribution. The lower the flow velocity, the larger the coefficient of pressure pulsation. Since the level of the pressure pulsation coefficient near the leading edges was very low in all cases, it could be considered that the change of the leaflet fully opening angle had little effect on pressure pulsation and the potential vibration of leaflets leading edges.

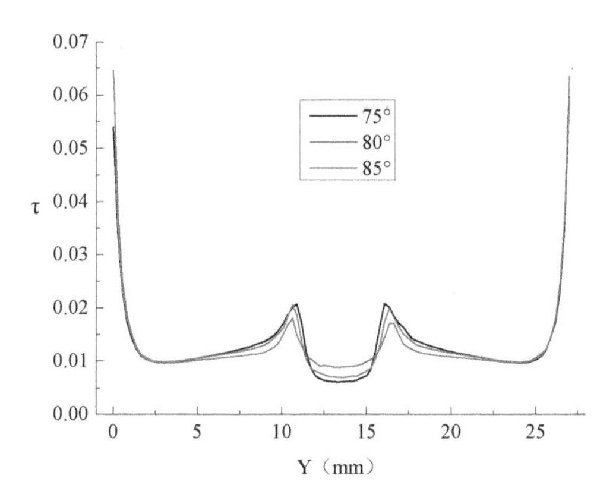

Figure 8. Coefficient distribution of pressure pulsation at the leading edge monitoring lines under different fully opening angle conditions at a flow rate of 5 L/min.

Figure 9 is the coefficient distribution of pressure pulsation at the trailing edge monitoring lines at different fully opening leaflet angles from 75° to 85°. In all three cases, the distribution of the pressure pulsation coefficient at the trailing edge monitoring lines were similar. The pressure pulsation coefficient near the trailing edge was relatively higher. At the main flow region, the pressure pulsation coefficients in the central and lateral orifices were close to zero. As the fully opening angle increased from 75° to 85°, the level of the pressure pulsation coefficient near the trailing edges gradually increased, and the maximum value of the pressure pulsation coefficient increased from 6.6 to 9.5. Additionally, in comparison to another two cases, the pressure pulsation coefficient in the central orifice showed a nonuniform distribution with the under 75° opening angle condition. The results showed that at the same flow rate condition, the level of the pressure pulsation coefficient near the trailing edges gradually increased as the leaflet fully opening angle increased. It could be inferred that the main influencing factor on the pressure pulsation coefficient level near the trailing edges was the different vortex structure distribution. The larger leaflet fully opening angle led to the larger vortex core region and a higher level of pressure pulsation coefficient. The nonuniform distribution of the pressure pulsation coefficient indicated that the nonuniform velocity distribution might be induced by the narrow central orifice under a relatively lower leaflet opening angle condition. Thus, considering the pressure pulsation and the flow uniformity, the recommended setting of leaflet fully opening angle was about 80°.

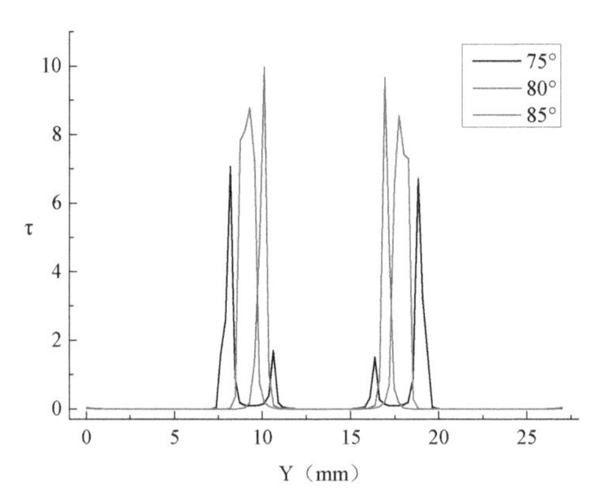

Figure 9. Coefficient distribution of pressure pulsation at the trailing edge monitoring lines under different fully opening angle conditions at a flow rate of 5 L/min.

4. Conclusions

In this study, the pressure pulsation characteristics induced by unsteady blood flow in the BMHV under different flow rates and leaflet fully opening angle conditions were analyzed. The conclusions can be listed as follows.

In regard to blood flow through a BMHV, the main influencing factor of the pressure pulsation coefficient level was the different velocity distribution. In all cases, the vortex core region and low-velocity region were mainly formed in the leading and trailing edges of leaflets. The lower the flow velocity on the edges of leaflets, the larger the coefficient of pressure pulsation. As the flow rate increased, the flow velocity near the edges of the leaflets increased and the coefficient of pressure pulsation decreased; the vortices then fell off downstream, and the influence of pressure pulsation decreased.

The level of the pressure pulsation coefficient near the trailing edges was much larger than the data obtained near the leading edges of leaflets, which indicated that one of the main reasons for leaflet vibration was the pressure pulsation located at the trailing edges of leaflets under low-velocity conditions. At the same flow rate and different leaflet fully opening angle conditions, the level of the pressure pulsation coefficient near the leading edges was close to 0, which indicated that the change in leaflet fully opening angle had little effect on the pressure pulsation and the potential vibration of leaflet leading edges.

The range of the vortex core region increased with the increased leaflet fully opening angle. The relatively lower the velocity in the vortex core region, the larger the level of the pressure pulsation coefficient. Meanwhile, the nonuniform velocity distribution could have been induced by the narrow central orifice under the relatively lower leaflet opening angle condition. Thus, considering the pressure pulsation and the flow uniformity, the recommended setting of leaflet fully opening angle was about 80°.

In this study, because the influence of the change of pressure and velocity boundary conditions to the phenomenon of leaflet fluttering are not well understood, a model of leaflets in fixed positions was adopted to simplify the calculation, which may have led to deviation between the simulation results and the actual situation. In addition, only one of the velocity profile monitoring lines was compared and validated by the experimental results. Considering the periodic boundary conditions, the simulation of the whole cardiac cycle and more detailed data are required for full validation in future work.

Author Contributions: Conceptualization, X.-g.-X. and S.-x.-L.; methodology, X.-g.-X.; software, X.-g.-X. and L.Z.; validation, T.-y.-L. and X.-g.-X.; formal analysis, T.-y.-L.; investigation, T.-y.-L. and X.-g.-X.; resources, S.-x.-L.; data curation, T.-y.-L. and C.L.; writing—original draft preparation, X.-g.-X. and C.L.; writing—review and editing,

X.-g.-X. and C.L.; visualization, X.-g.-X.; supervision, S.-x.-L.; project administration, X.-g.-X.; funding acquisition, S.-x.-L.

Funding: This research was funded by National Natural Science Foundation of China, grant number 51569012, and the Fundamental Research Funds for Colleges and Universities in Gansu province, grant number 20146302.

Conflicts of Interest: The authors declare no conflict of interest.

References

1. Yoganathan, A.P.; Chandran, K.B.; Sotiropoulos, F. Flow in prosthetic heart valves: State-of-the-art and future directions. *Ann. Biomed. Eng.* **2005**, *33*, 1689–1694. [CrossRef] [PubMed]
2. Smadi, O.; Hassan, I.; Pibarot, P.; Kadem, L. Numerical and experimental investigations of pulsatile blood flow pattern through a dysfunctional mechanical heart valve. *J. Biomech.* **2010**, *43*, 1565–1572. [CrossRef]
3. Alemu, Y.; Bluestein, D. Flow-induced platelet activation and damage accumulation in a mechanical heart valve: Numerical studies. *Artif. Organs* **2007**, *31*, 677–688. [CrossRef] [PubMed]
4. Zhou, F.; Cui, Y.Y.; Wu, L.L.; Yang, J.; Liu, L. Analysis of Flow Field in Mechanical Aortic Bileaflet Heart Valves Using Finite Volume Method. *J. Med. Biol. Eng.* **2016**, *36*, 110–120. [CrossRef]
5. Cheon, G.J.; Chandran, K.B. Dynamic behavior analysis of mechanical monoleaflet heart valve prostheses in the opening phase. *J. Biomech. Eng.* **1993**, *115*, 389–395. [CrossRef]
6. Chu, Y.P.; Cheng, J.L.; Liu, A.J. Evaluation on opening status of the mechanical heart valve in vitro under pulsatile flow. *Chin. J. Tissue Eng.* **2012**, *16*, 7480–7485. [CrossRef]
7. Lee, S.H.; Ryu, S.M.; Jeong, W.B. Vibration analysis of compressor piping system with fluid pulsation. *J. Mech. Sci. Technol.* **2012**, *26*, 3903–3909. [CrossRef]
8. Zhang, N.; Yang, M.; Gao, B.; Li, Z.; Ni, D. Experimental and numerical analysis of unsteady pressure pulsation in a centrifugal pump with slope volute. *J. Mech. Sci. Technol.* **2015**, *29*, 4231–4238. [CrossRef]
9. Nobili, M.; Morbiducci, U.; Ponzini, R.; Gaudio, C.; Balducci, A.; Grigioni, M.; Montevecchi, M.; Redaelli, A. Numerical simulation of the dynamics of a bileaflet prosthetic heart valve using a fluid-structure interaction approach. *J. Biomech.* **2008**, *41*, 2539–2550. [CrossRef]
10. Borazjani, I.; Ge, L.; Sotiropoulos, F. High-Resolution Fluid–Structure Interaction Simulations of Flow through a Bi-Leaflet Mechanical Heart Valve in an Anatomic Aorta. *Ann. Biomed. Eng.* **2010**, *38*, 326–344. [CrossRef] [PubMed]
11. Ge, L.; Dasi, L.P.; Sotiropoulos, F.; Yoganathan, A. Characterization of Hemodynamic Forces Induced by Mechanical Heart Valves: Reynolds vs. Viscous Stresses. *Ann. Biomed. Eng.* **2008**, *36*, 276–297. [CrossRef]
12. Redaelli, A.; Bothorel, H.; Votta, E.; Soncini, M.; Morbiducci, U.; Del Gaudio, C.; Balducci, A.; Grigioni, M. 3-D simulation of the SJM bileaflet valve opening process: Fluid–structure interaction study and experimental validation. *J. Heart Valve Dis* **2004**, *13*, 804–813. [CrossRef]
13. Cheng, R.; Lai, Y.G.; Chandran, K.B. Three-Dimensional Fluid-Structure Interaction Simulation of Bileaflet Mechanical Heart Valve Flow Dynamics. *Ann. Biomed. Eng.* **2004**, *32*, 1471–1483. [CrossRef]
14. Bluestein, D.; Rambod, E.; Gharib, M. Vortex shedding as a mechanism for free emboli formation in mechanical heart valves. *J. Biomech. Eng.* **2000**, *122*, 125–134. [CrossRef]
15. Shahriari, S.; Maleki, H.; Hassan, I.; Kadem, L. Evaluation of shear stress accumulation on blood components in normal and dysfunctional bileaflet mechanical heart valves using smoothed particle hydrodynamics. *J. Biomech.* **2012**, *45*, 2637–2644. [CrossRef]
16. Kwon, Y.J. Numerical analysis for the structural strength comparison of st. jude medical and edwards mira bileaflet mechanical heart valve prostheses. *J. Mech. Sci. Technol.* **2010**, *24*, 461–469. [CrossRef]
17. Qian, J.Y.; Li, X.J.; Gao, Z.X.; Jin, Z.J. Mixing Efficiency Analysis on Droplet Formation Process in Microchannels by Numerical Methods. *Processes* **2019**, *7*, 33. [CrossRef]
18. Qu, J.; Yan, T.; Sun, X.; Li, Z.; Li, W. Numerical Simulation of the Effects of the Helical Angle on the Decaying Swirl Flow of the Hole Cleaning Device. *Processes* **2019**, *7*, 109. [CrossRef]
19. MinYun, B.; McElhinney, D.B.; Shiva, A.; Lucia, M.; Aidun, C.K.; Yoganathan, A.P. Computational simulations of flow dynamics and blood damage through a bileaflet mechanical heart valve scaled to pediatric size and flow. *J. Biomech.* **2014**, *47*, 3169–3177. [CrossRef]

20. Kuan, Y.H.; Kabinejadian, F.; Nguyen, V.T.; Su, B.; Yoganathan, A.P.; Leo, H.L. Comparison of hinge microflow fields of bileaflet mechanical heart valves implanted in different sinus shape and downstream geometry. *Comput. Method. Biomech.* **2015**, *18*, 1785–1796. [CrossRef] [PubMed]
21. Kim, C.N.; Hong, T. The effects of the tilt angle of a bileaflet mechanical heart valve on blood flow and leaflet motion. *J. Mech. Sci. Technol.* **2012**, *26*, 819–825. [CrossRef]
22. Ge, L.; Leo, H.L.; Sotiropoulos, F.; Yoganathan, A.P. Flow in a mechanical bileaflet heart valve at laminar and near-peak systole flow rates: CFD simulations and experiments. *J. Biomech. Eng.* **2005**, *127*, 782–797. [CrossRef]
23. Jin, Z.J.; Gao, Z.X.; Qian, J.Y.; Wu, Z.; Sunden, B. A parametric study of hydrodynamic cavitation inside globe valves. *ASME J. Fluids Eng.* **2018**, *140*, 031208. [CrossRef]
24. Qian, J.Y.; Gao, Z.X.; Liu, B.Z.; Jin, Z.J. Parametric study on fluid dynamics of pilot-control angle globe valve. *ASME J. Fluids. Eng.* **2018**, *140*, 111103. [CrossRef]
25. Jin, Z.J.; Gao, Z.X.; Zhang, M.; Liu, B.Z.; Qian, J.Y. Computational fluid dynamics analysis on orifice structure inside valve core of pilot-control angle globe valve. *IMeChE Part C J. Mech. Eng. Sci.* **2018**, *232*, 2419–2429. [CrossRef]
26. Qian, J.Y.; Chen, M.R.; Liu, X.L.; Jin, Z.J. A numerical investigation of the flow of nanofluids through a micro Tesla valve. *J. Zhejiang Univ. Sci. A.* **2019**, *20*, 50–60. [CrossRef]

MDPI
St. Alban-Anlage 66
4052 Basel
Switzerland
Tel. +41 61 683 77 34
Fax +41 61 302 89 18
www.mdpi.com

Processes Editorial Office
E-mail: processes@mdpi.com
www.mdpi.com/journal/processes

CPSIA information can be obtained
at www.ICGtesting.com
Printed in the USA
LVHW070436010920
664725LV00035B/2141